Analytical Elements of Mechanisms

Mechanisms are fundamental components of machines. They are used to transmit forces and moments and to manipulate objects in industrial machinery, robots, automobiles, aircraft, mechatronics devices, and biomechanical systems. A knowledge of the kinematic and dynamic properties of mechanisms is essential for their design and control.

This book describes methods and algorithms for the analysis of kinematic systems. Beginning with basic concepts, the book then discusses a variety of problem-solving approaches and computational techniques. Its distinctive feature is its focus on the contour equation as a powerful, computationally efficient tool that will help the reader to design complex spatial mechanisms.

This handy text will be useful for senior or graduate students, researchers, and practicing engineers working in robotics, vehicle dynamics, mechatronics, and machine design.

Dan B. Marghitu is Associate Professor, Department of Mechanical Engineering, Auburn University.

Malcolm J. Crocker is Distinguished University Professor, Department of Mechanical Engineering, Auburn University.

Analytical Elements of Mechanisms

DAN B. MARGHITU
Auburn University

MALCOLM J. CROCKER
Auburn University

PUBLISHED BY THE PRESS SYNDICATE OF THE UNIVERSITY OF CAMBRIDGE
The Pitt Building, Trumpington Street, Cambridge, United Kingdom

CAMBRIDGE UNIVERSITY PRESS
The Edinburgh Building, Cambridge CB2 2RU, UK
40 West 20th Street, New York, NY 10011-4211, USA
10 Stamford Road, Oakleigh, VIC 3166, Australia
Ruiz de Alarcón 13, 28014 Madrid, Spain
Dock House, The Waterfront, Cape Town 8001, South Africa

http://www.cambridge.org

First published 2001

Printed in the United States of America

Typefaces Times Ten 10/13.5 pt. and Franklin Gothic *System* LaTeX 2ε [TB]

A catalog record for this book is available from the British Library.

Library of Congress Cataloging in Publication Data

Marghitu, Dan B.

Analytical elements of mechanisms / Dan B. Marghitu, Malcolm J. Crocker.

p. cm.

ISBN 0-521-62383-9

1. Mechanical movements. 2. Machinery, Dynamics of. 3. Machinery, Kinematics of. I. Crocker, Malcolm J. II. Title.

TJ181 .M25 2001

621.8′11 – dc21 00-064137

ISBN 0 521 62383 9 hardback

Contents

Preface

Mechanisms have been and continue to be fundamental components of all machines. Mechanisms are used to transmit forces and moments and to manipulate objects. A knowledge of the kinematics and dynamics properties of these mechanisms is crucial for machine design and control.

With this book, readers can become familiar with the solution of kinematics and dynamics problems of mechanisms, methods of solution, and their software implementation. Readers can also study and compare the available methods of analysis from a unified viewpoint.

The unique feature of this book and, from the authors' point of view, its main advantage, is that it presents the contour equation method for kinematics and kinetostatics analysis of mechanisms with bars and gears. This will enable the reader to design complex mechanisms in a more advantageous way with the use of computers.

The general approach involves the presentation of a systematic explanation of the basic concepts of the kinematics and dynamics of mechanical systems. The treatment of kinematics methods begins with the traditional vector approach and proceeds to the contour equation method as the one that is the most suitable and computationally efficient. The other feature of the approach used here is the methodical review of the dynamics methods of solution. Special attention is devoted to symbolical software such as *Mathematica*.

The book will assist all those interested in the design of mechanisms such as robots, manipulators, building machines, textile machines, vehicles, aircraft, satellites, ships, biomechanical systems (vehicle simulators, barrier tests, human motion studies, etc.), controlled mechanical systems, mechatronical devices, and many others. It is appropriate for use as a text for senior level or graduate courses in mechanical engineering dealing with the subjects of the analysis and design of mechanisms, fundamentals of robotics, vehicle dynamics, mechatronics, and multibody modeling. It is possible to cover all essential contents of the book in one semester. A basic knowledge of mechanics and calculus is assumed. The book may also be useful for practicing engineers and researchers in the fields of machine design and dynamics, and also biomechanics and mechatronics. These engineers will obtain a basic knowledge of methods for the solution of everyday problems and explanations of the algorithms on which commercial software packages are based. An understanding of these methods of analysis for mechanisms is important in order to use the software packages correctly and to avoid the possibility of obtaining erroneous design conclusions.

1 Introduction

1.1 Degrees of Freedom

The *number of degrees of freedom* (DOF) of a system is equal to the number of independent parameters (measurements) that are needed to uniquely define its position in space at any instant of time. The number of DOF is defined with respect to a reference frame.

Figure 1.1 shows a rigid body (RB) lying in a plane. The rigid body is assumed to be incapable of deformation, and the distance between two particles on the rigid body is constant at any time. If this rigid body always remains in the plane, three parameters (three DOF) are required to completely define its position: two linear coordinates (x, y) to define the position of any one point on the rigid body, and one angular coordinate θ to define the angle of the body with respect to the axes. The minimum number of measurements needed to define its position are shown in the figure as x, y, and θ. A rigid body in a plane then has three degrees of freedom. Note that the particular parameters chosen to define its position are not unique. Any alternative set of three parameters could be used. There is an infinity of sets of parameters possible, but in this case there must always be three parameters per set, such as two lengths and an angle, to define the position because a rigid body in plane motion has three DOF.

Six parameters are needed to define the position of a free rigid body in a three-dimensional (3-D) space. One possible set of parameters that could be used are three lengths (x, y, z), plus three angles $(\theta_x, \theta_y, \theta_z)$. Any free rigid body in three-dimensional space has six degrees of freedom.

1.2 Motion

A rigid body free to move in a reference frame will, in the general case, have complex motion, which is simultaneously a combination of rotation and

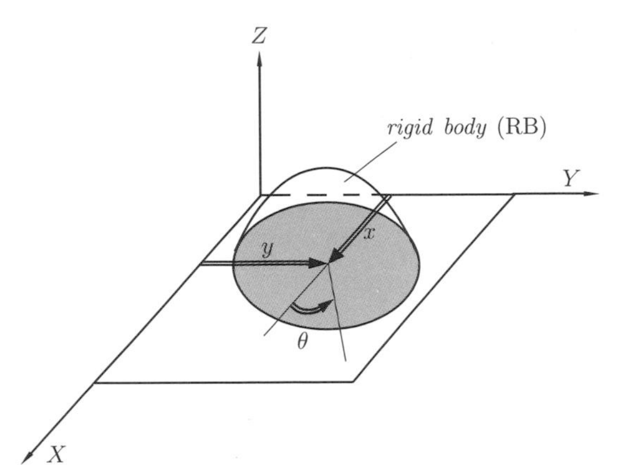

Figure 1.1. RB in planar motion with three DOF: translation along the x axis, translation along the y axis, and rotation, θ, about the z.

translation. For simplicity, only the two-dimensional (2-D) or planar case will be presented. For planar motion the following terms will be defined (see Fig. 1.2).

- Pure rotation is that in which the body possesses one point (center of rotation) that has no motion with respect to a "fixed" reference frame; see Fig. 1.2(a). All other points on the body describe arcs about that center.
- Pure translation is that in which all points on the body describe parallel paths; see Fig. 1.2(b).
- Complex motion is that which exhibits a simultaneous combination of rotation and translation; see Fig. 1.2(c). With general plane motion, points on the body will travel nonparallel paths, and there will be, at every instant, a center of rotation, which will continuously change location.

Translation and rotation represent independent motions of the body. Each can exist without the other. For a 2-D coordinate system, as shown in Fig. 1.1, the x and y terms represent the translation components of motion, and the θ term represents the rotation component.

1.3 Links and Joints

Linkages are basic elements of all mechanisms. Linkages are made up of links and joints. A *link*, sometimes known as an *element* or a *member*, is an (assumed) rigid body that possesses nodes. *Nodes* are defined as points at which links can be attached. A link connected to its neighboring elements by s nodes is an element of *degree* s. A link of degree 1 is also called unary, as in Fig. 1.3(a); that of degree 2, binary, as in Fig. 1.3(b); that of degree 3, ternary, as in Fig. 1.3(c); and so on.

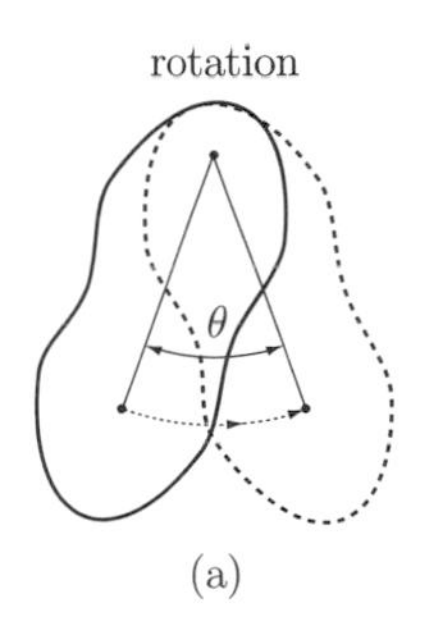

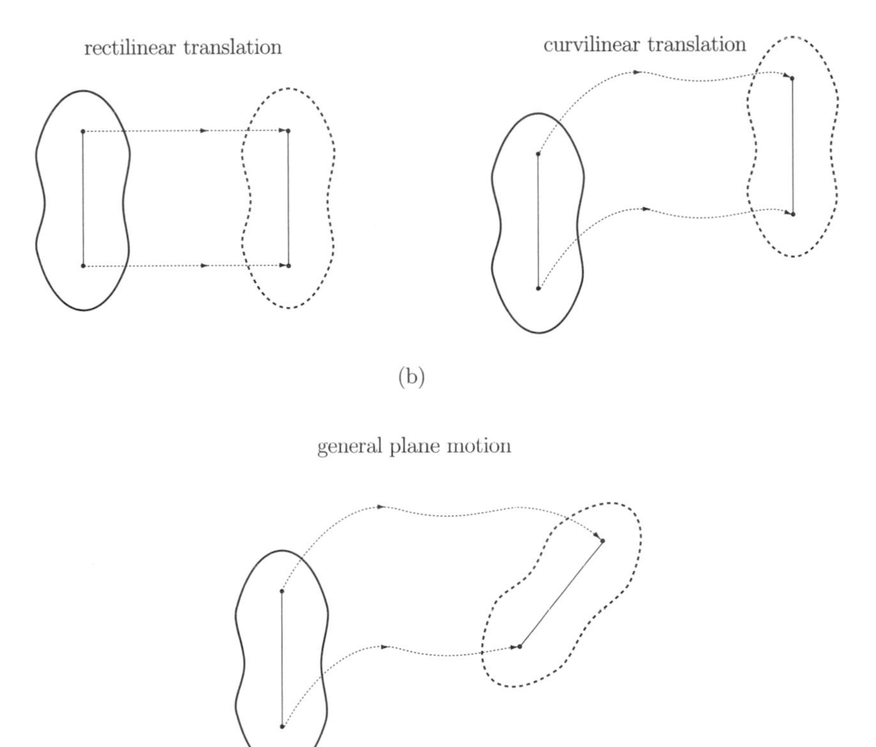

Figure 1.2. RB in motion: (a) pure rotation, (b) pure translation, and (c) general motion.

A *joint* is a connection between two or more links (at their nodes). A joint allows some relative motion between the connected links. Joints are also called *kinematic pairs*.

The number of independent coordinates that uniquely determine the relative position of two constrained links is termed the *degree of freedom* of a given joint. Alternatively the term *joint class* is introduced. A kinematic pair is of the jth class if it diminishes the relative motion of linked bodies by j

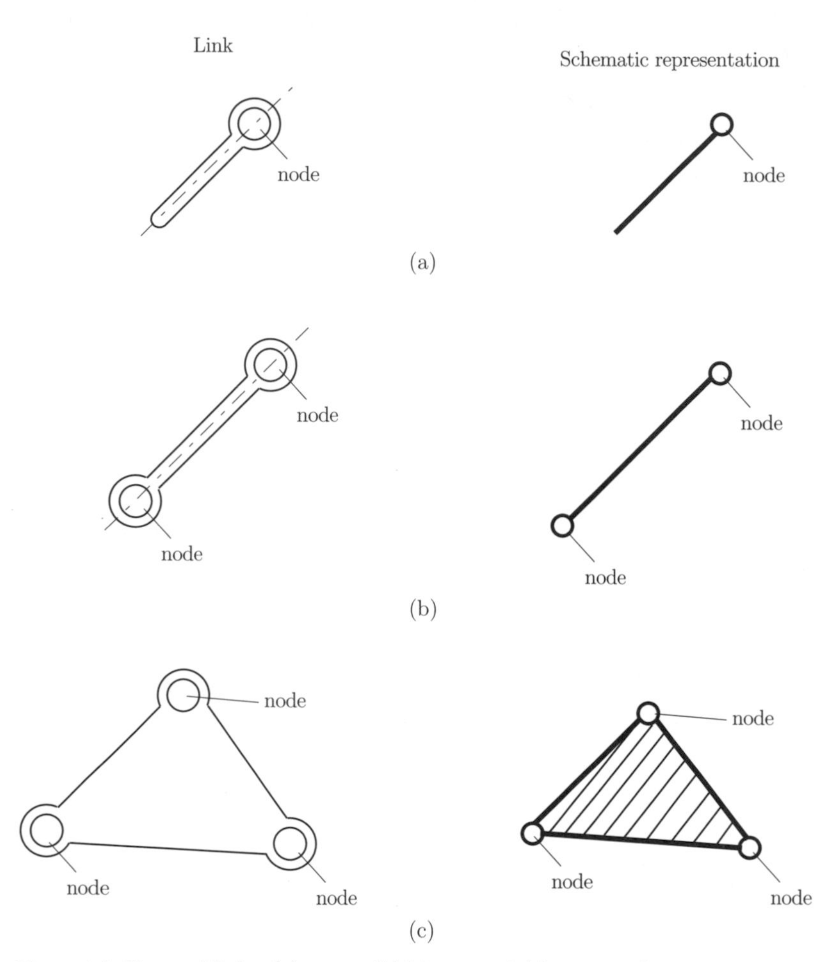

Figure 1.3. Types of links: (a) unary, (b) binary, and (c) ternary elements.

degrees of freedom; that is, j scalar constraint conditions correspond to the given kinematic pair. It follows that such a joint has $(6j)$ independent coordinates. The number of degrees of freedom is the fundamental characteristic quantity of joints. One of the links of a system is usually considered to be the reference link, and the position of other RBs is determined in relation to this reference body. If the reference link is stationary, the term *frame* or *ground* is used.

The coordinates in the definition of degree of freedom can be linear or angular. Also, the coordinates used can be absolute (measured with regard to the frame) or relative. Figures 1.4–1.9 show examples of joints commonly found in mechanisms. Figures 1.4(a) and 1.4(b) show two forms of a planar, one degree of freedom joint, namely a rotating pin joint and a translating slider

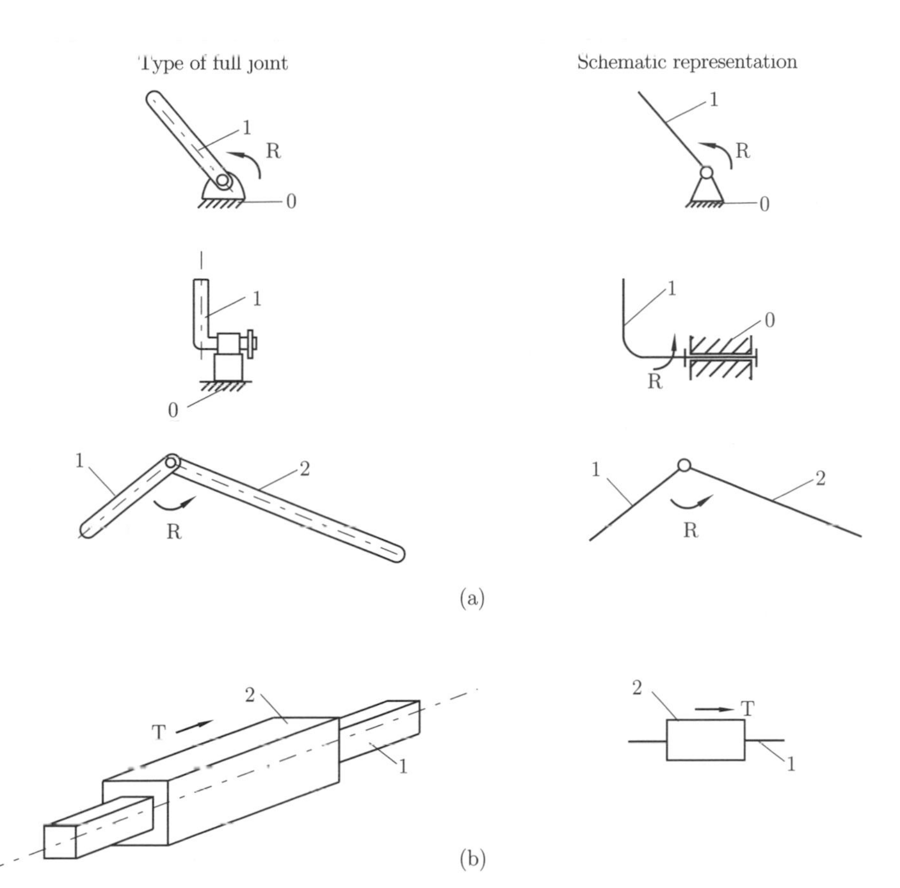

Figure 1.4. One degree of freedom joint, full joint (fifth class): (a) pin joint and (b) slider joint.

joint. These are both typically referred to as *full joints* and are of the fifth class. The pin joint allows one rotational (R) DOF, and the slider joint allows one translational (T) DOF between the joined links. These are both special cases of another common, one degree of freedom joint, the screw and nut, shown in Fig. 1.5(a). Motion of either the nut or the screw relative to the other results in helical motion. If the helix angle is made zero, Fig. 1.5(b), the nut rotates without advancing and it becomes a pin joint. If the helix angle is made 90°, the nut will translate along the axis of the screw, and it becomes a slider joint.

Figure 1.6 shows examples of two degrees of freedom joints, which simultaneously allow two independent, relative motions, namely translation (T) and rotation (R), between the joined links. A two degrees of freedom joint is usually referred to as a *half-joint* and is of the fourth class. A half-joint is sometimes also called a roll–slide joint because it allows both rotation (rolling) and translation (sliding).

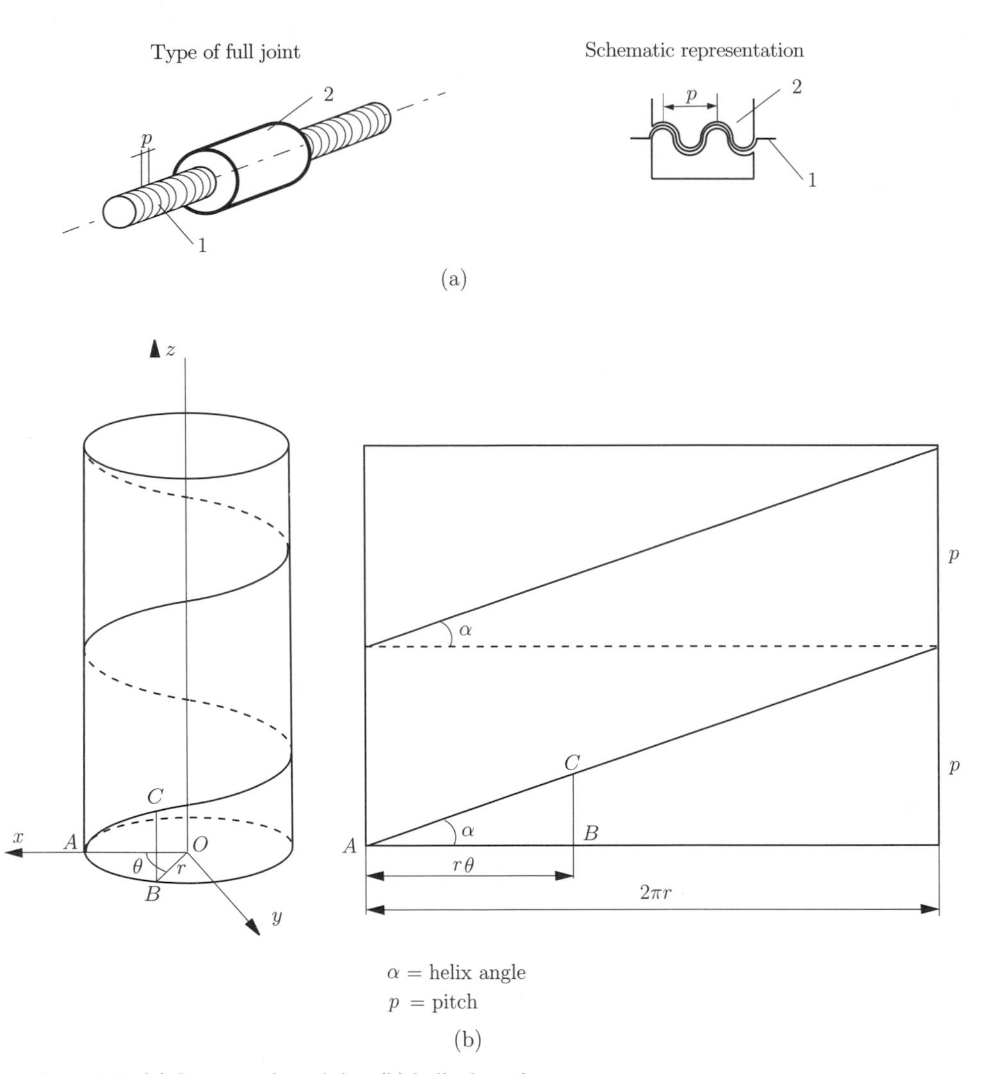

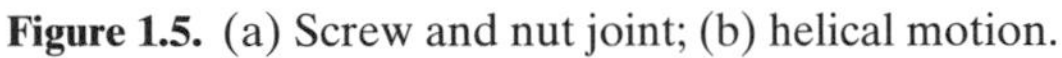

Figure 1.5. (a) Screw and nut joint; (b) helical motion.

A joystick, ball-and-socket joint, or sphere joint, shown in Fig. 1.7(a), is an example of a three degrees of freedom joint (third class), which allows three independent angular motions between the two links that are joined. This ball joint would typically be used in a 3-D mechanism, one example being the ball joints used in automotive suspension systems. A plane joint, Fig. 1.7(b), is also an example of a three degrees of freedom joint, which allows two translations and one rotation.

Note that in order to visualize the degree of freedom of a joint in a mechanism, it is helpful to "mentally disconnect" the two links that create the joint from the rest of the mechanism. It is easier to see how many degrees of

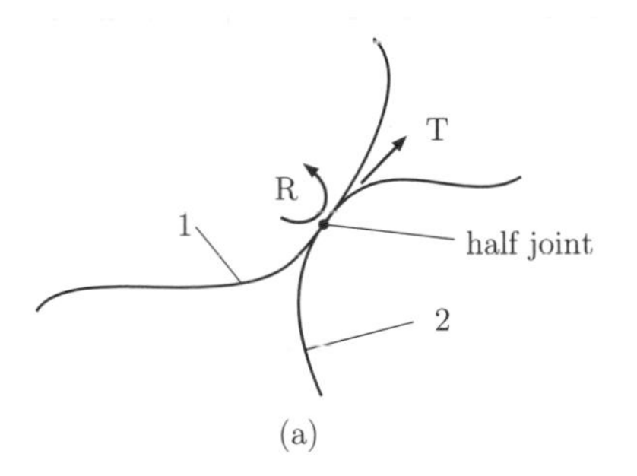

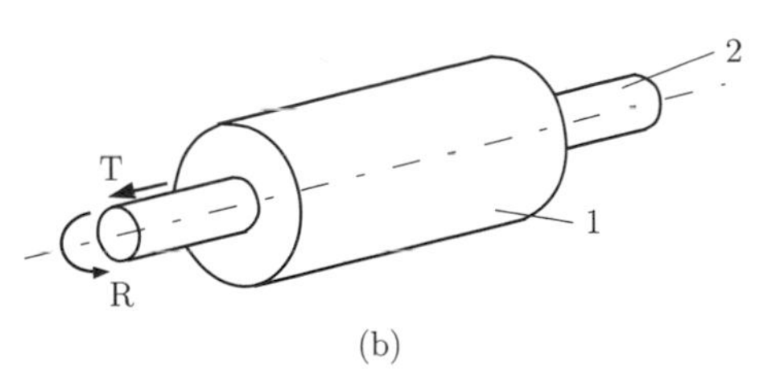

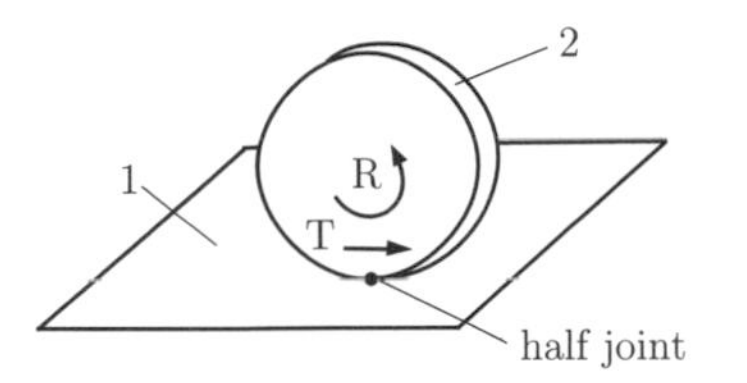

2
cam
1
follower
half joint
(d)

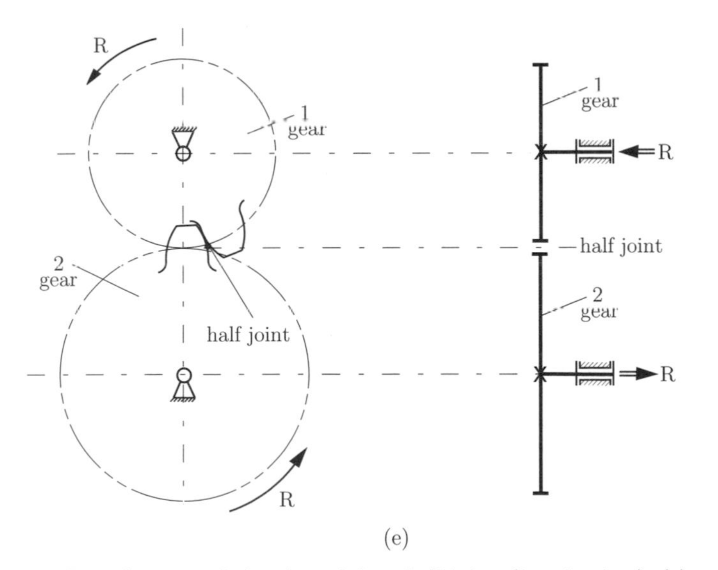

Figure 1.6. Two degrees of freedom joint, half-joint (fourth class): (a) general joint, (b) cylinder joint, (c) roll–slide disk, (d) cam-follower joint, and (e) gear joint.

freedoms the two joined links have with respect to one another. Figure 1.8 shows an example of a second class joint (cylinder on plane), and Fig. 1.9 represents a first class joint (sphere on plane).

The type of contact between the elements can be point (P), curve (C), or surface (S). The term *lower joint* was coined by Reuleaux to describe joints with

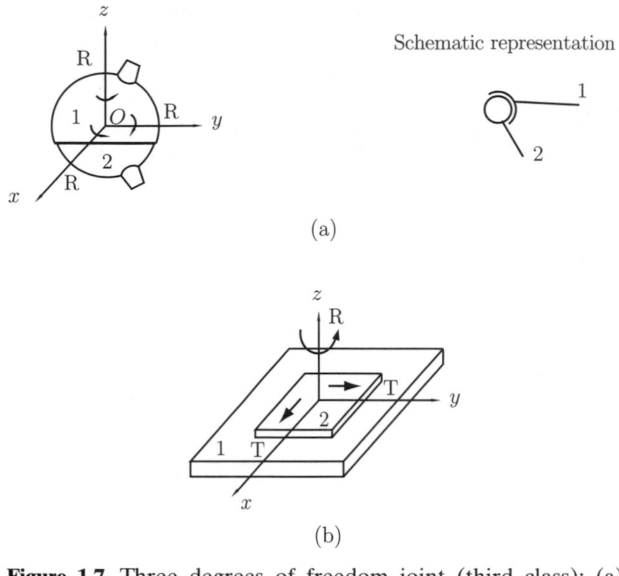

Figure 1.7. Three degrees of freedom joint (third class): (a) ball and socket joint, and (b) plane joint.

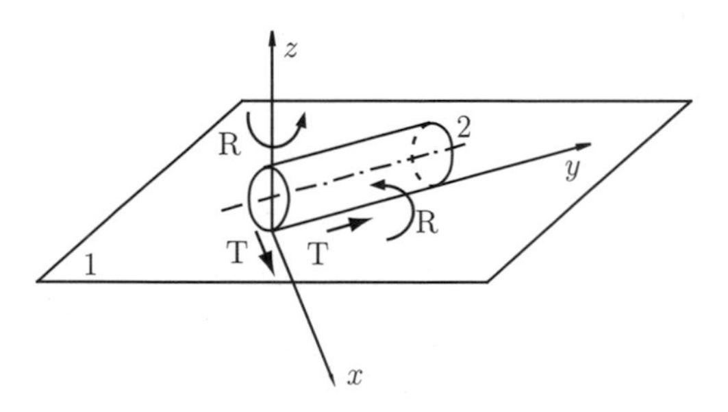

Figure 1.8. Four degrees of freedom joint (second class) cylinder on a plane.

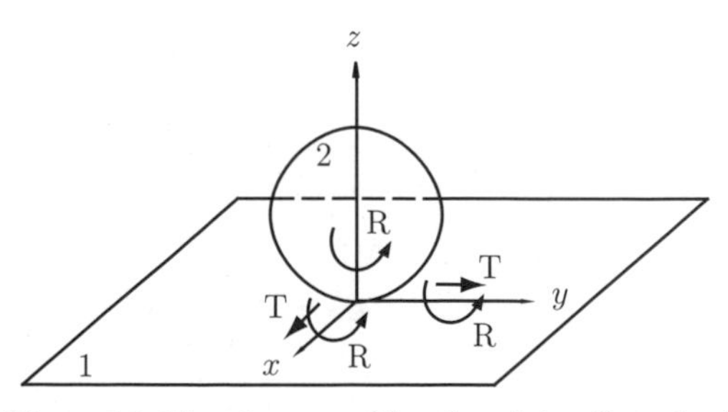

Figure 1.9. Five degrees of freedom joint (first class) sphere on a plane.

surface contact. He used the term *higher joint* to describe joints with point or curve contact. The main practical advantage of lower joints over higher joints is their better ability to trap lubricant between their enveloping surfaces. This is especially true for the rotating pin joint.

A *closed joint* is a joint that is kept together or closed by its geometry. A pin in a hole and a slider in a two-sided slot are forms of closed joints. A *force closed joint*, such as a pin in a half-bearing or a slider on a surface, requires some external force to keep it together or closed. This force could be supplied by gravity, by a spring, or by some external means. In linkages, closed joints are usually preferred, and they are easy to accomplish. For cam-follower systems, force closure is often preferred.

The *order of a joint* is defined as the number of links joined minus one. The simplest joint combination of two links has order one and it is a single joint, shown in Fig. 1.10(a). As additional links are placed on the same joint, the order is increased on a one for one basis, as shown in Fig. 1.10(b). Joint order

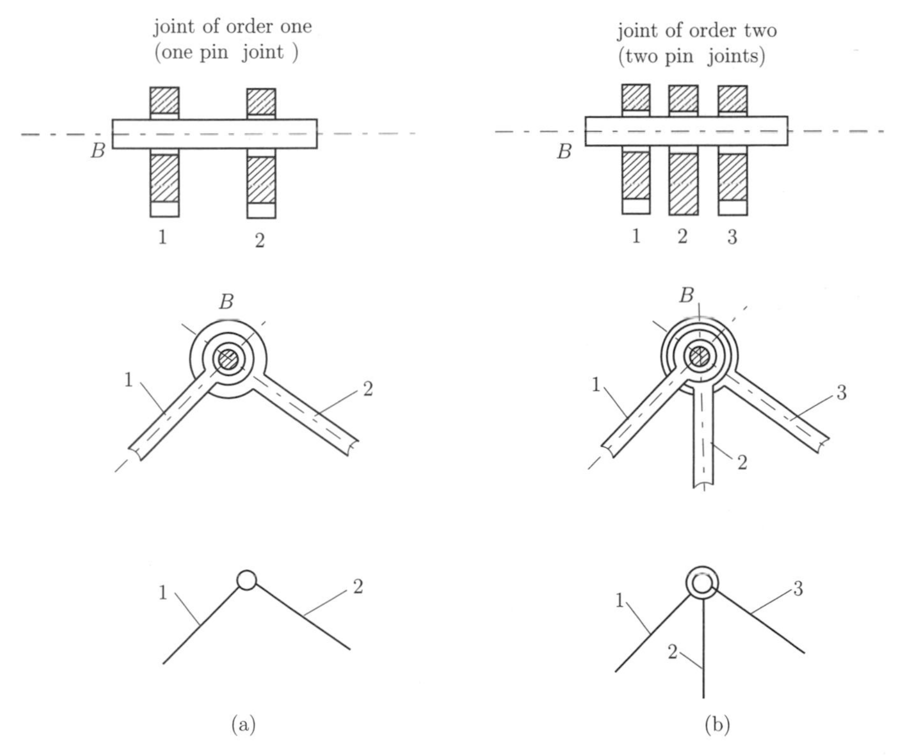

Figure 1.10. Order of a joint: (a) joint of order one, and (b) joint of order two (multiple joints).

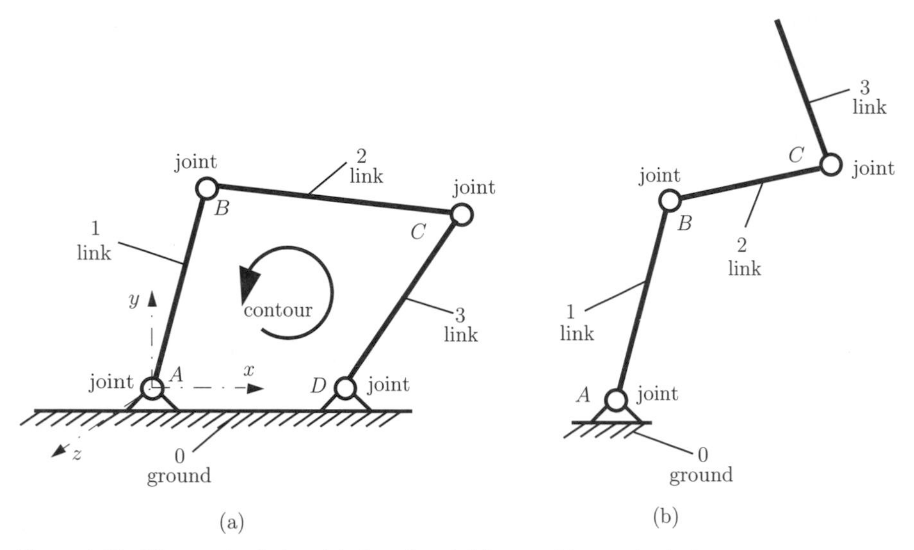

Figure 1.11. Kinematic chains: (a) closed and (b) open kinematic chains.

has significance in the proper determination of overall degrees of freedom for an assembly.

Bodies linked by joints form a *kinematic chain*. Simple kinematic chains are shown in Fig. 1.11.

A *contour* or *loop* is a configuration described by a polygon, as shown in Fig. 1.11(a).

The presence of loops in a mechanical structure can be used to define the following types of chains.

- *Closed kinematic chains* have one or more loops so that each link and each joint is contained in at least one of the loops, as shown in Fig. 1.11(a). A closed kinematic chain has no open attachment point.
- *Open kinematic chains* contain no loops, as shown in Fig. 1.11(b). A common example of an open kinematic chain is an industrial robot.
- *Mixed kinematic chains* are a combination of closed and open kinematic chains.

Another classification is also useful.

- *Simple chains* contain only binary elements.
- *Complex chains* contain at least one element of degree 3 or higher.

A *mechanism* is defined as a kinematic chain in which at least one link has been "grounded" or attached to the frame, as shown in Figs. 1.11(a) and 1.12. According to Reuleaux's definition, a *machine* is a collection of mechanisms

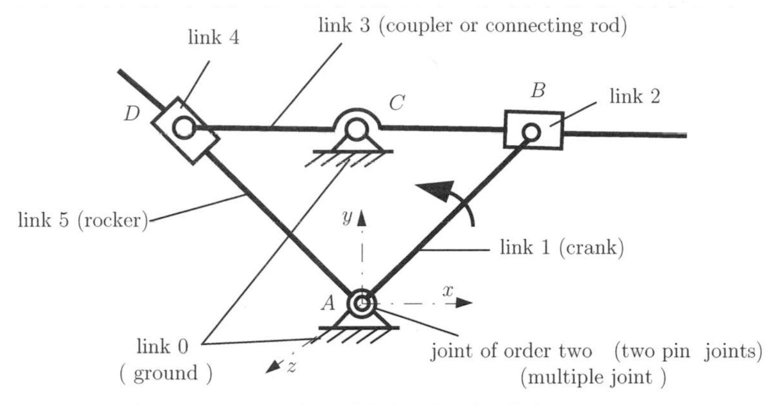

Figure 1.12. Complex mechanism with five moving links.

arranged to transmit forces and do work. He viewed all energy, or force-transmitting devices, as machines that utilize mechanisms, as their building blocks, to provide the necessary motion constraints.

The following terms can be defined (Fig. 1.12).

- A *crank* is a link that makes a complete revolution about a fixed grounded pivot.
- A *rocker* is a link that has oscillatory (back and forth) rotation and is fixed to a grounded pivot.
- A *coupler* or connecting rod is a link that has complex motion and is not fixed to ground.

Ground is defined as any link or links that are fixed (nonmoving) with respect to the reference frame. Note that the reference frame may in fact itself be in motion.

1.4 Number of Degrees of Freedom

The concept of *number of degrees of freedom* is fundamental to the analysis of mechanisms. It is usually necessary to be able to determine quickly the number of DOF of any collection of links and joints that may be used to solve a problem.

The number of degrees of freedom or the *mobility* of a system can be defined as

- the number of inputs that have to be provided in order to create a predictable system output, or
- the number of independent coordinates required to define the position of the system.

The *family* f of a mechanism is the number of DOF that are eliminated from all the links of the system.

Every free body in space has six degrees of freedom. A system of family f consisting of n movable links has $(6-f)n$ degrees of freedom. Each joint of class j diminishes the freedom of motion of the system by $j-f$ degrees of freedom. When the number of joints of class k is designated as c_k, it follows that the number of degrees of freedom of the particular system is

$$M = (6-f)n - \sum_{j=f+1}^{5} (j-f)c_j. \tag{1.1}$$

This is referred to in the literature on mechanisms as the Dobrovolski formula.

A *driver* link is that part of a mechanism that causes motion. An example is a crank. The number of driver links is equal to the number of DOF of the mechanism. A *driven* link or *follower* is that part of a mechanism whose motion is affected by the motion of the driver.

Mechanisms of Family $f = 1$

The family of a mechanism can be computed with the help of a mobility table (Table 1.1). Consider the mechanism, shown in Fig. 1.13, which can be used to measure the weight of postal envelopes. The translation along the i axis is denoted by T_i, and the rotation about the i axis is denoted by R_i, where $i = x, y, z$. Every link in the mechanism is analyzed in terms of its translation and rotation about the reference frame xyz. For example, link 0 (ground) has no translations, T_i = No, and no rotations, R_i = No. Link 1 has a rotation motion about the z axis, R_z = Yes. Link 2 has a planar motion (xy is the plane of motion) with a translation along the x axis, T_x = Yes, a translation along the y axis,

Table 1.1. Mobility Table for the Mechanism Shown in Fig. 1.13

Link	T_x	T_y	T_z	R_x	R_y	R_z
0	No	No	No	No	No	No
1	No	No	No	No	No	Yes
2	Yes	Yes	No	No	No	Yes
3	No	Yes	No	No	No	No
4	No	Yes	Yes	Yes	No	No
5	No	No	No	Yes	No	No
					No	

For all links R_y = No $\Longrightarrow f = 1$.

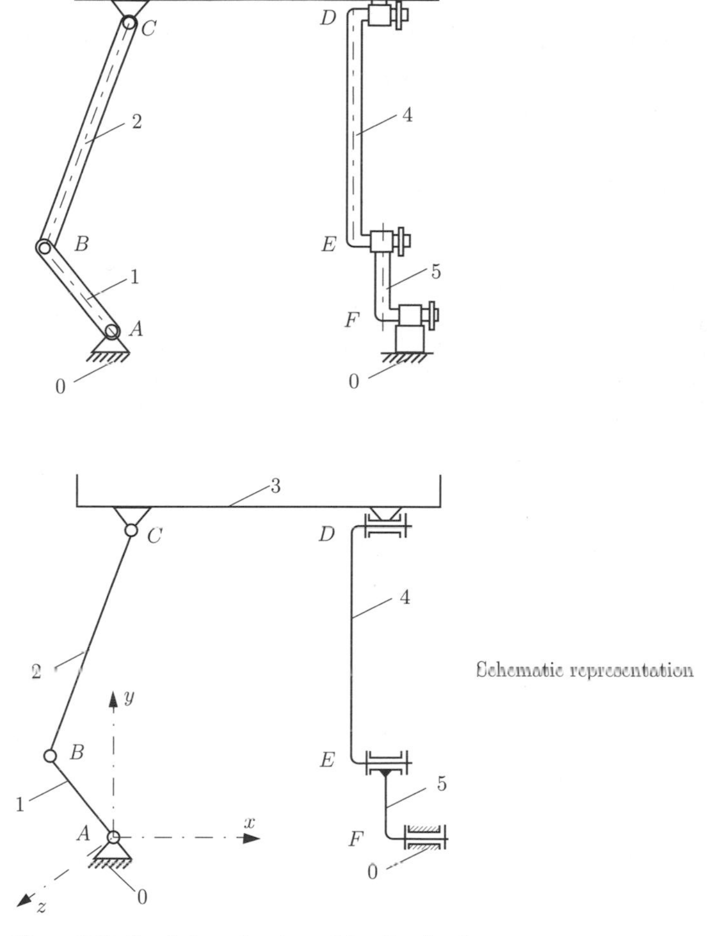

Figure 1.13. Spatial mechanism of family $f = 1$.

T_y = Yes, and a rotation about the z axis, R_z = Yes. Link 3 has a translation along y, T_y = Yes. Link 4 has a planar motion (yz the plane of motion) with a translation along y, T_y = Yes, a translation along z, T_z = Yes, and a rotation about x, R_x = Yes. Link 5 has a rotation about x axis, R_x = Yes. The results of this analysis are presented with the help of the mobility table give in Table 1.1. From the mobility table it can be seen that link i, $i = 0, 1, 2, 3, 4, 5$ has no rotation about the y axis; that is, there is no rotation about the y axis for any of the links of the mechanism (R_y = No). The family of the mechanism is $f = 1$ because there is one DOF, rotation about y, which is eliminated from all the links.

There are six joints of class 5 (rotational joints) in the system at A, B, C, D, E, and F.

The number of DOF for the mechanism in Fig. 1.13, which is of the $f = 1$ family, is given by

$$M = 5n - \sum_{j=2}^{5}(j-1)c_j = 5n - 4c_5 - 3c_4 - 2c_3 - c_2$$
$$= 5(5) - 4(6) = 1.$$

The mechanism has one DOF and one driver link.

Mechanisms of Family $f = 2$

A mobility table for a mechanism of family $f = 2$, Fig. 1.14, is given in Table 1.2.

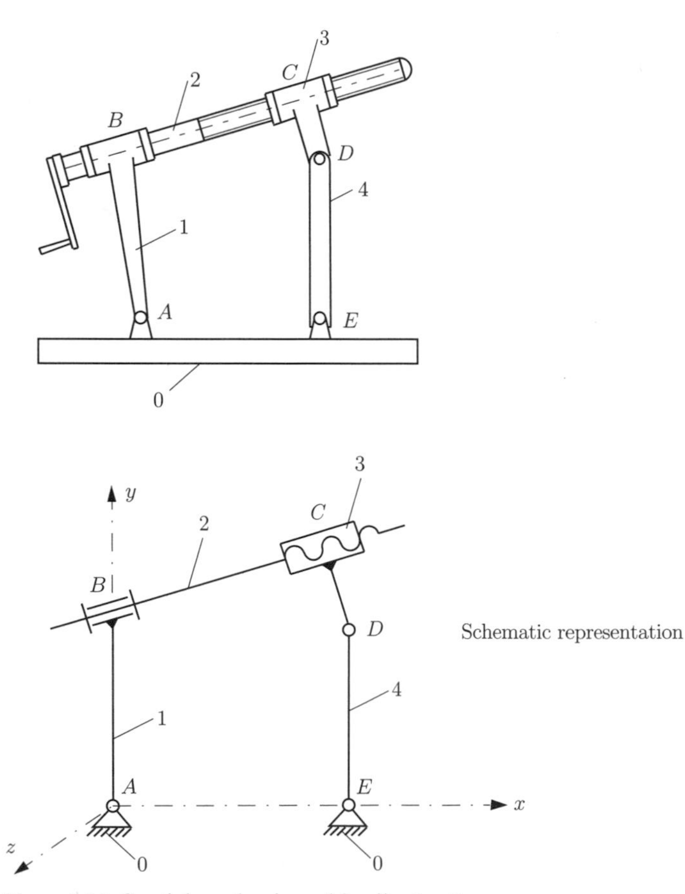

Figure 1.14. Spatial mechanism of family $f = 2$.

Table 1.2. Mobility Table for the Mechanism Shown in Fig. 1.14

Link	T_x	T_y	T_z	R_x	R_y	R_z
0	No	No	No	No	No	No
1	No	No	No	No	No	Yes
2	Yes	Yes	No	Yes	No	Yes
3	Yes	Yes	No	No	No	Yes
4	No	No	No	No	No	Yes
			No		No	

For all Links T_z = No and R_y = No $\Longrightarrow f = 2$.

The number of DOF for the $f = 2$ family mechanism is given by

$$M = 4n - \sum_{j=3}^{5}(j-2)c_j = 4n - 3c_5 - 2c_4 - c_3.$$

The mechanism in Fig. 1.14 has four moving links ($n = 4$), four rotational joints (A, B, D, E), and one screw and nut joint (C); that is, there are five joints of class 5 ($c_5 = 5$). The number of DOF for this mechanism is

$$M = 4n - 3c_5 - 2c_4 - c_3 = 4(4) - 3(5) = 1.$$

Mechanisms of Family $f = 3$

The number of DOF for mechanisms of family $f = 3$ is given by

$$M = 3n - \sum_{j=4}^{5}(j-3)c_j = 3n - 2c_5 - c_4.$$

For the mechanism in Fig. 1.11(a), the mobility table is given in Table 1.3.

Table 1.3. Mobility Table for the Mechanism Shown in Fig. 1.11(a)

Link	T_x	T_y	T_z	R_x	R_y	R_z
0	No	No	No	No	No	No
1	No	No	No	No	No	Yes
2	Yes	Yes	No	No	No	Yes
3	No	No	No	No	No	Yes
			No	No	No	

For all links T_z = No, R_x = No, and R_y = No $\Longrightarrow f = 3$.

Table 1.4. Mobility Table for the Mechanism Shown in Fig. 1.12

Link	T_x	T_y	T_z	R_x	R_y	R_z
0	No	No	No	No	No	No
1	No	No	No	No	No	Yes
2	Yes	Yes	No	No	No	Yes
3	No	No	No	No	No	Yes
4	Yes	Yes	No	No	No	Yes
5	No	No	No	No	No	Yes
			No	No	No	

For all links T_z = No, R_x = No, and R_y = No $\Longrightarrow f = 3$.

The mechanism in Fig. 1.11(a) has three moving links ($n = 3$) and four rotational joints at A, B, C, and D ($c_5 = 4$). The number of DOF for this mechanism is given by

$$M = 3n - 2c_5 - c_4 = 3(3) - 2(4) = 1.$$

The mobility table for the mechanism shown in Fig. 1.12 is given in Table 1.4. There are seven joints of class 5 ($c_5 = 7$) in the system.

- At A there is one rotational joint between link 0 and link 1.
- At B there is one rotational joint between link 1 and link 2.
- At B there is one translational joint between link 2 and link 3.
- At C there is one rotational joint between link 0 and link 3.
- At D there is one rotational joint between link 3 and link 4.
- At D there is one translational joint between link 4 and link 5.
- At A there is one rotational joint between link 5 and link 0.

The number of moving links is five ($n = 5$). The number of DOF for this mechanism is given by

$$M = 3n - 2c_5 - c_4 = 3(5) - 2(7) = 1,$$

and this mechanism has one driver link.

Mechanisms of Family $f = 4$

The number of DOF for mechanisms of family $f = 4$ is given by

$$M = 2n - \sum_{j=5}^{5}(j - 4)c_j = 2n - c_5.$$

For the mechanism shown in Fig. 1.15 the mobility table is given in Table 1.5.

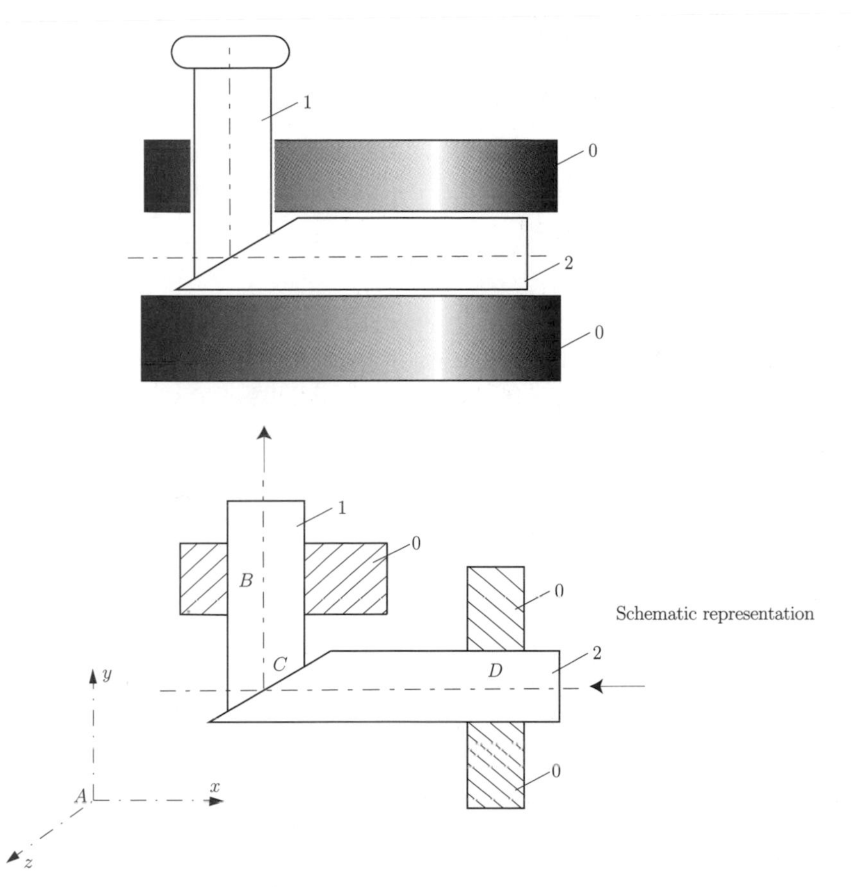

Figure 1.15. Spatial mechanism of family $f = 4$.

There are three translational joints of class 5 ($c_5 = 3$) in the system.

- At B there is one translational joint between link 0 and link 1.
- At C there is one translational joint between link 1 and link 2.
- At D there is one translational joint between link 2 and link 0.

Table 1.5. Mobility Table for the Mechanism Shown in Fig. 1.15

Link	T_x	T_y	T_z	R_x	R_y	R_z
0	No	No	No	No	No	No
1	No	Yes	No	No	No	No
2	Yes	No	No	No	No	No
			No	No	No	No

For all links T_z = No, R_x = No, R_y = No, and R_z = No $\Longrightarrow f = 4$.

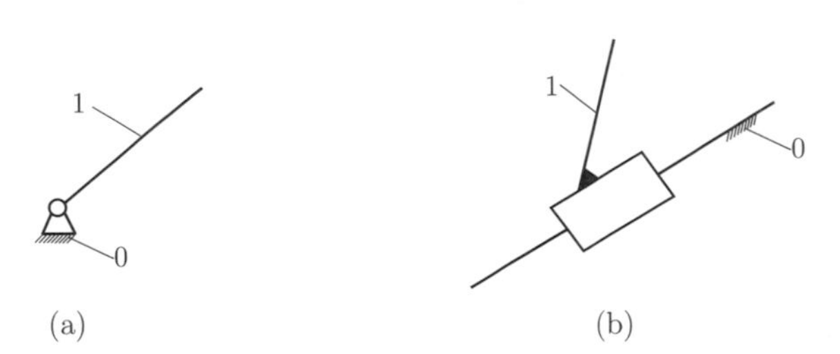

Figure 1.16. Spatial mechanism of family $f = 5$: driver link with (a) rotational and (b) translational motions.

The number of DOF for this mechanism with two moving links ($n = 2$) is given by

$$M = 2n - c_5 = 2(2) - (3) = 1.$$

Mechanisms of Family $f = 5$

The number of DOF for mechanisms of family $f = 5$ is equal to the number of moving links:

$$M = n.$$

The driver link with rotational motion, Fig. 1.16(a), and the driver link with translational motion, Fig. 1.16(b), are in the $f = 5$ category.

1.5 Planar Mechanisms

For the special case of planar mechanisms ($f = 3$), Eq. (1.1) has the form

$$M = 3n - 2c_5 - c_4, \tag{1.2}$$

where c_5 is the number of full joints and c_4 is the number of half-joints.

Kinematic chains that do not change their freedom of motion (mobility) after being connected to an arbitrary system have special significance. Kinematic chains that can be defined in this way are called *system groups*. If the system groups are connecting to or disconnecting from a given system, it enables such a system to be modified or new structural systems to be created, but at the same time maintains the original freedom of motion. The term *system group* has been used by Assur for the classification of planar mechanisms and investigated further by Artobolevskij. Consider planar systems containing only kinematic pairs of class 2; from Eq. (1.2) the following relationship can be obtained:

$$3n - 2c_5 = 0. \tag{1.3}$$

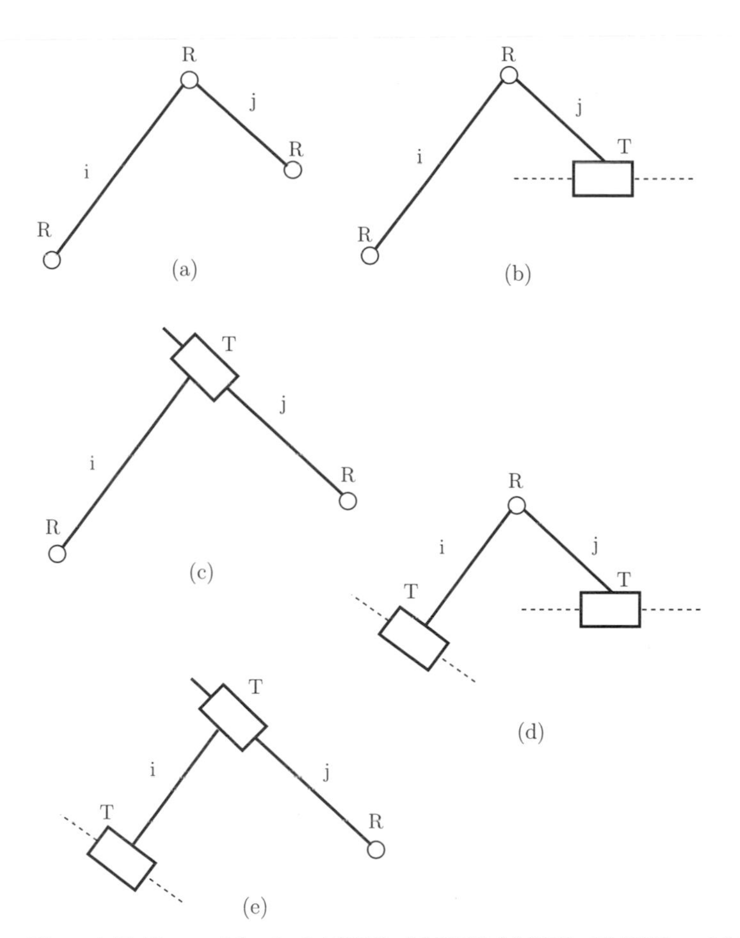

Figure 1.17. Types of dyads: (a) RRR, (b) RRT, (c) RTR, (d) TRT, and (e) TTR.

According to Eq. (1.3), the number of system group links, n, is always even. The simplest such system is the binary group with two links ($n = 2$) and three full joints ($c_5 = 3$). The binary group is called also a *dyad*. The sets of links shown in Fig. 1.17 are dyads, and one can distinguished the following types:

- rotation rotation rotation type (dyad RRR), Fig. 1.17(a);
- rotation rotation translation type (dyad RRT), Fig. 1.17(b);
- rotation translation rotation type (dyad RTR), Fig. 1.17(c);
- translation rotation translation type (dyad TRT), Fig. 1.17(d);
- translation translation rotation type (dyad TTR), Fig. 1.17(e).

The advantage of classifying systems into groups is the resulting simplicity. The solution of the whole system can then be obtained by forming solutions of its parts.

The mechanism shown in Fig. 1.11(a) has one driver, link 1, with rotational motion and one dyad, RRR, composed of link 2 and link 3.

The mechanism shown in Fig. 1.12 is obtained by forming a system with the following:

- the driver link 1, which has a rotational motion;
- the dyad RTR composed of links 2 and 3, and joints B having rotation (B_R), B having translation (B_T), and C having rotation (C_R);
- the dyad RTR composed of links 4 and 5, and joints D having rotation (D_R), D having translation (D_T), and A having rotation (A_R).

2 Position Analysis

2.1 Absolute Cartesian Method

The position analysis of a kinematic chain requires the determination of the joint positions and/or the position of the center of gravity of the link. The position, velocity, and acceleration of a point are specified, in general, relative to an arbitrary reference frame. Newton's second law can be expressed in terms of a reference frame that is fixed relative to the earth. A reference frame in which Newton's second law can be used is said to be *Newtonian*, or an *inertial* reference frame. Normally an inertial reference frame is assumed in this book.

A planar link with end nodes A and B is considered in Fig. 2.1. Let (x_A, y_A) be the coordinates of joint A with respect to the reference frame xOy, and let (x_B, y_B) be the coordinates of joint B with the same reference frame. With the use of the Pythagoras theorem, the following relation can be written:

$$(x_B - x_A)^2 + (y_B - y_A)^2 = AB^2 = L_{AB}^2, \tag{2.1}$$

where L_{AB} is the length of link AB.

Let ϕ be the angle of link AB with the horizontal axis Ox. Then, the slope m of link AB is defined as

$$m = \tan\phi = (y_B - y_A)/(x_B - x_A). \tag{2.2}$$

Let n be the intercept of AB with the vertical axis Oy. With the use of slope m and the intercept n, the equation of the straight link, in the plane, is

$$y = mx + n, \tag{2.3}$$

where x and y are the coordinates of any point on this link.

For a link with a translational joint, the sliding direction (Δ) is given by the equation (Fig. 2.2)

$$x\cos\alpha - y\sin\alpha - p = 0, \tag{2.4}$$

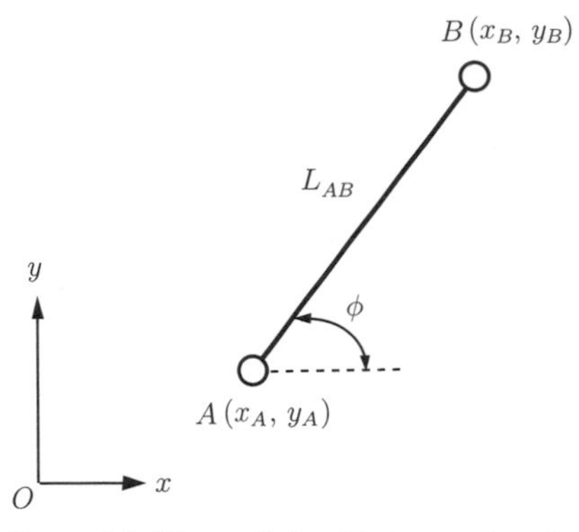

Figure 2.1. Planar link with two end nodes A and B.

where p is the distance from the origin O to the sliding line (Δ). The position function for joint $A(x_A, y_A)$ is

$$x_A \cos\alpha - y_A \sin\alpha - p = \pm d, \tag{2.5}$$

where d is the distance from A to the sliding line. The relation between joint A and a point B on the sliding direction, $B \in (\Delta)$, where the symbol $\in$ means belongs to, is

$$(x_A - x_B)\sin\beta - (y_A - y_B)\cos\beta = \pm d, \tag{2.6}$$

where $\beta = \alpha + (\pi/2)$.

If $Ax + By + C = 0$ is the linear equation of the line (Δ), then the distance d is (Fig. 2.2)

$$d = \frac{|Ax_A + By_A + C|}{\sqrt{A^2 + B^2}}. \tag{2.7}$$

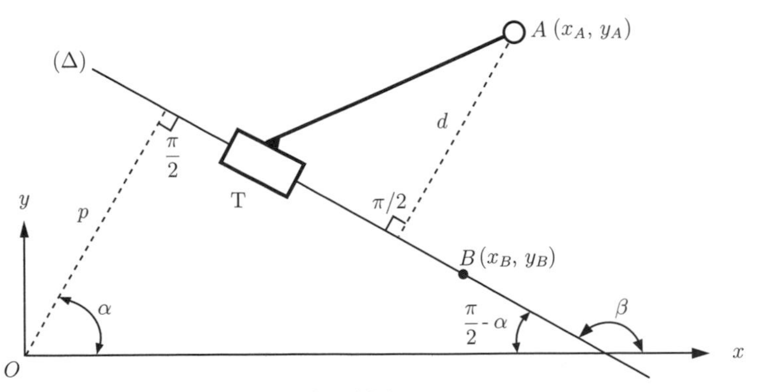

Figure 2.2. Link with a translational joint.

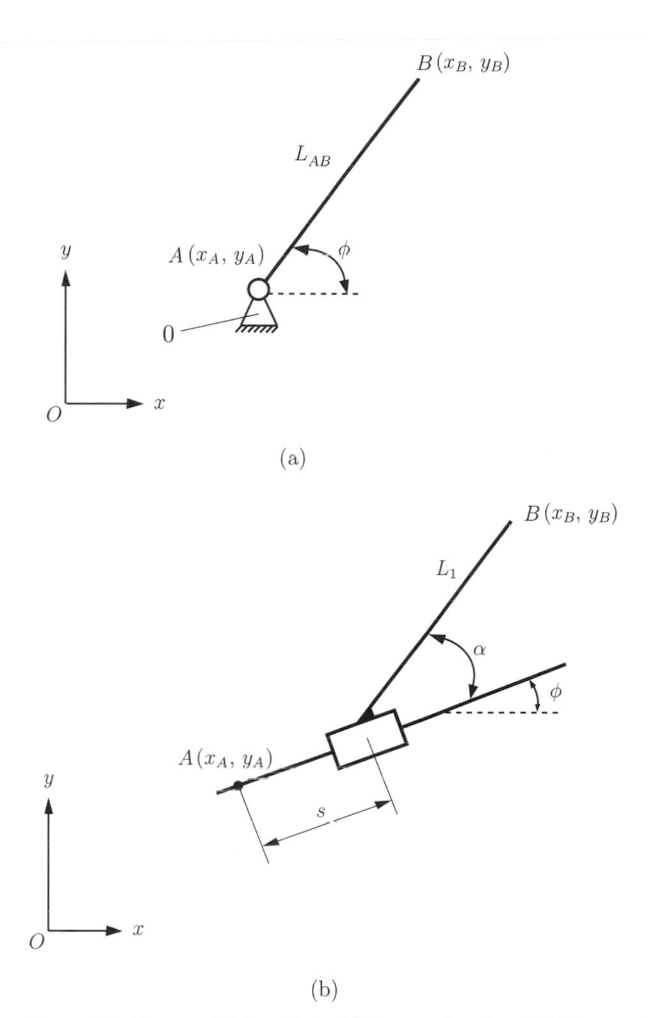

Figure 2.3. Types of driver link: (a) in rotational and (b) in translational motion.

For a driver link in rotational motion, shown in Fig. 2.3(a), the following relations can be written:

$$x_B = x_A + L_{AB} \cos\phi,$$
$$y_B = y_A + L_{AB} \sin\phi. \tag{2.8}$$

From Fig. 2.3(b), for a driver link in translational motion, one can write the equations

$$x_B = x_A + s\cos\phi + L_1 \cos(\phi + \alpha),$$
$$y_B = y_A + s\sin\phi + L_1 \sin(\phi + \alpha). \tag{2.9}$$

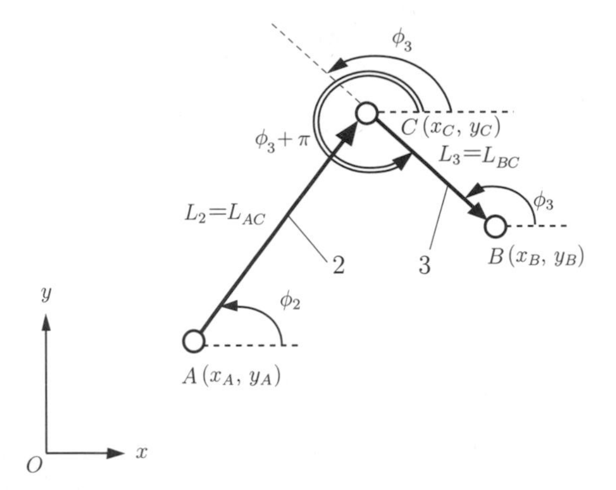

Figure 2.4. RRR dyad.

For the RRR dyad there are two quadratic equations of the form (Fig. 2.4)

$$(x_A - x_C)^2 + (y_A - y_C)^2 = AC^2 = L_{AC}^2 = L_2^2,$$
$$(x_B - x_C)^2 + (y_B - y_C)^2 = BC^2 = L_{BC}^2 = L_3^2, \tag{2.10}$$

where the coordinates of the joint C, x_C and y_C, are the unknowns. With x_C and y_C determined, the angles ϕ_2 and ϕ_3 are computed from the relations

$$y_C - y_A - (x_C - x_A)\tan\phi_2 = 0,$$
$$y_C - y_B - (x_C - x_B)\tan\phi_3 = 0. \tag{2.11}$$

The following relations can be written for the RRT dyad, shown in Fig. 2.5(a):

$$(x_A - x_C)^2 + (y_A - y_C)^2 = AC^2 = L_{AC}^2 = L_2^2,$$
$$(x_C - x_B)\sin\alpha - (y_C - y_B)\cos\alpha = \pm h. \tag{2.12}$$

From these two equations the two unknowns, x_C and y_C, are computed. Figure 2.5(b) depicts the particular case for the RRT dyad when $L_3 = h = 0$.

For the RTR dyad, Fig. 2.6(a), the known data are: the positions of joints A and B, x_A, y_A, x_B, and y_B; the angle α; and the length L_2 ($h = L_2 \sin\alpha$). There are four unknowns in the position of $C(x_C, y_C)$ and in the equation for the sliding line (Δ): $y = mx + n$. The unknowns in the sliding line m and n can be computed from the relations

$$L_2 \sin\alpha = \frac{|mx_A - y_A + n|}{\sqrt{m^2+1}},$$
$$y_B = mx_B + n. \tag{2.13}$$

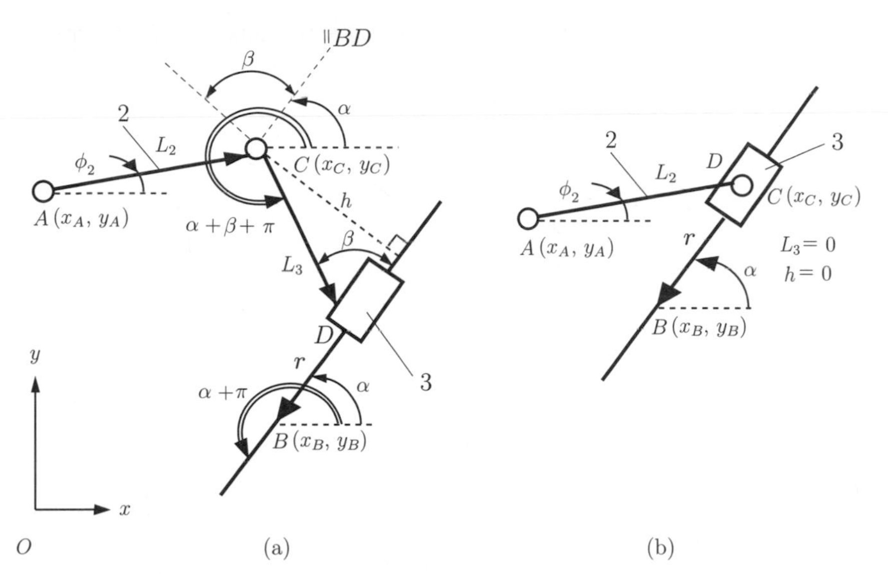

Figure 2.5. (a) RRT dyad; (b) RRT dyad, particular case $L_3 = h = 0$.

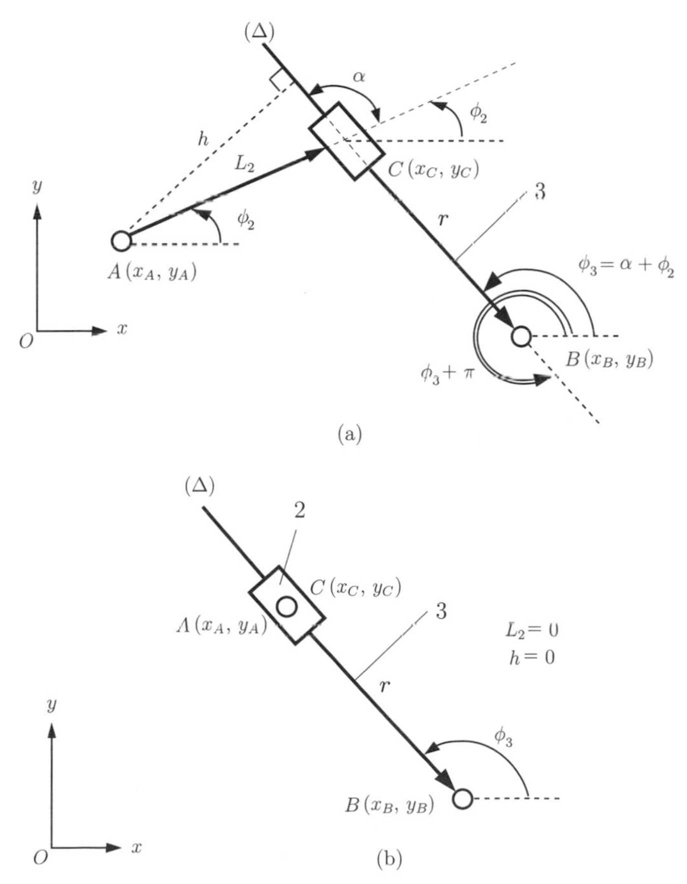

Figure 2.6. (a) RTR dyad; (b) RTR dyad, particular case $L_2 = h = 0$.

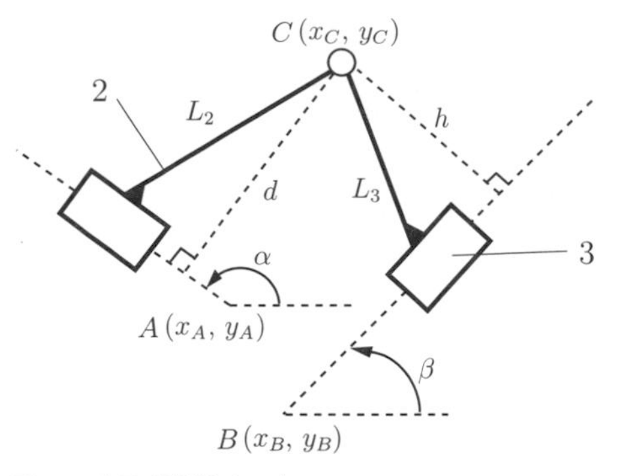

Figure 2.7. TRT dyad.

The coordinates of joint C can be obtained by using the equations

$$(x_A - x_C)^2 + (y_A - y_C)^2 = L_2^2,$$
$$y_C = m x_C + n. \qquad (2.14)$$

In Fig. 2.6(b) the particular case, when $L_2 = h = 0$, is shown.

To compute the coordinates of joint C for the TRT dyad, shown in Fig. 2.7, two equations can be written:

$$(x_C - x_A)\sin\alpha - (y_C - y_A)\cos\alpha = \pm d,$$
$$(x_C - x_B)\sin\beta - (y_C - y_B)\cos\beta = \pm h. \qquad (2.15)$$

The input data are $x_A, y_A, x_B, y_B, \alpha, \beta, d$, and h, and the output data are x_C, y_C.

EXAMPLE

Determine the positions of the joints of the mechanism shown in Fig. 2.8. The known elements are $AB = l_1, CD = l_3, CE = l_4$, and $AD = d_1$, and h is the distance from slider 5 to the horizontal axis Ax.

The origin of the system is at A, $A \equiv O$; that is, $x_A = y_A = 0$. The coordinates of the R joint at B are

$$x_B = l_1 \cos\phi, \qquad y_B = l_1 \sin\phi.$$

For the dyad DBB (RTR), the following equations can be written with respect to the sliding line CD:

$$m x_B - y_B + n = 0, \qquad y_D = m x_D + n.$$

With $x_D = d_1$ and $y_D = 0$ from the above system, slope m of link CD and intercept n can be calculated:

$$m = \frac{l_1 \sin\phi}{l_1 \cos\phi - d_1}, \qquad n = \frac{d_1 l_1 \sin\phi}{d_1 - l_1 \cos\phi}.$$

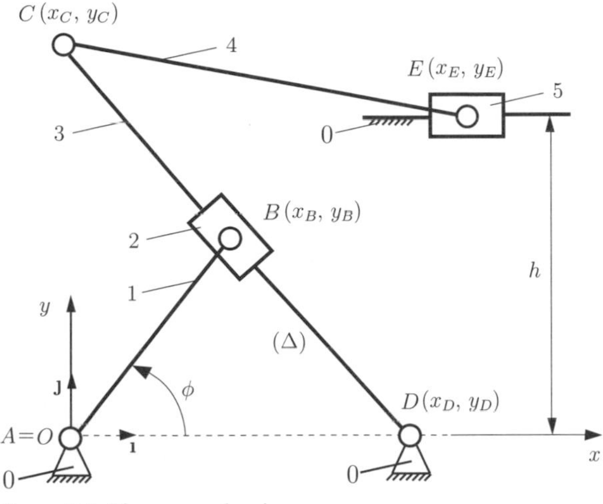

Figure 2.8. Planar mechanism.

The coordinates x_C and y_C of the center of the R joint C result from the system of two equations:

$$y_C = \frac{l_1 \sin\phi}{l_1 \cos\phi - d_1} x_C + \frac{d_1 l_1 \sin\phi}{d_1 - l_1 \cos\phi}, \quad (x_C - x_D)^2 + (y_C - y_D)^2 = l_3^2.$$

Because of the quadratic equation, two solutions are obtained for x_C and y_C. For continuous motion of the mechanism, there are constraint relations for the choice of the correct solution; that is, $x_C < x_B < x_D$ and $y_C > 0$.

For the last dyad CEE (RRT), a position function can be written for joint E:

$$(x_C - x_E)^2 + (y_C - h)^2 = l_4^2.$$

The equation produces values for x_{E1} and x_{E2}, and the solution $x_E > x_C$ is selected for continuous motion of the mechanism.

2.2 Vector Loop Method

First, the independent closed loops are identified. A vector equation corresponding to each independent loop is established. The vector equation gives rise to two scalar equations, one for the horizontal axis x, and another for the vertical axis y.

For an open kinematic chain, shown in Fig. 2.9, with general joints (pin joints, slider joints, etc.), a vector loop equation can be considered:

$$\mathbf{r}_A + \mathbf{r}_1 + \cdots + \mathbf{r}_n = \mathbf{r}_B, \tag{2.16}$$

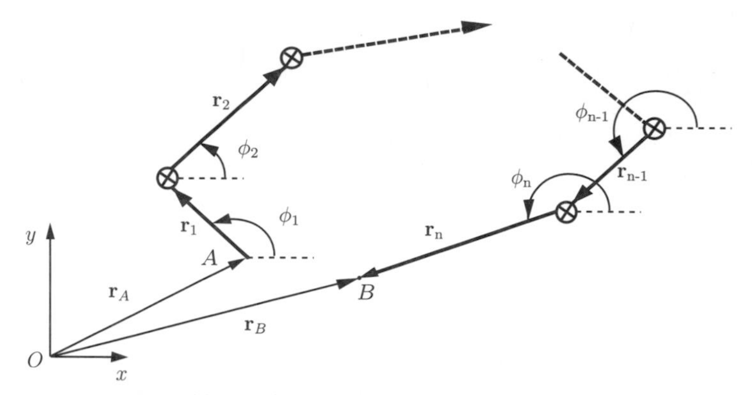

Figure 2.9. Open kinematic chain with general joints.

or

$$\sum_{k=1}^{n} \mathbf{r}_k = \mathbf{r}_B - \mathbf{r}_A. \tag{2.17}$$

The vectorial Eq. (2.17) can be projected on the reference frame xOy:

$$\sum_{k=1}^{n} r_k \cos\phi_k = x_B - x_A,$$

$$\sum_{k=1}^{n} r_k \sin\phi_k = y_B - y_A. \tag{2.18}$$

RRR Dyad

The input data are as follows: the position of A is (x_A, y_A), the position of B is (x_B, y_B), the length of AC is $L_{AC} = L_2$, and the length of BC is $L_{BC} = L_3$; see Fig. 2.4. The unknown data are as follows: the position of $C(x_C, y_C)$, and the angles ϕ_2 and ϕ_3. The position equation for the RRR dyad is $\mathbf{AC} + \mathbf{CB} = \mathbf{r}_B - \mathbf{r}_A$, or

$$L_2 \cos\phi_2 + L_3 \cos(\phi_3 + \pi) = x_B - x_A,$$
$$L_2 \sin\phi_2 + L_3 \sin(\phi_3 + \pi) = y_B - y_A. \tag{2.19}$$

The angles ϕ_2 and ϕ_3 can be computed from Eq. (2.19). The position of C can be computed by using the known angle ϕ_2:

$$x_C = x_A + L_2 \cos\phi_2,$$
$$y_C = y_A + L_2 \sin\phi_2. \tag{2.20}$$

RRT Dyad

The input data are as follows: the position of A is (x_A, y_A), the position of B is (x_B, y_B), the length of AC is L_2, the length of CB is L_3, and the angles α and β are constants; see Fig. 2.5(a). The unknown data are as follows: the position of $C(x_C, y_C)$, the angle ϕ_2, and the distance $r = DB$.

The vectorial equation for this kinematic chain is $\mathbf{AC} + \mathbf{CD} + \mathbf{DB} = \mathbf{r}_B - \mathbf{r}_A$, or

$$L_2 \cos\phi_2 + L_3 \cos(\alpha + \beta + \pi) + r \cos(\alpha + \pi) = x_B - x_A,$$
$$L_2 \sin\phi_2 + L_3 \sin(\alpha + \beta + \pi) + r \sin(\alpha + \pi) = y_B - y_A. \tag{2.21}$$

One can compute r and ϕ_2 from Eq. (2.21). The position of C can be found from Eq. (2.20).

PARTICULAR CASE $L_3 = 0$

See Fig. 2.5(b). In this case Eq. (2.21) can be written as

$$L_2 \cos\phi_2 + r \cos(\alpha + \pi) = x_B - x_A,$$
$$L_2 \sin\phi_2 + r \sin(\alpha + \pi) = y_B - y_A. \tag{2.22}$$

RTR Dyad

The input data are as follows: the position of A is (x_A, y_A), the position of B is (x_B, y_B), the length of AC is L_2, and the angle α is constant, as shown in Fig. 2.6(a). The unknown data are as follows: the distance $r = CB$ and the angles ϕ_2 and ϕ_3.

The vectorial loop equation can be written as $\mathbf{AC} + \mathbf{CB} = \mathbf{r}_B - \mathbf{r}_A$, or

$$L_2 \cos\phi_2 + r \cos(\alpha + \phi_2 + \pi) = x_B - x_A,$$
$$L_2 \sin\phi_2 + r \sin(\alpha + \phi_2 + \pi) = y_B - y_A. \tag{2.23}$$

One can compute r and ϕ_2 from Eq. (2.23). The angle ϕ_3 can be written as

$$\phi_3 = \phi_2 + \alpha. \tag{2.24}$$

PARTICULAR CASE $L_2 = 0$

See Fig. 2.6(b). In this case, from Eqs. (2.23) and (2.24), one can obtain

$$r \cos(\phi_3) = x_B - x_A,$$
$$r \sin(\phi_3) = y_B - y_A. \tag{2.25}$$

The method is illustrated by using some examples as follows.

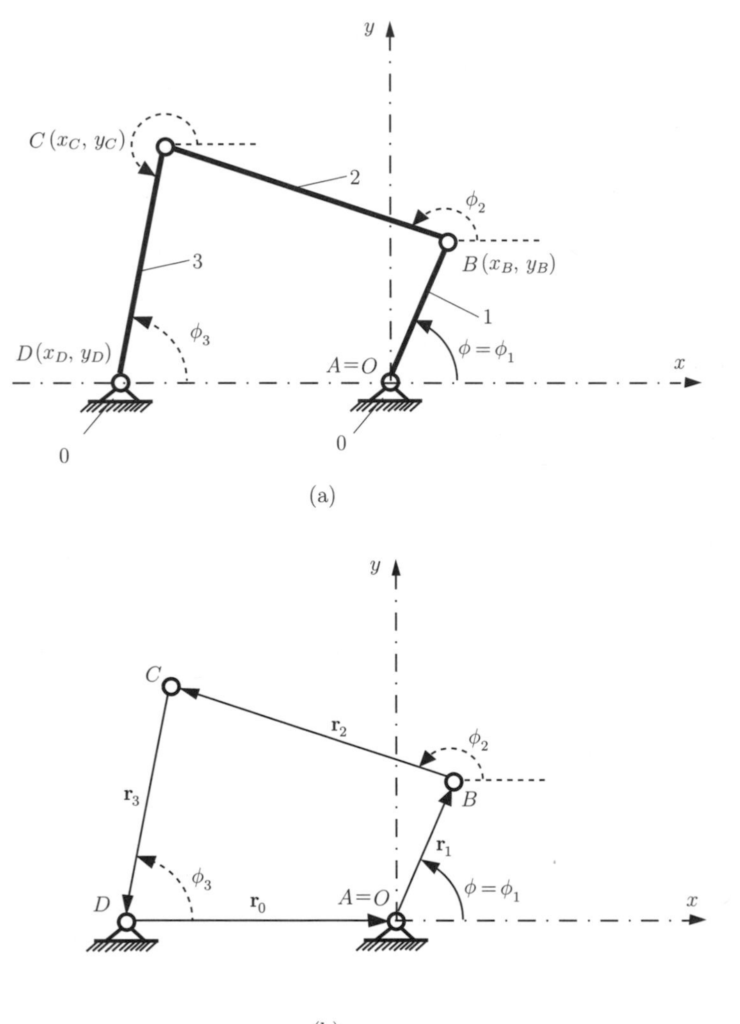

Figure 2.10. (a) 4R mechanism and (b) closed-loop $ABCD$.

■ EXAMPLE 1

Figure 2.10(a) shows a 4R linkage with link lengths r_0, r_1, r_2, and r_3. Find angles ϕ_2 and ϕ_3 as functions of the driver link angle $\phi = \phi_1$.

The links are denoted as vectors $\mathbf{r}_0$, $\mathbf{r}_1$, $\mathbf{r}_2$, and $\mathbf{r}_3$, ($|\mathbf{r}_i| = r_i, \quad i = 0, 1, 2, 3$), and the angles are measured counterclockwise from the x axis, as shown in Fig. 2.10(b). For the closed loop $ABCD$, a vectorial equation can be written:

$$\mathbf{r}_0 + \mathbf{r}_1 + \mathbf{r}_2 + \mathbf{r}_3 = \mathbf{0}. \tag{2.26}$$

When the above vectorial equation is projected onto the x and y axes, two scalar equations are obtained:

$$r_0 + r_1 \cos\phi_1 + r_2 \cos\phi_2 - r_3 \cos\phi_3 = 0, \tag{2.27}$$

$$r_1 \sin\phi_1 + r_2 \sin\phi_2 - r_3 \sin\phi_3 = 0. \tag{2.28}$$

Equations (2.27) and (2.28) represent a set of nonlinear equations in two unknowns, ϕ_2 and ϕ_3. The solution of these two equations solves the position analysis.

Rearranging Eqs. (2.27) and (2.28) gives

$$r_2 \cos\phi_2 = (r_3 \cos\phi_3 - r_0) - r_1 \cos\phi_1, \tag{2.29}$$

$$r_2 \sin\phi_2 = r_3 \sin\phi_3 - r_1 \sin\phi_1. \tag{2.30}$$

Squaring both sides of the above equations and summing gives

$$r_2^2 = r_0^2 + r_1^2 + r_3^2 - 2r_3 \cos\phi_3(r_0 + r_1 \cos\phi_1)$$
$$- 2r_1r_3 \sin\phi_1 \sin\phi_3 + 2r_0r_1 \cos\phi_1,$$

or

$$a \sin\phi_3 + b \cos\phi_3 = c, \tag{2.31}$$

where

$$a = \sin\phi_1, \qquad b = \cos\phi_1 + (r_0/r_1),$$
$$c = (r_0/r_3)\cos\phi_1 + \left[\left(r_0^2 + r_1^2 + r_3^2 - r_2^2\right)/(2r_1r_3)\right]. \tag{2.32}$$

With the use of the relations

$$\sin\phi_3 = 2\tan(\phi_3/2)[1 + \tan^2(\phi_3/2)],$$
$$\cos\phi_3 = [1 - \tan^2(\phi_3/2)]/1 + \tan^2(\phi_3/2)], \tag{2.33}$$

in Eq. (2.31), the following relation is obtained:

$$(b + c)\tan^2(\phi_3/2) - 2a\tan(\phi_3/2) + (c - b) = 0,$$

which gives

$$\tan(\phi_3/2) = (a \pm \sqrt{a^2 + b^2 - c^2})/(b + c). \tag{2.34}$$

Thus, for each given value of ϕ_1 and the length of the links, two distinct values of angle ϕ_3 are obtained:

$$\phi_{3(1)} = 2\tan^{-1}[(a + \sqrt{a^2 + b^2 - c^2})/(b + c)],$$
$$\phi_{3(2)} = 2\tan^{-1}[(a - \sqrt{a^2 + b^2 - c^2})/(b + c)]. \tag{2.35}$$

The two values of ϕ_3 correspond to the two different positions of the mechanism.

The angle ϕ_2 can be eliminated from Eqs. (2.27) and (2.28) to give ϕ_1 in a similar way to that just described.

EXAMPLE 2

Figure 2.11(a) shows a quick-return shaper mechanism. Given the lengths $AB = 0.20$ m, $AD = 0.40$ m, $CD = 0.70$ m, and $CE = 0.30$ m, and the input angle $\phi = \phi_1 = 45°$, obtain the positions of all the other joints. The distance from slider 5 to horizontal axis Ax is $y_E = 0.35$ m.

The coordinates of joint B are

$$y_B = AB \sin\phi = 0.20 \sin 45° = 0.141 \text{ m},$$
$$x_B = AB \cos\phi = 0.20 \cos 45° = 0.141 \text{ m}.$$

The vector diagram of Fig. 2.11(b) is drawn by representing the RTR (BBD) dyad. The vector equation, corresponding to this loop, is written as

$$\mathbf{r}_B + \mathbf{r} - \mathbf{r}_D = \mathbf{0},$$

or

$$\mathbf{r} = \mathbf{r}_D - \mathbf{r}_B,$$

where $\mathbf{r} = \mathbf{BD}$ and $|\mathbf{r}| = r$. When the above vectorial equation is projected on the x and y axes, two scalar equations are obtained:

$$r\cos(\pi + \phi_3) = x_D - x_B = -0.141 \text{ m},$$
$$r\sin(\pi + \phi_3) = y_D - y_B = -0.541 \text{ m}.$$

Angle ϕ_3 is obtained by solving the system of the two previous scalar equations:

$$\tan\phi_3 = \frac{0.541}{0.141} \Longrightarrow \phi_3 = 75.36°.$$

The distance r is

$$r = \frac{x_D - x_B}{\cos(\pi + \phi_3)} = 0.56 \text{ m}.$$

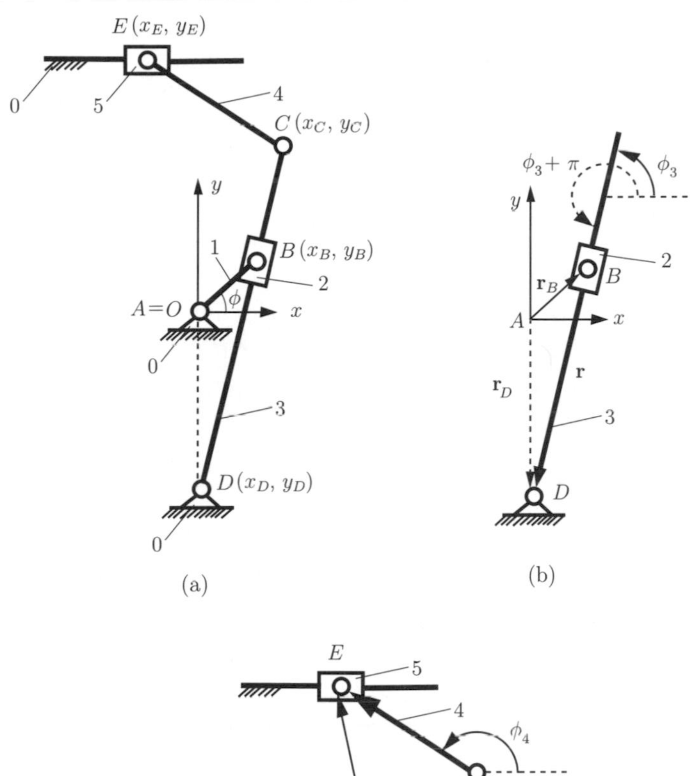

Figure 2.11. (a) Quick-return shaper mechanism, (b) vector diagram representing the RTR (BBD) dyad, and (c) vector diagram representing the RRT (CEE) dyad.

The coordinates of joint C are

$$x_C = CD \cos \phi_3 = 0.17 \text{ m},$$
$$y_C = CD \sin \phi_3 - AD = 0.27 \text{ m}.$$

For the next dyad RRT (CEE), Fig. 2.11(c), one can write

$$CE \cos(\pi - \phi_4) = x_E - x_C,$$
$$CE \sin(\pi - \phi_4) = y_E - y_C.$$

When this system of equations is solved, the unknowns ϕ_4 and x_E are obtained:

$$\phi_4 = 165.9^\circ, \qquad x_E = -0.114 \text{ m}.$$

2.3 R–RRR–RRT Mechanism

The planar R–RRR–RRT mechanism considered is shown in Fig. 2.12. The driver link is the rigid link 1 (the element AB). The following data are given: $AB = 0.150$ m, $BC = 0.400$ m, $CD = 0.370$ m, $CE = 0.230$ m, $EF = CE$, $L_a = 0.300$ m, $L_b = 0.450$ m, and $L_c = CD$. The angle of driver link 1 with the horizontal axis is $\phi = \phi_1 = 45^\circ$.

Position Analysis of the Mechanism

POSITION OF JOINT *A*

A Cartesian reference frame $xOyz$ with the versors $[\mathbf{\imath}, \mathbf{\jmath}, \mathbf{k}]$ is selected, as shown in Fig. 2.12. Since joint A is at the origin of the reference system $A \equiv O$, one can write

$$x_A = 0, \qquad y_A = 0.$$

POSITION OF JOINT *B*

The unknowns are the coordinates of joint B, x_B and y_B. Because joint A is fixed and angle ϕ is known, the coordinates of joint B are computed from the

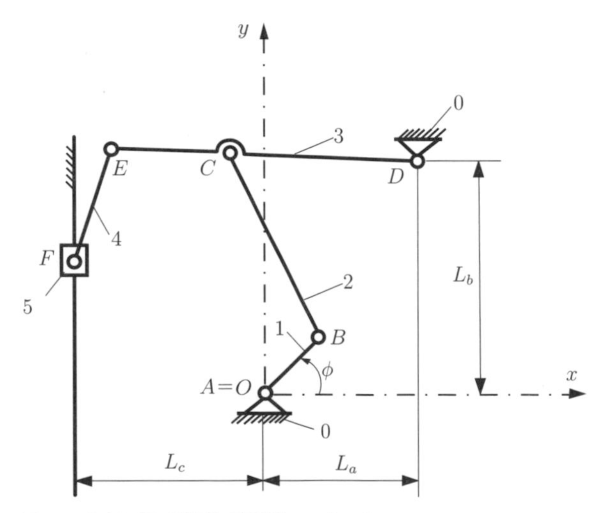

Figure 2.12. R–RRR–RRT mechanism.

following expressions:

$$x_B = AB\cos\phi = 0.106 \text{ m}, \qquad y_B = AB\sin\phi = 0.106 \text{ m}. \tag{2.36}$$

POSITION OF JOINT D

The coordinates of joint D are

$$x_D = L_a, \qquad y_D = L_b.$$

POSITION OF JOINT C

The unknowns are the coordinates of joint C, x_C and y_C. When the positions of joints B and D are known, the position of joint C can be computed, using the fact that the lengths of links BC and CD are constants:

$$\begin{aligned}(x_C - x_B)^2 + (y_C - y_B)^2 &= BC^2,\\ (x_C - x_D)^2 + (y_C - y_D)^2 &= CD^2.\end{aligned} \tag{2.37}$$

Equations (2.37) consist of two quadratic equations. When this system of equations is solved, two sets of solutions are found for the position of joint C. These solutions are

$$\begin{aligned}x_{C_1} &= -0.069 \text{ m}, & y_{C_1} &= 0.465 \text{ m},\\ x_{C_2} &= 0.504 \text{ m}, & y_{C_2} &= 0.141 \text{ m}.\end{aligned} \tag{2.38}$$

Points C_1 and C_2 are the intersections of the circle of radius BC (with its center at B) with the circle of radius CD (with its center at D), as shown in Fig. 2.13. For the position of joint C to be determined for this mechanism, an additional constraint condition is needed: $x_C < x_D$. Because $x_D = 0.300$ m, the coordinates of joint C have the following numerical values:

$$x_C = x_{C_1} = -0.069 \text{ m}, \qquad y_C = y_{C_1} = 0.465 \text{ m}. \tag{2.39}$$

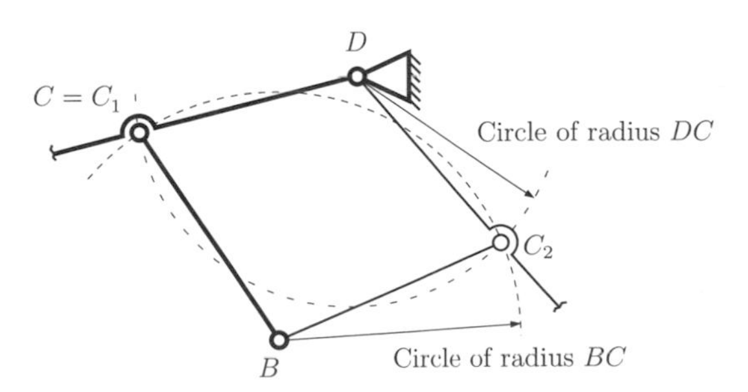

Figure 2.13. Position of joint C.

POSITION OF JOINT *E*

The unknowns are the coordinates of joint E, x_E and y_E. The position of joint E is determined from the equation

$$(x_E - x_C)^2 + (y_E - y_C)^2 = CE^2. \tag{2.40}$$

Joints D, C, and E are located on the same straight link DE. For these joints, the following equations can be written:

$$\begin{aligned} y_D &= m x_D + n, \\ y_C &= m x_C + n, \\ y_E &= m x_E + n, \end{aligned} \tag{2.41}$$

where m is the slope of link DE and n is the intercept of link DE with the y axis. With the use of the first two equations of Eqs. (2.41), slope m and intercept n are found to be

$$\begin{aligned} m &= \frac{y_C - y_D}{x_C - x_D} = -0.041, \\ n &= y_D - m x_D = 0.462 \text{ m}. \end{aligned} \tag{2.42}$$

Equation (2.40) and the last equation of Eqs. (2.41) form a system of equations from which the coordinates of joint E can be computed. Two solutions are obtained, shown in Fig. 2.14, and the numerical values of the coordinates for E_1 and E_2 are

$$\begin{aligned} x_{E_1} &= -0.299 \text{ m}, \qquad y_{E_1} = 0.474 \text{ m}, \\ x_{E_2} &= 0.160 \text{ m}, \qquad y_{E_2} = 0.455 \text{ m}. \end{aligned} \tag{2.43}$$

For continuous motion of the mechanism, a constraint condition is needed, $x_E < x_C$. When this condition is used, the coordinates of joint E are

$$x_E = x_{E_1} = -0.300 \text{ m}, \qquad y_E = y_{E_1} = 0.475 \text{ m}. \tag{2.44}$$

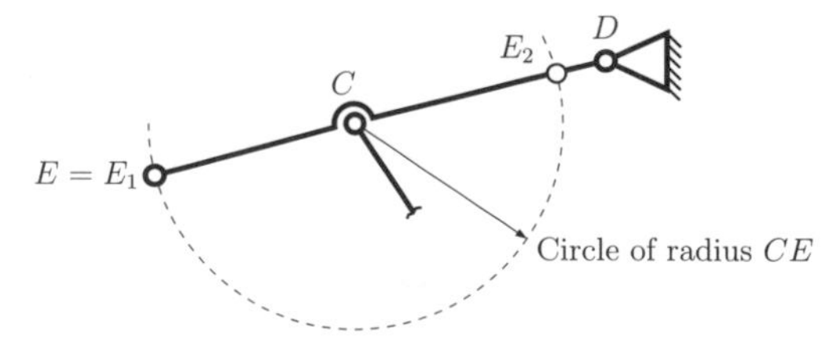

Figure 2.14. Position of joint E.

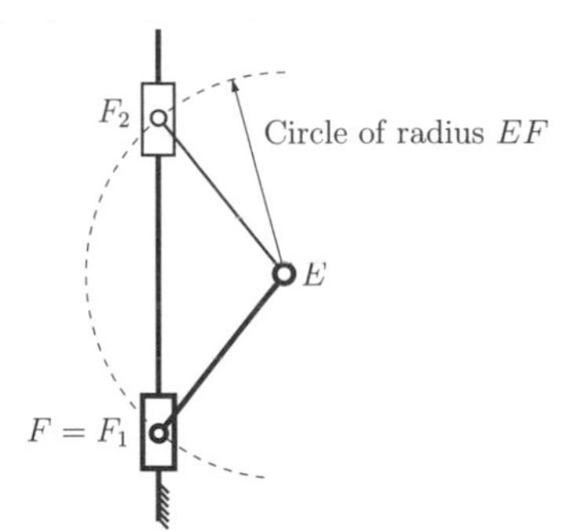

Figure 2.15. Position of joint F.

POSITION OF JOINT F

Joint F is restricted to move in a vertical direction; that is,

$$x_F = -L_c. \tag{2.45}$$

Coordinate y_F of the joint F can be calculated from the following quadratic equation:

$$(x_F - x_E)^2 + (y_F - y_E)^2 = EF^2. \tag{2.46}$$

The solutions of Eq. (2.46) are

$$y_{F_1} = 0.256 \text{ m}, \qquad y_{F_2} = 0.693 \text{ m}. \tag{2.47}$$

Points F_1 and F_2 are located at the intersections of the circle of radius EF (centered at E) and the vertical line with $x = x_F$, as shown in Fig. 2.15. For the mechanism depicted in Fig. 2.13, the y coordinate of joint F should be smaller than the y coordinate of joint E, $y_F < y_E$. The y coordinate of joint F is

$$y_F = y_{F_1} = 0.256 \text{ m}. \tag{2.48}$$

For a complete rotation of driver link $\phi \in [0, 360°]$, the position analysis can be carried out by using *Mathematica*; the program is given in Appendix 1.

A graph showing successive instantaneous locations of the mechanism for a complete rotation of the driver link is shown in Fig. 2.16. The planar displacement of the center of mass of link 2 is shown in Fig. 2.17.

The R–TRR–RRT mechanism is shown in Fig. 2.18. The following data are given: $AC = 0.100$ m, $BC = 0.300$ m, $BD = 0.900$ m, and $L_a = 0.100$ m. The angle of the driver element (link AB) with the horizontal axis is $\phi = 45°$.

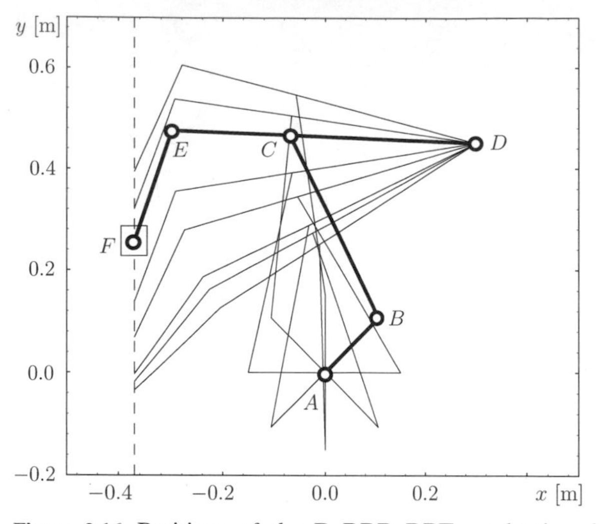

Figure 2.16. Positions of the R–RRR–RRT mechanism for a complete rotation of the driver link.

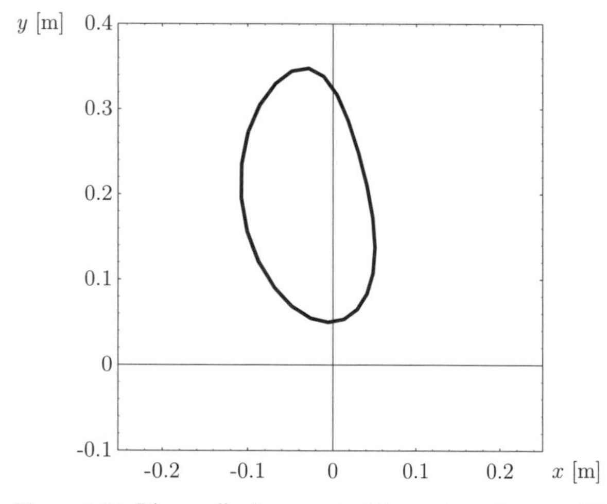

Figure 2.17. Planar displacement of the center of mass of link 2.

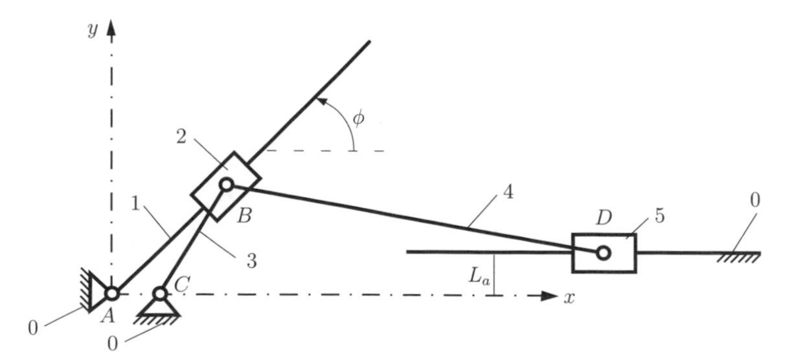

Figure 2.18. R–TRR–RRT mechanism.

Position Analysis of the Mechanism

POSITION OF JOINT *A*

A Cartesian reference frame with the origin at A is selected. The coordinates of joint A are

$$x_A = 0, \qquad y_A = 0.$$

POSITION OF JOINT *C*

The coordinates of joint C are

$$x_C = AC, \qquad y_C = 0.$$

POSITION OF JOINT *B*

The unknowns are the coordinates of joint B, x_B and y_B. Because the joint A is fixed, one can obtain

$$x_B = AB\cos\phi, \qquad y_B = AB\sin\phi. \tag{2.49}$$

In Eq. (2.49) the variable length AB is unknown. An extra equation is necessary:

$$(x_B - x_C)^2 + (y_B - y_C)^2 = BC^2. \tag{2.50}$$

Equations (2.49) and (2.50) form a system of three equations with the unknowns x_B, y_B, and AB. The following numerical results can be obtained:

$$\begin{aligned} &AB_1 = -0.221 \text{ m}, \qquad x_{B_1} = -0.156 \text{ m}, \qquad y_{B_1} = -0.156 \text{ m},\\ &AB_2 = 0.362 \text{ m}, \qquad x_{B_2} = 0.256 \text{ m}, \qquad y_{B_2} = 0.256 \text{ m}. \end{aligned} \tag{2.51}$$

It is evident from physical considerations that the solution for which the length AB is positive has to be chosen, that is, $AB = 0.362$ m. Thus, the coordinates of joint B are

$$x_B = x_{B_2} = 0.256 \text{ m}, \qquad y_B = y_{B_2} = 0.256 \text{ m}. \tag{2.52}$$

POSITION OF JOINT *D*

Slider 5 has a translational motion in the horizontal direction and $y_D = L_a$. There is only one unknown, x_D, for joint D. The following expression can be written:

$$(x_B - x_D)^2 + (y_B - y_D)^2 = BD^2. \tag{2.53}$$

When Eq. (2.53), is solved, two numerical values are obtained:

$$x_{D_1} = -0.630 \text{ m}, \qquad x_{D_2} = 1.142 \text{ m}. \tag{2.54}$$

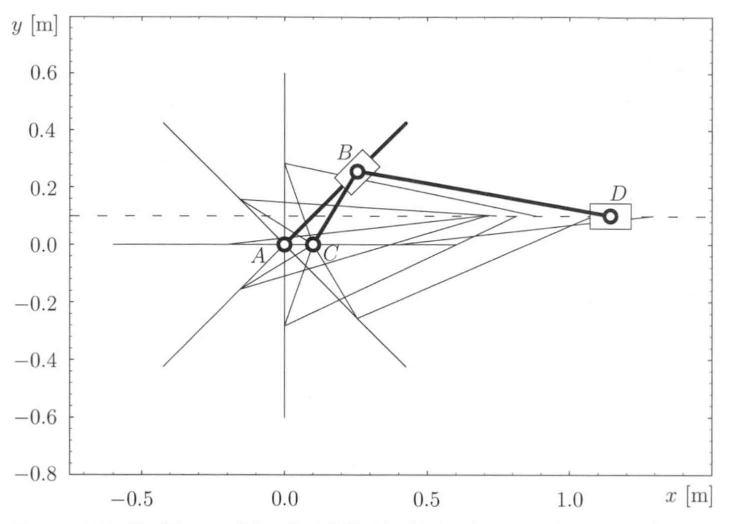

Figure 2.19. Positions of the R–TRR–RRT mechanism for a complete rotation of the driver link.

For continuous motion of the mechanism, a geometric constraint $x_D > x_B$ has to be selected. With the use of this relation, the coordinates of joint D are

$$x_D = 1.142 \text{ m}, \qquad y_D = 0.100 \text{ m}. \tag{2.55}$$

Successive instantaneous positions of the mechanism are shown in Fig. 2.19, for a complete rotation of the driver link. The *Mathematica* computer program used to obtained the mechanism positions is given in Appendix 2.

2.4 R–RRR–RTT Mechanism

The R–RRR–RTT mechanism is shown in Fig. 2.20. The following data are given: $AB = 0.080$ m, $BC = 0.350$ m, $EC = 0.200$ m, $CD = 0.150$ m, $L_a = 0.200$ m, $L_b = 0.350$ m, and $L_c = 0.040$ m. The angle of the driver element (link AB) with the horizontal axis is $\phi = 135°$.

Position Analysis of the Mechanism

POSITION OF JOINT *A*

Joint A is located at the origin of a Cartesian axis reference system. The coordinates of joint A are

$$x_A = y_A = 0.$$

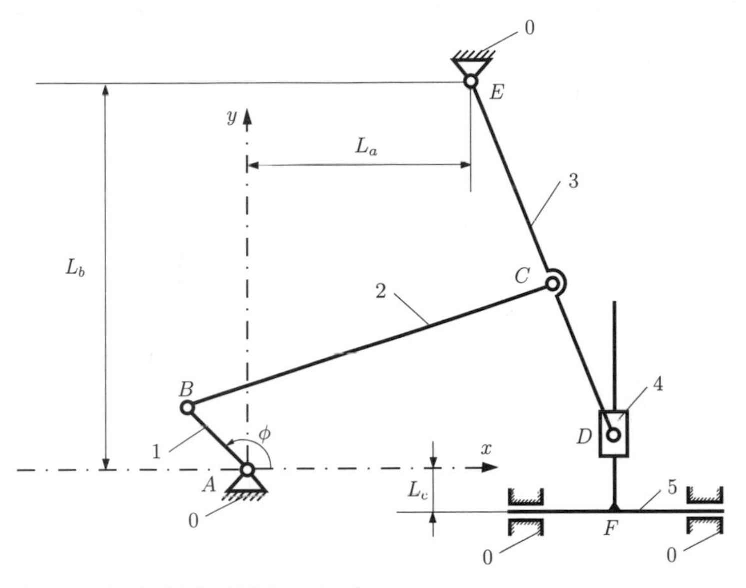

Figure 2.20. R–RRR–RTT mechanism.

POSITION OF JOINT *E*

The coordinates of joint E are given as

$$x_E = L_a, \quad y_E = L_b.$$

POSITION OF JOINT *B*

The coordinates of joint B, x_B and y_B, are

$$x_B = AB\cos\phi = -0.056 \text{ m}, \qquad y_B = AB\sin\phi = 0.056 \text{ m}. \tag{2.56}$$

POSITION OF JOINT *C*

For joint $C(x_C, y_C)$, the following equations can be written:

$$\begin{aligned}(x_C - x_B)^2 + (y_C - y_B)^2 &= BC^2,\\ (x_C - x_E)^2 + (y_C - y_E)^2 &= EC^2.\end{aligned} \tag{2.57}$$

Equations (2.57) consist of two quadratic equations, and two sets of solutions C_1 and C_2 are found for the position of joint C:

$$\begin{aligned}x_{C_1} &= 0.006 \text{ m}, & y_{C_1} &= 0.400 \text{ m},\\ x_{C_2} &= 0.276 \text{ m}, & y_{C_2} &= 0.165 \text{ m}.\end{aligned} \tag{2.58}$$

To determine the correct position of joint C, a constraint condition is needed: $x_C > x_E$. The coordinates of joint C have the following numerical values:

$$x_C = x_{C_2} = 0.276 \text{ m}, \qquad y_C = y_{C_2} = 0.165 \text{ m}. \tag{2.59}$$

POSITION OF JOINT *D*

For joint D the following equation can be written:

$$(x_D - x_E)^2 + (y_D - y_E)^2 = DE^2. \tag{2.60}$$

Because joints E, C, and D are on the same straight line ED, one can write

$$\begin{aligned} y_E &= m x_E + n, \\ y_C &= m x_C + n, \\ y_D &= m x_D + n. \end{aligned} \tag{2.61}$$

where m is the slope of link ED and n is the intercept of link ED with the y axis. With the use of the first two equations in Eqs. (2.61), slope m and intercept n can be determined:

$$\begin{aligned} m &= \frac{y_E - y_C}{x_E - x_C} = -2.427, \\ n &= y_C - m x_C = 0.835 \text{ m}. \end{aligned} \tag{2.62}$$

Once m and n have been determined, Eq. (2.60) and the last equation in Eqs. (2.61) form a system of equations from which the coordinates of joint

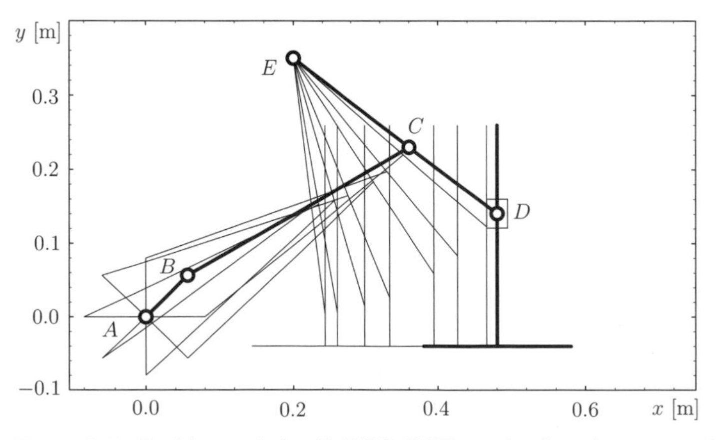

Figure 2.21. Positions of the R–RRR–RTT mechanism for a complete rotation of the driver link.

D can be found. The following numerical values are obtained:

$$x_{D_1} = 0.219 \text{ m}, \qquad y_{D_1} = 0.304 \text{ m},$$
$$x_{D_2} = 0.333 \text{ m}, \qquad y_{D_2} = 0.026 \text{ m}. \tag{2.63}$$

With the use of the constraint condition $y_D < y_C$, the coordinates of joint D are

$$x_D = 0.333 \text{ m}, \qquad y_D = 0.026 \text{ m}. \tag{2.64}$$

POSITION OF JOINT *F*

The coordinates of point F are given by

$$x_F = x_D = 0.333 \text{ m}, \qquad y_F = -L_c = 0.040 \text{ m}. \tag{2.65}$$

The graph showing successive instantaneous locations of the mechanism for a complete rotation of the driver link is shown in Fig. 2.21.

3 Velocity and Acceleration Analysis

The classical method for obtaining the velocities and/or accelerations of links and joints is to compute the derivatives of the positions and/or velocities with respect to time.

3.1 Driver Link

For a driver link in rotational motion, Fig. 2.3(a), the following position relation can be written:

$$x_B(t) = x_A + L_{AB}\cos\phi(t),$$
$$y_B(t) = y_A + L_{AB}\sin\phi(t). \tag{3.1}$$

When Eq. (3.1) is differentiated with respect to time, the following expressions are obtained for the x and y components, υ_{Bx} and υ_{By}, of the linear velocity of joint B:

$$\upsilon_{Bx} = \dot{x}_B = \frac{dx_B(t)}{dt} = -L_{AB}\,\dot{\phi}\,\sin\phi,$$
$$\upsilon_{By} = \dot{y}_B = \frac{dy_B(t)}{dt} = L_{AB}\,\dot{\phi}\,\cos\phi. \tag{3.2}$$

The angular velocity of the driver link is $\omega = \dot{\phi}$.

The time derivative of Eq. (3.2) yields the x and y components, a_{Bx} and a_{By}, of the linear acceleration of joint B:

$$a_{Bx} = \ddot{x}_B = \frac{d\upsilon_B(t)}{dt} = -L_{AB}\dot{\phi}^2\cos\phi - L_{AB}\ddot{\phi}\sin\phi,$$
$$a_{By} = \ddot{y}_B = \frac{d\upsilon_B(t)}{dt} = -L_{AB}\dot{\phi}^2\sin\phi + L_{AB}\ddot{\phi}\cos\phi, \tag{3.3}$$

where $\alpha = \ddot{\phi}$ is the angular acceleration of driver link AB.

3.2 RRR Dyad

For the RRR dyad, Fig. 2.4, there are two quadratic equations of the form

$$[x_C(t) - x_A]^2 + [y_C(t) - y_A]^2 = L_{AC} = L_2^2,$$
$$[x_C(t) - x_B]^2 + [y_C(t) - y_B]^2 = L_{BC} = L_3^2. \tag{3.4}$$

When the above system of quadratic equations is solved, the coordinates of joint C, $x_C(t)$ and $y_C(t)$, are obtained.

The derivative of Eq. (3.4) with respect to time yields

$$(x_C - x_A)(\dot{x}_C - \dot{x}_A) + (y_C - y_A)(\dot{y}_C - \dot{y}_A) = 0,$$
$$(x_C - x_B)(\dot{x}_C - \dot{x}_B) + (y_C - y_B)(\dot{y}_C - \dot{y}_B) = 0. \tag{3.5}$$

The velocity vector $\mathbf{v}_C = [\dot{x}_C, \dot{y}_C]^T$ of the above system of equations can be written in matrix form:

$$\mathbf{v}_C = \mathbf{M}_1 \cdot \mathbf{v}, \tag{3.6}$$

where

$$\mathbf{v} = [\dot{x}_A, \dot{y}_A, \dot{x}_B, \dot{y}_B]^T,$$
$$\mathbf{M}_1 = \mathbf{A}_1^{-1} \cdot \mathbf{A}_2,$$
$$\mathbf{A}_1 = \begin{bmatrix} x_C - x_A & y_C - y_A \\ x_C - x_B & y_C - y_B \end{bmatrix},$$
$$\mathbf{A}_2 = \begin{bmatrix} x_C - x_A & y_C - y_A & 0 & 0 \\ 0 & 0 & x_C - x_B & y_C - y_B \end{bmatrix}.$$

Similarly, by differentiation of Eq. (3.5), the following equations are obtained, which include the x and y components, $\ddot{x}_C$ and $\ddot{y}_C$, of the acceleration vector of joint C:

$$(\dot{x}_C - \dot{x}_A)^2 + (x_C - x_A)(\ddot{x}_C - \ddot{x}_A) + (\dot{y}_C - \dot{y}_A)^2 + (y_C - y_A)(\ddot{y}_C - \ddot{y}_A) = 0,$$
$$(\dot{x}_C - \dot{x}_B)^2 + (x_C - x_B)(\ddot{x}_C - \ddot{x}_B) + (\dot{y}_C - \dot{y}_B)^2 + (y_C - y_B)(\ddot{y}_C - \ddot{y}_B) = 0. \tag{3.7}$$

The acceleration vector $\mathbf{a}_C$ of joint C is obtained from the above system of equations:

$$\mathbf{a}_C = [\ddot{x}_C, \ddot{y}_C]^T = \mathbf{M}_1 \cdot \mathbf{a} + \mathbf{M}_2, \tag{3.8}$$

where

$$\mathbf{a} = [\ddot{x}_A, \ddot{y}_A, \ddot{x}_B, \ddot{y}_B]^T,$$
$$\mathbf{M}_2 = -\mathbf{A}_1^{-1} \cdot \mathbf{A}_3,$$
$$\mathbf{A}_3 = \begin{bmatrix} (\dot{x}_C - \dot{x}_A)^2 + (\dot{y}_C - \dot{y}_A)^2 \\ (\dot{x}_C - \dot{x}_B)^2 + (\dot{y}_C - \dot{y}_B)^2 \end{bmatrix}.$$

As a way to compute the angular velocity and acceleration of the RRR dyad, the following equations can be written:

$$\begin{aligned} y_C(t) - y_A + [x_C(t) - x_A] \tan \phi_2(t) &= 0, \\ y_C(t) - y_B + [x_C(t) - x_B] \tan \phi_3(t) &= 0. \end{aligned} \tag{3.9}$$

The derivative with respect to time of Eq. (3.9) yields

$$\begin{aligned} \dot{y}_C - \dot{y}_A - (\dot{x}_C - \dot{x}_A) \tan \phi_2 - (x_C - x_A) \frac{1}{\cos^2 \phi_2} \dot{\phi}_2 &= 0, \\ \dot{y}_C - \dot{y}_B - (\dot{x}_C - \dot{x}_B) \tan \phi_3 - (x_C - x_B) \frac{1}{\cos^2 \phi_3} \dot{\phi}_3 &= 0. \end{aligned} \tag{3.10}$$

The angular velocity vector ω is computed as

$$\omega = [\dot{\phi}_2, \dot{\phi}_3]^T = \mathbf{\Omega}_1 \cdot \mathbf{v} + \mathbf{\Omega}_2 \cdot \mathbf{v}_C, \tag{3.11}$$

where

$$\mathbf{\Omega}_1 = \begin{bmatrix} \frac{x_C - x_A}{L_2^2} & -\frac{x_C - x_A}{L_2^2} & 0 & 0 \\ 0 & 0 & \frac{x_C - x_B}{L_3^2} & -\frac{x_C - x_B}{L_3^2} \end{bmatrix},$$

$$\mathbf{\Omega}_2 = \begin{bmatrix} -\frac{x_C - x_A}{L_2^2} & \frac{x_C - x_A}{L_2^2} \\ -\frac{x_C - x_B}{L_3^2} & \frac{x_C - x_B}{L_3^2} \end{bmatrix}.$$

Differentiating Eq. (3.11), the angular acceleration vector, $\alpha = \dot{\omega} = [\ddot{\phi}_2, \ddot{\phi}_3]^T$ is

$$\alpha = \dot{\mathbf{\Omega}}_1 \cdot \mathbf{v} + \dot{\mathbf{\Omega}}_2 \cdot \mathbf{v}_C + \mathbf{\Omega}_1 \cdot \mathbf{a} + \mathbf{\Omega}_2 \cdot \mathbf{a}_C. \tag{3.12}$$

3.3 RRT Dyad

For the RRT dyad, Fig. 2.5(a), the following equations can be written for position analysis:

$$[x_C(t) - x_A]^2 + [y_C(t) - y_A]^2 = AC^2 = L_{AC}^2 = L_2^2,$$
$$[x_C(t) - x_B]\sin\alpha - [y_C(t) - y_B]\cos\alpha = \pm h. \tag{3.13}$$

From the above system of equations, the x and y coordinates of joint C, $x_C(t)$ and $y_C(t)$, can be computed.

The time derivative of Eq. (3.13) yields

$$(x_C - x_A)(\dot{x}_C - \dot{x}_A) + (y_C - y_A)(\dot{y}_C - \dot{y}_A) = 0,$$
$$(\dot{x}_C - \dot{x}_B)\sin\alpha - (\dot{y}_C - \dot{y}_B)\cos\alpha = 0. \tag{3.14}$$

The solution for velocity vector $\mathbf{v}_C$ of joint C from Eq. (3.14) is

$$\mathbf{v}_C = [\dot{x}_C, \dot{y}_C]^T = \mathbf{M}_3 \cdot \mathbf{v}, \tag{3.15}$$

where

$$\mathbf{M}_3 = \mathbf{A}_4^{-1} \cdot \mathbf{A}_5,$$

$$\mathbf{A}_4 = \begin{bmatrix} x_C - x_A & y_C - y_A \\ \sin\alpha & -\cos\alpha \end{bmatrix},$$

$$\mathbf{A}_5 = \begin{bmatrix} x_C - x_A & y_C - y_A & 0 & 0 \\ 0 & 0 & \sin\alpha & \cos\alpha \end{bmatrix}.$$

Differentiating Eq. (3.14) with respect to time,

$$(\dot{x}_C - \dot{x}_A)^2 + (x_C - x_A)(\ddot{x}_C - \ddot{x}_A) + (\dot{y}_C - \dot{y}_A)^2$$
$$+ (y_C - y_A)(\ddot{y}_C - \ddot{y}_A) = 0,$$
$$(\ddot{x}_C - \ddot{x}_B)\sin\alpha - (\ddot{y}_C - \ddot{y}_B)\cos\alpha = 0, \tag{3.16}$$

giving the acceleration vector $\mathbf{a}_C$ of joint C

$$\mathbf{a}_C = [\ddot{x}_C, \ddot{y}_C]^T = \mathbf{M}_3 \cdot \mathbf{a} + \mathbf{M}_4, \tag{3.17}$$

where

$$\mathbf{M}_4 = -\mathbf{A}_4^{-1} \cdot \mathbf{A}_6,$$

$$\mathbf{A}_6 = \begin{bmatrix} (\dot{x}_C - \dot{x}_A)^2 + (\dot{y}_C - \dot{y}_A)^2 \\ 0 \end{bmatrix}. \tag{3.18}$$

The angular position of link 2 is described by the following equation:

$$y_C(t) - y_A - [x_C(t) - x_A] \tan\phi_2(t) = 0. \tag{3.19}$$

The time derivative of Eq. (3.19) yields

$$\dot{y}_C - \dot{y}_A - (\dot{x}_C - \dot{x}_A) \tan\phi_2 - (x_C - x_A)\frac{1}{\cos^2\phi_2}\dot{\phi}_2 = 0, \tag{3.20}$$

and the angular velocity, ω_2, of link 2 is

$$\omega_2 = \frac{x_C - x_A}{L_2^2}[(\dot{y}_C - \dot{y}_A) - (\dot{x}_C - x_A) \tan\phi_2]. \tag{3.21}$$

The angular acceleration, α_2, of link 2 is $\alpha_2 = \dot{\omega}_2$.

3.4 RTR Dyad

For the RTR dyad, Fig. 2.6(a), the following relations can be written:

$$\begin{aligned}
&[x_C(t) - x_A]^2 + [y_C(t) - y_A]^2 = L_2^2,\\
&\tan\alpha = \frac{(y_C - y_B)/(x_C - x_B) - (y_C - y_A)/(x_C - x_A)}{1 + (y_C - y_B)/(x_C - x_B) \times (y_C - y_A)/(x_C - x_A)}\\
&\qquad = \frac{(y_C - y_B)(x_C - x_A) - (y_C - y_A)(x_C - x_B)}{(x_C - x_B)(x_C - x_A) + (y_C - y_B)(y_C - y_A)}.
\end{aligned} \tag{3.22}$$

The time derivative of Eq. (3.22) yields

$$\begin{aligned}
&(x_C - x_A)(\dot{x}_C - \dot{x}_A) + (y_C - y_A)(\dot{y}_C - \dot{y}_A) = 0,\\
&\tan\alpha[(\dot{x}_C - \dot{x}_B)(x_C - x_A) + (x_C - x_B)(\dot{x}_C - \dot{x}_A)]\\
&\quad + \tan\alpha[(\dot{y}_C - \dot{y}_A)(y_C - y_B) + (y_C - y_A)(\dot{y}_C - \dot{y}_B)]\\
&\quad + (\dot{y}_C - \dot{y}_A)(x_C - x_B) + (y_C - y_A)(\dot{x}_C - \dot{x}_B)\\
&\quad - (\dot{y}_C - \dot{y}_B)(x_C - x_A) - (y_C - y_B)(\dot{x}_C - \dot{x}_A) = 0,
\end{aligned} \tag{3.23}$$

or in a matrix form

$$\mathbf{A}_7 \cdot \mathbf{v}_C = \mathbf{A}_8 \cdot \mathbf{v}, \tag{3.24}$$

where

$$\mathbf{A}_7 = \begin{bmatrix} x_C - x_A & y_C - y_A \\ \gamma_1 & \gamma_2 \end{bmatrix},$$

$$\mathbf{A}_8 = \begin{bmatrix} x_C - x_A & y_C - y_A & 0 & 0 \\ \gamma_3 & \gamma_4 & \gamma_5 & \gamma_6 \end{bmatrix}.$$

In addition,

$$\gamma_1 = [(x_C - x_B) + (x_C - x_A)] \tan\alpha - (y_C - y_B) + (y_C - y_A),$$
$$\gamma_2 = [(y_C - y_A) + (y_C - y_B)] \tan\alpha - (x_C - x_A) + (x_C - x_B),$$
$$\gamma_3 = (x_C - x_B) \tan\alpha + (y_C - y_B),$$
$$\gamma_4 = (x_C - x_A) \tan\alpha + (y_C - y_A),$$
$$\gamma_5 = (y_C - y_B) \tan\alpha + (x_C - x_B),$$
$$\gamma_6 = (y_C - y_A) \tan\alpha - (x_C - x_A).$$

The solution for the velocity vector, $\mathbf{v}_C$, of joint C, from Eq. (3.24) is

$$\mathbf{v}_C = \mathbf{M}_5 \cdot \mathbf{v}, \tag{3.25}$$

where

$$\mathbf{M}_5 = \mathbf{A}_7^{-1} \cdot \mathbf{A}_8.$$

When Eq. (3.24) is differentiated, the following relation is obtained:

$$\mathbf{A}_7 \cdot \mathbf{a}_C = \mathbf{A}_8 \cdot \mathbf{a} - \mathbf{A}_9, \tag{3.26}$$

where

$$\mathbf{A}_9 = \begin{bmatrix} (\dot{x}_C - \dot{x}_A)^2 + (\dot{y}_C - \dot{y}_A)^2 \\ \gamma_7 \end{bmatrix},$$
$$\gamma_7 = 2(\dot{x}_C - \dot{x}_B)(\dot{x}_C - \dot{x}_A) \tan\alpha + 2(\dot{y}_C - \dot{y}_B)(\dot{y}_C - \dot{y}_A) \tan\alpha$$
$$- 2(\dot{y}_C - \dot{y}_B)(\dot{x}_C - \dot{x}_A) + 2(\dot{y}_C - \dot{y}_A)(\dot{x}_C - \dot{x}_B).$$

Acceleration vector $\mathbf{a}_C$ of joint C is

$$\mathbf{a}_C = \mathbf{M}_5 \cdot \mathbf{a} - \mathbf{M}_6, \tag{3.27}$$

where

$$\mathbf{M}_6 = \mathbf{A}_7^{-1} \cdot \mathbf{A}_9.$$

As a way to compute the angular velocities for the RTR dyad, the following equations can be written:

$$y_C(t) - y_A = [x_C(t) - x_A] \tan\phi_2,$$
$$\phi_3 = \phi_2 + \alpha. \tag{3.28}$$

The time derivative of Eq. (3.28) yields

$$(\dot{y}_C - \dot{y}_A) = (\dot{x}_C - \dot{x}_A)\tan\phi_2 + (x_C - x_A)\frac{1}{\cos^2\phi_2}\dot{\phi}_2,$$
$$\dot{\phi}_3 = \dot{\phi}_2. \tag{3.29}$$

The angular velocities ω_2 and ω_3 of links 2 and 3 are

$$\omega_2 = \omega_3 = \frac{\cos^2\phi_2}{x_C - x_A}[(\dot{y}_C - \dot{y}_A) - (\dot{x}_C - \dot{x}_A)\tan\phi_2]. \tag{3.30}$$

The angular accelerations α_2 and α_3 of links 2 and 3 are found to be

$$\alpha_2 = \alpha_3 = \dot{\omega}_2 = \dot{\omega}_3. \tag{3.31}$$

3.5 TRT Dyad

For the TRT dyad, Fig. 2.7, two equations can be written:

$$[x_C(t) - x_A]\sin\alpha - [y_C(t) - y_A]\cos\alpha = \pm d,$$
$$[x_C(t) - x_B]\sin\beta - [y_C(t) - y_B]\cos\beta = \pm h. \tag{3.32}$$

The derivative with respect to time of Eq. (3.32) yields

$$(\dot{x}_C - \dot{x}_A)\sin\alpha - (\dot{y}_C - \dot{y}_A)\cos\alpha + (x_C - x_A)\dot{\alpha}\cos\alpha$$
$$+ (y_C - y_A)\dot{\alpha}\sin\alpha = 0,$$
$$(\dot{x}_C - \dot{x}_B)\sin\beta - (\dot{y}_C - \dot{y}_B)\cos\beta + (x_C - x_B)\dot{\beta}\cos\beta$$
$$+ (y_C - y_B)\dot{\beta}\sin\beta = 0, \tag{3.33}$$

which can be rewritten as

$$\mathbf{A}_{10} \cdot \mathbf{v}_C = \mathbf{A}_{11} \cdot \mathbf{v}_1, \tag{3.34}$$

where

$$\mathbf{v}_1 = [\dot{x}_A, \dot{y}_A, \dot{\alpha}, \dot{x}_B, \dot{y}_B, \dot{\beta}]^T,$$
$$\mathbf{A}_{10} = \begin{bmatrix} -\sin\alpha & -\cos\alpha \\ \sin\beta & -\cos\beta \end{bmatrix},$$
$$\mathbf{A}_{11} = \begin{bmatrix} \sin\alpha & -\cos\alpha & \xi_1 & 0 & 0 & 0 \\ 0 & 0 & 0 & \sin\beta & -\cos\beta & \xi_2 \end{bmatrix},$$
$$\xi_1 = (x_A - x_C)\cos\alpha + (y_A - y_C)\sin\alpha,$$
$$\xi_2 = (x_B - x_C)\cos\beta + (y_B - y_C)\sin\beta.$$

The solution of Eq. (3.34) gives the velocity vector $\mathbf{v}_C$ of joint C:

$$\mathbf{v}_C = \mathbf{M}_7 \cdot \mathbf{v}_1, \tag{3.35}$$

where

$$\mathbf{M}_7 = \mathbf{A}_{10}^{-1} \cdot \mathbf{A}_{11}.$$

Differentiating Eq. (3.34) with respect to time gives

$$\mathbf{A}_{10} \cdot \mathbf{a}_C = \mathbf{A}_{11} \cdot \mathbf{a}_1 - \mathbf{A}_{12}, \tag{3.36}$$

where

$$\mathbf{a}_1 = [\ddot{x}_A, \ddot{y}_A, \ddot{\alpha}, \ddot{x}_B, \ddot{y}_B, \ddot{\beta}]^T,$$

$$\mathbf{A}_{12} = \begin{bmatrix} \xi_3 \\ \xi_4 \end{bmatrix},$$

$$\xi_3 = 2(\dot{x}_C - \dot{x}_A)\dot{\alpha} \cos\alpha + 2(\dot{y}_C - \dot{y}_A)\dot{\beta} \sin\alpha - (x_C - x_A)\dot{\alpha}^2 \sin\alpha + (y_C - y_A)\dot{\alpha}^2 \cos\alpha,$$

$$\xi_4 = 2(\dot{x}_C - \dot{x}_B)\dot{\beta} \cos\beta + 2(\dot{y}_C - \dot{y}_B)\dot{\beta} \sin\beta - (x_C - x_B)\dot{\beta}^2 \sin\beta + (y_C - y_B)\dot{\beta}^2 \cos\beta.$$

The solution of the above equations gives the acceleration vector $\mathbf{a}_C$ of joint C:

$$\mathbf{a}_C = \mathbf{M}_7 \cdot \mathbf{a} + \mathbf{M}_8, \tag{3.37}$$

where

$$\mathbf{M}_8 = \mathbf{A}_{10}^{-1} \cdot \mathbf{A}_{12}.$$

EXAMPLE: R–RTR–RRT Mechanism

The R–RTR–RRT mechanism considered is shown in Fig. 3.1. The motion of the mechanism is to be determined for the following data: $AB = 0.100$ m, $CD = 0.075$ m, $DE = 0.200$ m, $AC = 0.150$ m, driver link angle $\phi = \phi_1 = 45°$, and the angular speed of driver link 1 $\omega = \omega_1 = 4.712$ rad/s.

Velocity and Acceleration Analysis of the Mechanism

The origin of the fixed reference frame is at $C \equiv 0$. The position of the fixed joint A is

$$x_A = 0, \qquad y_A = AC. \tag{3.38}$$

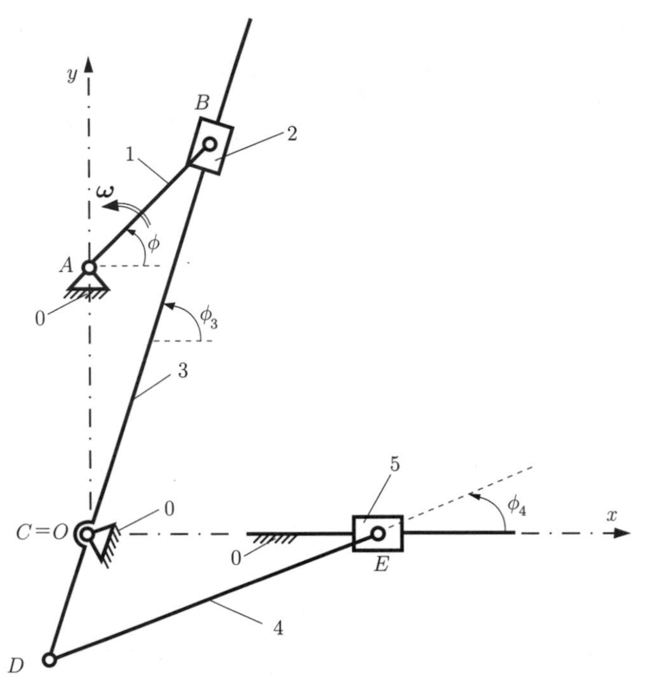

Figure 3.1. R–RTR–RRT mechanism.

Joint *B*

The position of joint B is

$$x_B(t) = x_A + AB\cos\phi(t), \qquad y_B(t) = y_A + AB\sin\phi(t), \tag{3.39}$$

and for $\phi = 45°$,

$$x_B = 0.000 + 0.100\cos 45° = 0.070 \text{ m},$$
$$y_B = 0.150 + 0.100\sin 45° = 0.220 \text{ m}.$$

VELOCITY ANALYSIS

The linear velocity vector of joint B is

$$\mathbf{v}_B = \dot{x}_B\mathbf{ı} + \dot{y}_B\mathbf{J},$$

where

$$\dot{x}_B = \frac{dx_B}{dt} = -AB\dot{\phi}\sin\phi, \qquad \dot{y}_B = \frac{dy_B}{dt} = AB\dot{\phi}\cos\phi. \tag{3.40}$$

With $\phi = 45^\circ$ and $\dot{\phi} = \omega = 4.712$ rad/s,

$$\begin{aligned} \dot{x}_B &= -0.100(4.712)\sin 45^\circ = -0.333 \text{ m/s}, \\ \dot{y}_B &= 0.100(4.712)\cos 45^\circ = 0.333 \text{ m/s}, \\ v_B &= |\mathbf{v}_B| = \sqrt{\dot{x}_B^2 + \dot{y}_B^2} = \sqrt{(-0.333)^2 + 0.333^2} = 0.471 \text{ m/s}. \end{aligned} \tag{3.41}$$

ACCELERATION ANALYSIS

The linear acceleration vector of joint B is

$$\mathbf{a}_B = \ddot{x}_B\mathbf{\imath} + \ddot{y}_B\mathbf{\jmath},$$

where

$$\begin{aligned} \ddot{x}_B &= \frac{d\dot{x}_B}{dt} = -AB\dot{\phi}^2\cos\phi - AB\ddot{\phi}\sin\phi, \\ \ddot{y}_B &= \frac{d\dot{y}_B}{dt} = -AB\dot{\phi}^2\sin\phi + AB\ddot{\phi}\cos\phi. \end{aligned} \tag{3.42}$$

The angular acceleration of link 1 is $\ddot{\phi} = \dot{\omega} = 0$. The numerical values are

$$\begin{aligned} \ddot{x}_B &= -0.100(4.712)^2\cos 45^\circ = -1.569 \text{ m/s}^2, \\ \ddot{y}_B &= -0.100(4.712)^2\sin 45^\circ = -1.569 \text{ m/s}^2, \\ a_B &= |\mathbf{a}_B| = \sqrt{\ddot{x}_B^2 + \ddot{y}_B^2} = \sqrt{(-1.569)^2 + (-1.569)^2} = 2.220 \text{ m/s}^2. \end{aligned} \tag{3.43}$$

Link 3

Joints B, C, and D are located on the same link BD. Thus the following equation can be written:

$$y_B(t) - y_C - [x_B(t) - x_C]\tan\phi_3(t) = 0. \tag{3.44}$$

Angle $\phi_3 = \phi_2$ can be computed as follows:

$$\phi_3 = \phi_2 = \arctan(y_B - y_C)/(x_B - x_C), \tag{3.45}$$

and for $\phi = 45^\circ$ one can obtain

$$\phi_3 = \arctan(0.22)/(0.07) = 72.235^\circ.$$

VELOCITY ANALYSIS

The derivative of Eq. (3.44) yields

$$\dot{y}_B - \dot{y}_C - (\dot{x}_B - \dot{x}_C)\tan\phi_3 - (x_B - x_C)\frac{1}{\cos^2\phi_3}\dot{\phi}_3 = 0. \tag{3.46}$$

The angular velocity of link 3, $\omega_3 = \omega_2 = \dot{\phi}_3$, can be computed as follows:

$$\omega_3 = \omega_2 = \frac{\cos^2\phi_3[\dot{y}_B - \dot{y}_C - (\dot{x}_B - \dot{x}_C)\tan\phi_3]}{x_B - x_C}, \tag{3.47}$$

$$\omega_3 = \frac{\cos^2 72^\circ(0.333 + 0.333\ \tan\ 72.235^\circ)}{0.07} = 1.807\ \text{rad/s}.$$

ACCELERATION ANALYSIS

The angular acceleration of link 3, $\alpha_3 = \alpha_2 = \ddot{\phi}_3$, can be computed from the time derivative of Eq. (3.46):

$$\begin{aligned}&\ddot{y}_B - \ddot{y}_C - (\ddot{x}_B - \ddot{x}_C)\tan\phi_3 - 2(\dot{x}_B - \dot{x}_C)\frac{1}{\cos^2\phi_3}\dot{\phi}_3\\&\quad - 2(x_B - x_C)\frac{\sin\phi_3}{\cos^3\phi_3}\dot{\phi}_3^2 - (x_B - x_C)\frac{1}{\cos^2\phi_3}\ddot{\phi}_3 = 0.\end{aligned} \tag{3.48}$$

The solution of the previous equation is

$$\begin{aligned}\alpha_3 = \alpha_2 = \Bigg[&\ddot{y}_B - \ddot{y}_C - (\ddot{x}_B - \ddot{x}_C)\tan\phi_3 - 2(\dot{x}_B - \dot{x}_C)\frac{1}{\cos^2\phi_3}\dot{\phi}_3\\&- 2(x_B - x_C)\frac{\sin\phi_3}{\cos^3\phi_3}\dot{\phi}_3^2\Bigg]\frac{\cos^2\phi_3}{x_B - x_C},\end{aligned} \tag{3.49}$$

and for the given numerical data the angular acceleration α_3 of link 3 is

$$\begin{aligned}\alpha_3 = \alpha_2 = \Bigg[&-1.569 + 1.569\ \tan\ 72.235^\circ + 2(0.333)\frac{1}{\cos^2\ 72.235^\circ}1.807\\&- 2(0.07)\frac{\sin\ 72.235^\circ}{\cos^3\ 72.235^\circ}(1.807)^2\Bigg]\frac{\cos^2\ 72.235^\circ}{0.07} = 1.020\ \text{rad/s}^2.\end{aligned} \tag{3.50}$$

Joint *D*

For the position analysis of joint D, the following quadratic equations can be written:

$$[x_D(t) - x_C]^2 + [y_D(t) - y_C]^2 = BC^2, \tag{3.51}$$

$$[x_D(t) - x_C]\sin\phi_3(t) - [y_D(t) - y_C]\cos\phi_3(t) = 0. \tag{3.52}$$

The previous equations can be rewritten as follows:

$$x_D^2(t) + y_D^2(t) = CD^2,$$
$$x_D(t)\sin\phi_3(t) - y_D(t)\cos\phi_3(t) = 0. \tag{3.53}$$

The solution of the above system of equations is

$$x_D = \pm\frac{CD}{\sqrt{1+\tan^2\phi_3}} = \pm\frac{0.075}{\sqrt{1+\tan^2\ 72.235^\circ}} = -0.023 \text{ m},$$
$$y_D = x_D \tan\phi_3 = -0.023\ \tan 72.235^\circ = -0.071 \text{ m}. \tag{3.54}$$

The negative value for x_D was selected for this position of the mechanism.

VELOCITY ANALYSIS

The velocity analysis can be carried out by differentiating Eq. (3.53):

$$x_D\dot{x}_D + y_D\dot{y}_D = 0,$$
$$\dot{x}_D\sin\phi_3 + x_D\cos\phi_3\dot{\phi}_3 - \dot{y}_D\cos\phi_3 + y_D\sin\phi_3\dot{\phi}_3 = 0, \tag{3.55}$$

which can be rewritten as

$$-0.023\dot{x}_D - 0.07\dot{y}_D = 0,$$
$$0.95\dot{x}_D - 0.023(0.3)(1.807) - 0.3\dot{y}_D - 0.07(0.95)(1.807) = 0. \tag{3.56}$$

The solution is

$$\dot{x}_D = 0.129 \text{ m/s}, \qquad \dot{y}_D = -0.041 \text{ m/s}.$$

The magnitude of the velocity of joint D is

$$v_D = |\mathbf{v}_D| = \sqrt{\dot{x}_D^2 + \dot{y}_D^2} = \sqrt{0.129^2 + (-0.041)^2} = 0.135 \text{ m/s}.$$

ACCELERATION ANALYSIS

The acceleration analysis can be obtained by using the derivative of the velocity given by Eq. (3.55):

$$\dot{x}_D^2 + x_D\ddot{x}_D + \dot{y}_D^2 + y_D\ddot{y}_D = 0,$$
$$\ddot{x}_D\sin\phi_3 + 2\dot{x}_D\dot{\phi}_3\cos\phi_3 - x_D\dot{\phi}_3^2\sin\phi_3 + x_D\ddot{\phi}_3\cos\phi_3 - \ddot{y}_D\cos\phi_3$$
$$+ 2\dot{y}_D\dot{\phi}_3\sin\phi_3 + y_D\dot{\phi}_3^2\cos\phi_3 + y_D\ddot{\phi}_3^2\sin\phi_3 = 0. \tag{3.57}$$

The solution of the above system is

$$\ddot{x}_D = 0.147 \text{ m/s}^2, \qquad \ddot{y}_D = 0.210 \text{ m/s}^2.$$

The absolute acceleration of joint D is

$$a_D = |\mathbf{a}_D| = \sqrt{\ddot{x}_D^2 + \ddot{y}_D^2} = \sqrt{(0.150)^2 + (0.212)^2} = 0.256 \text{ m/s}^2.$$

Joint *E*

The position of joint E is determined from the following equation:

$$[x_E(t) - x_D(t)]^2 + [y_E(t) - y_D(t)]^2 = DE^2. \tag{3.58}$$

For joint E, with the coordinate $y_E = 0$, Eq. (3.58) becomes

$$[x_E(t) - x_D(t)]^2 + y_D^2(t) = DE^2, \tag{3.59}$$

or

$$(x_E + 0.023)^2 + (0.071)^2 = 0.2^2,$$

with the correct solution $x_E = 0.164$ m.

VELOCITY ANALYSIS

The velocity of joint E is determined by differentiating Eq. (3.59) as follows:

$$2(\dot{x}_E - \dot{x}_D)(x_E - x_D) + 2y_D\dot{y}_D = 0, \tag{3.60}$$

which can be rewritten

$$\dot{x}_E - \dot{x}_D = (y_D\dot{y}_D)/(x_E - x_D).$$

The solution of the above equation is

$$\dot{x}_E = 0.129 - \frac{(-0.071)(-0.041)}{0.164 + 0.023} = 0.113 \text{ m/s}.$$

ACCELERATION ANALYSIS

The derivative of Eq. (3.60) yields

$$(\ddot{x}_E - \ddot{x}_D)(x_E - x_D) + (\dot{x}_E - \dot{x}_D)^2 + \dot{y}_D^2 + y_D\ddot{y}_D = 0, \tag{3.61}$$

with the solution

$$\ddot{x}_E = \ddot{x}_D - \frac{\dot{y}_D^2 + y_D\ddot{y}_D + (\dot{x}_E - \dot{x}_D)^2}{x_E - x_D},$$

or with numerical values

$$\ddot{x}_E = 0.150 - \frac{(-0.041)^2 + (-0.07)(0.21) + (0.112 - 0.129)^2}{0.164 + 0.023}$$
$$= 0.217 \text{ m/s}^2.$$

Link 4

Angle ϕ_4 is determined from the following equation:

$$y_E - y_D(t) - [x_E(t) - x_D(t)] \tan \phi_4(t) = 0, \quad (3.62)$$

where $y_E = 0$. The above equation can be rewritten:

$$-y_D(t) - [x_E(t) - x_D(t)] \tan \phi_4(t) = 0, \quad (3.63)$$

and the solution is

$$\phi_4 = \arctan\left(\frac{-y_D}{x_E - x_D}\right) = \arctan\left(\frac{0.071}{0.164 + 0.023}\right) = 20.923^\circ.$$

VELOCITY ANALYSIS

The derivative of Eq. (3.63) yields

$$-\dot{y}_D - (\dot{x}_E - \dot{x}_D) \tan \phi_4 - (x_E - x_D)\frac{1}{\cos^2 \phi_4}\dot{\phi}_4 = 0. \quad (3.64)$$

Hence

$$\omega_4 = \dot{\phi}_4 = -\frac{\cos^2 \phi_4[\dot{y}_d + (\dot{x}_e - \dot{x}_d) \tan \phi_4]}{x_e - x_d} = 0.221 \text{ rad/s}.$$

ACCELERATION ANALYSIS

The angular acceleration of link 4 is determined by differentiating Eq. (3.64) as follows:

$$-\ddot{y}_D - (\ddot{x}_E - \ddot{x}_D) \tan \phi_4 - 2(\dot{x}_E - \dot{x}_D)\frac{1}{\cos^2 \phi_4}\dot{\phi}_4 - 2(x_E - x_D)\frac{\sin \phi_4}{\cos^3 \phi_4}\dot{\phi}_4^2$$
$$- (x_E - x_D)\frac{1}{\cos^2 \phi_4}\ddot{\phi}_4 = 0.$$

The solution of the above equation gives the angular acceleration α_4 of link 4:

$$\alpha_4 = \ddot{\phi}_4 = -1.105 \text{ rad/s}^2.$$

The *Mathematica* program for the position analysis for a complete rotation of the driver link is given in Appendix 3, and the velocity and acceleration program for $\phi = 45^\circ$ is given in Appendix 4.

The *Mathematica* program for the velocity and acceleration analysis for the mechanism considered in Section 2.3 is presented in Appendix 5.

4 Contour Equations

4.1 Introduction

This chapter aims at providing an algebraic method to compute the velocities and accelerations of any closed kinematic chain. The classical method for obtaining the velocities and accelerations involves the computation of the derivatives with respect to time of the position vectors, as described in Chapter 3. The method of contour equations avoids this task and utilizes only algebraic equations. With the use of this approach, the numerical implementation that results is much simpler and more efficient. The contour equations method described here can be applied to both planar and spatial mechanisms.

Instantaneously Coincident Points

Two rigid links (j) and (k) are connected by a joint (kinematic pair) at A, shown in Fig. 4.1. The point A_j of the rigid body (j) is guided along a path prescribed in the body (k). The point A_j belonging to body (j) and the point A_k belonging to body (k) are coincident at the instant of motion under consideration. The following relation exists between the velocity $\mathbf{v}_{Aj}$ of point A_j and the velocity $\mathbf{v}_{Ak}$ of point A_k:

$$\mathbf{v}_{Aj} = \mathbf{v}_{Ak} + \mathbf{v}^r_{Ajk}, \tag{4.1}$$

where $\mathbf{v}^r_{Ajk}$ indicates the velocity of A_j as seen by an observer at A_k attached to body k or the relative velocity of A_j with respect to A_k, allowed at joint A. The direction of $\mathbf{v}^r_{Ajk}$ is obviously tangent to the path prescribed in body (k).

From Eq. (4.1) the accelerations of A_j and A_k are expressed as

$$\mathbf{a}_{Aj} = \mathbf{a}_{Ak} + \mathbf{a}^r_{Ajk} + \mathbf{a}^c_{Ajk}, \tag{4.2}$$

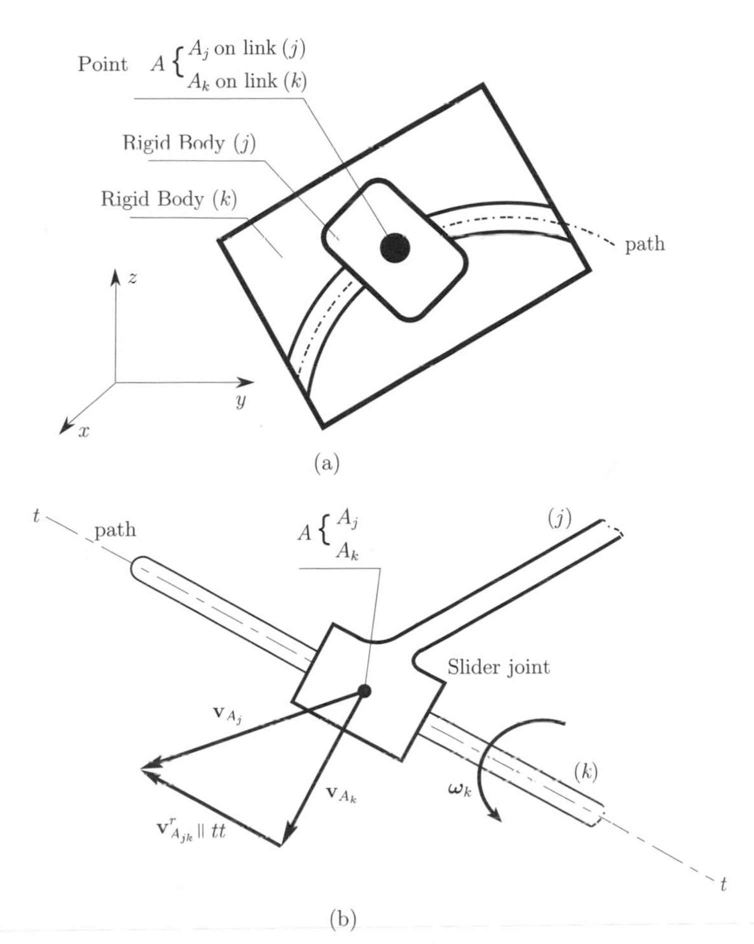

Figure 4.1. Two rigid links (j) and (k) connected by a joint at A: (a) general case and (b) a slider joint in general motion.

where $\mathbf{a}^c_{Ajk}$ is known as the *Coriolis acceleration* and is given by

$$\mathbf{a}^c_{Ajk} = 2\,\omega_k \times \mathbf{v}^r_{Ajk}, \tag{4.3}$$

where ω_k is the angular velocity of the body (k).

Equations (4.1) and (4.2) are useful even for coincident points belonging to two links that may not be directly connected. A graphical representation of Eq. (4.1) is shown in Fig. 4.1(b) for a rotating slider joint.

4.2 Formulation of the Problem

Figure 4.2 shows a monocontour closed kinematic chain with n rigid links. The joint A_i, $i = 0, 1, 2, \ldots, n$ is the connection between links (i) and ($i - 1$).

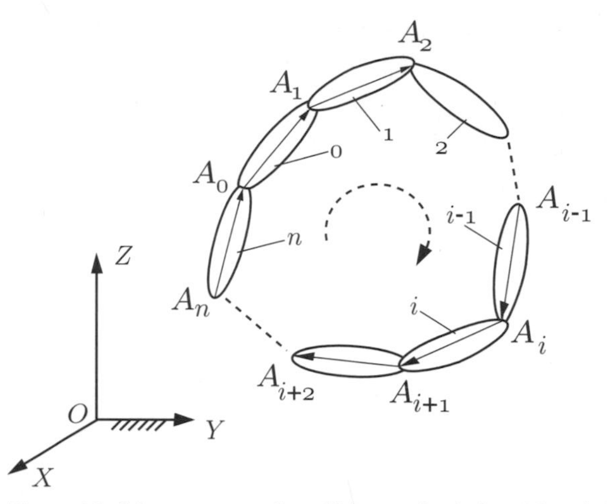

Figure 4.2. Monocontour closed kinematic chain with n rigid links.

The last link n is connected with the first link 0 of the chain. For the closed kinematic chain, a path is chosen from link 0 to link n. At joint A_i there are two instantaneously coincident points: (1) point $A_{i,i}$ belonging to link (i), $A_{i,i} \in (i)$, and (2) point $A_{i,i-1}$ belonging to body $(i-1)$, $A_{i,i-1} \in (i-1)$.

4.3 Contour Velocity Equations

The following relation exists between the velocity $\mathbf{v}_{Ai,i}$ of point $A_{i,i}$ and the velocity $\mathbf{v}_{Ai,i-1}$ of point $A_{i,i-1}$:

$$\mathbf{v}_{Ai,i} = \mathbf{v}_{Ai,i-1} + \mathbf{v}^r_{Ai,i-1}, \tag{4.4}$$

where $\mathbf{v}^r_{Ai,i-1}$ is the relative velocity of $A_{i,i}$ on link (i) with respect to $A_{i,i-1}$ on link $(i-1)$. With the use of the velocity relation for two points on rigid body (i), the following relation exists:

$$\mathbf{v}_{Ai+1,i} = \mathbf{v}_{Ai,i} + \boldsymbol{\omega}_i \times \mathbf{A}_i\mathbf{A}_{i+1}, \tag{4.5}$$

where $\boldsymbol{\omega}_i$ is the absolute angular velocity of link (i) in the reference frame $OXYZ$, and $\mathbf{A}_i\mathbf{A}_{i+1}$ is the distance vector from A_i to A_{i+1}. With the use of Eqs. (4.4) and (4.5), the velocity of point $A_{i+1,i} \in (i+1)$ can be written as

$$\mathbf{v}_{Ai+1,i} = \mathbf{v}_{Ai,i-1} + \boldsymbol{\omega}_i \times \mathbf{A}_i\mathbf{A}_{i+1} + \mathbf{v}^r_{Ai,i-1}. \tag{4.6}$$

The following expressions can be obtained for the n link closed kinematic chain:

$$
\begin{aligned}
\mathbf{v}_{A3,2} &= \mathbf{v}_{A2,1} + \boldsymbol{\omega}_2 \times \mathbf{A}_2\mathbf{A}_3 + \mathbf{v}^r_{A2,1} \\
\mathbf{v}_{A4,3} &= \mathbf{v}_{A3,2} + \boldsymbol{\omega}_3 \times \mathbf{A}_3\mathbf{A}_4 + \mathbf{v}^r_{A3,2} \\
&\cdots\cdots \\
\mathbf{v}_{Ai+1,i} &= \mathbf{v}_{Ai,i-1} + \boldsymbol{\omega}_i \times \mathbf{A}_i\mathbf{A}_{i+1} + \mathbf{v}^r_{Ai,i-1} \\
&\cdots\cdots \\
\mathbf{v}_{A1,0} &= \mathbf{v}_{A0,n} + \boldsymbol{\omega}_0 \times \mathbf{A}_0\mathbf{A}_1 + \mathbf{v}^r_{A0,n} \\
\mathbf{v}_{A2,1} &= \mathbf{v}_{A1,0} + \boldsymbol{\omega}_1 \times \mathbf{A}_1\mathbf{A}_2 + \mathbf{v}^r_{A1,0}.
\end{aligned}
\tag{4.7}
$$

Summing the relations in Eq. (4.7), one can obtain

$$
\begin{aligned}
&[\boldsymbol{\omega}_1 \times \mathbf{A}_1\mathbf{A}_2 + \boldsymbol{\omega}_2 \times \mathbf{A}_2\mathbf{A}_3 + \cdots + \boldsymbol{\omega}_i \times \mathbf{A}_i\mathbf{A}_{i+1} + \cdots + \boldsymbol{\omega}_0 \times \mathbf{A}_0\mathbf{A}_1] \\
&\quad + \left[\mathbf{v}^r_{A2,1} + \mathbf{v}^r_{A3,2} + \cdots + \mathbf{v}^r_{Ai,i-1} + \cdots + \mathbf{v}^r_{A0,n} + \mathbf{v}^r_{A1,0}\right] = \mathbf{0}.
\end{aligned}
\tag{4.8}
$$

Because the reference system $OXYZ$ is considered "fixed" in space, vector $\mathbf{A}_{i-1}\mathbf{A}_i$ can be written in terms of the position vectors of points A_{i-1} and A_i:

$$
\mathbf{A}_{i-1}\mathbf{A}_i = \mathbf{OA}_i - \mathbf{OA}_{i-1}, \tag{4.9}
$$

and Eq. (4.8) becomes

$$
\begin{aligned}
&[\mathbf{OA}_1 \times (\boldsymbol{\omega}_1 - \boldsymbol{\omega}_0) + \mathbf{OA}_2 \times (\boldsymbol{\omega}_2 - \boldsymbol{\omega}_1) + \cdots + \mathbf{OA}_0 \times (\boldsymbol{\omega}_0 - \boldsymbol{\omega}_n)] \\
&\quad + \left[\mathbf{v}^r_{A1,0} + \mathbf{v}^r_{A2,1} + \cdots + \mathbf{v}^r_{Ai,i-1} + \cdots + \mathbf{v}^r_{A0,n}\right] = \mathbf{0}.
\end{aligned}
\tag{4.10}
$$

The following relations exist between the absolute angular velocity $\boldsymbol{\omega}_i$ of rigid body (i) and the relative angular velocity $\boldsymbol{\omega}_{i,i-1}$ of rigid body (i) with respect to rigid body $(i-1)$:

$$
\begin{aligned}
\boldsymbol{\omega}_1 &= \boldsymbol{\omega}_0 + \boldsymbol{\omega}_{1,0} \\
\boldsymbol{\omega}_2 &= \boldsymbol{\omega}_1 + \boldsymbol{\omega}_{2,1} \\
&\cdots\cdots \\
\boldsymbol{\omega}_i &= \boldsymbol{\omega}_{i-1} + \boldsymbol{\omega}_{i,i-1} \\
&\cdots\cdots \\
\boldsymbol{\omega}_0 &= \boldsymbol{\omega}_n + \boldsymbol{\omega}_{0,n}.
\end{aligned}
\tag{4.11}
$$

With the use of Eq. (4.11), Eq. (4.10) becomes

$$
\begin{aligned}
&[\mathbf{OA}_1 \times \boldsymbol{\omega}_{1,0} + \mathbf{OA}_2 \times \boldsymbol{\omega}_{2,1} + \cdots + \mathbf{OA}_0 \times \boldsymbol{\omega}_{0,n}] \\
&\quad + \left[\mathbf{v}^r_{A1,0} + \mathbf{v}^r_{A2,1} + \cdots + \mathbf{v}^r_{A0,n}\right] = \mathbf{0}.
\end{aligned}
\tag{4.12}
$$

The previous equation can be written as

$$
\sum_{(i)} \mathbf{OA}_i \times \boldsymbol{\omega}_{i,i-1} + \sum_{(i)} \mathbf{v}^r_{Ai,i-1} = \mathbf{0}. \tag{4.13}
$$

Summing the expressions given in Eq. (4.11), one can obtain

$$\omega_{1,0} + \omega_{2,1} + \cdots + \omega_{0,n} = \mathbf{0}, \tag{4.14}$$

which may be rewritten as

$$\sum_{(i)} \omega_{i,i-1} = \mathbf{0}. \tag{4.15}$$

Equations (4.13) and (4.15) represent the velocity equations for a simple closed kinematic chain.

4.4 Contour Acceleration Equations

Using the acceleration distributions of the relative motion of two rigid bodies (i) and $(i-1)$, one can write

$$\mathbf{a}_{Ai,i} = \mathbf{a}_{Ai,i-1} + \mathbf{a}^{r}_{Ai,i-1} + \mathbf{a}^{c}_{Ai,i-1}, \tag{4.16}$$

where $\mathbf{a}_{Ai,i}$ and $\mathbf{a}_{Ai,i-1}$ are the linear accelerations of points $A_{i,i}$ and $A_{i,i-1}$, and $\mathbf{a}^{r}_{Ai,i-1}$ is the relative acceleration between $A_{i,i}$ on link (i) and $A_{i,i-1}$ on link $(i-1)$. Finally, $\mathbf{a}^{c}_{Ai,i-1}$ is the Coriolis acceleration defined as

$$\mathbf{a}^{c}_{Ai,i-1} = 2\omega_{i-1} \times \mathbf{v}^{r}_{Ai,i-1}. \tag{4.17}$$

Using the acceleration distribution relations for two points on a rigid body, one can write

$$\mathbf{a}_{Ai+1,i} = \mathbf{a}_{Ai,i} + \alpha_i \times \mathbf{A}_i\mathbf{A}_{i+1} + \omega_i \times (\omega_i \times \mathbf{A}_i\mathbf{A}_{i+1}), \tag{4.18}$$

where α_i is the angular acceleration of link (i). From Eqs. (4.16) and (4.18) one can obtain

$$\begin{aligned}\mathbf{a}_{Ai+1,i} = {} & \mathbf{a}_{Ai,i-1} + \mathbf{a}^{r}_{Ai,i-1} + \mathbf{a}^{c}_{Ai,i-1} + \alpha_i \times \mathbf{A}_i\mathbf{A}_{i+1} \\ & + \omega_i \times (\omega_i \times \mathbf{A}_i\mathbf{A}_{i+1}).\end{aligned} \tag{4.19}$$

When similar equations are written for all the links of the kinematic chain, the following relations are obtained:

$$\begin{aligned}\mathbf{a}_{A3,2} &= \mathbf{a}_{A2,1} + \mathbf{a}^{r}_{A2,1} + \mathbf{a}^{c}_{A2,1} + \alpha_2 \times \mathbf{A}_2\mathbf{A}_3 + \omega_2 \times (\omega_2 \times \mathbf{A}_2\mathbf{A}_3), \\ \mathbf{a}_{A4,3} &= \mathbf{a}_{A3,2} + \mathbf{a}^{r}_{A3,2} + \mathbf{a}^{c}_{A3,2} + \alpha_3 \times \mathbf{A}_3\mathbf{A}_4 + \omega_3 \times (\omega_3 \times \mathbf{A}_3\mathbf{A}_4), \\ & \cdots\cdots\cdots\cdots\cdots\cdots\cdots\cdots\cdots\cdots \\ \mathbf{a}_{A_{1,0}} &= \mathbf{a}_{A_{0,n}} + \mathbf{a}^{r}_{A0,n} + \mathbf{a}^{c}_{A0,n} + \alpha_0 \times \mathbf{A}_0\mathbf{A}_1 + \omega_0 \times (\omega_0 \times \mathbf{A}_0\mathbf{A}_1), \\ \mathbf{a}_{A2,1} &= \mathbf{a}_{A1,0} + \mathbf{a}^{r}_{A1,0} + \mathbf{a}^{c}_{A1,0} + \alpha_1 \times \mathbf{A}_1\mathbf{A}_2 + \omega_1 \times (\omega_1 \times \mathbf{A}_1\mathbf{A}_2).\end{aligned} \tag{4.20}$$

Summing the expressions in Eq. (4.20), one can obtain

$$\left[\mathbf{a}^r_{A1,0} + \mathbf{a}^r_{A2,1} + \cdots + \mathbf{a}^r_{A0,n}\right] + \left[\mathbf{a}^c_{A1,0} + \mathbf{a}^c_{A2,1} + \cdots + \mathbf{a}^c_{A0,n}\right]$$
$$+ [\alpha_1 \times \mathbf{A}_1\mathbf{A}_2 + \alpha_2 \times \mathbf{A}_2\mathbf{A}_3 + \cdots + \alpha_0 \times \mathbf{A}_0\mathbf{A}_1]$$
$$+ \omega_1 \times (\omega_1 \times \mathbf{A}_1\mathbf{A}_2) + \omega_2 \times (\omega_2 \times \mathbf{A}_2\mathbf{A}_3) + \cdots$$
$$+ \omega_0 \times (\omega_0 \times \mathbf{A}_0\mathbf{A}_1) = \mathbf{0}. \tag{4.21}$$

The following vectorial relations exist between the absolute angular acceleration α_i of (i) and the relative angular acceleration $\alpha_{i,i-1}$ of link (i) with respect to link $(i-1)$:

$$\alpha_2 = \alpha_1 + \alpha_{2,1} + \omega_2 \times \omega_{2,1}$$
$$\alpha_3 = \alpha_2 + \alpha_{3,2} + \omega_3 \times \omega_{3,2}$$
$$\cdots\cdots\cdots\cdots\cdots\cdots\cdots\cdots\cdots$$
$$\alpha_i = \alpha_i + \alpha_{i,i-1} + \omega_i \times \omega_{i,i-1}$$
$$\cdots\cdots\cdots\cdots\cdots\cdots\cdots\cdots\cdots$$
$$\alpha_1 = \alpha_0 + \alpha_{1,0} + \omega_1 \times \omega_{1,0}. \tag{4.22}$$

Using the relation $\mathbf{A}_{i-1}\mathbf{A}_i = \mathbf{OA}_i - \mathbf{OA}_{i-1}$ in Eq. (4.21), one can obtain

$$\left[\mathbf{a}^r_{A1,0} + \mathbf{a}^r_{A2,1} + \cdots + \mathbf{a}^r_{A0,n}\right] + \left[\mathbf{a}^c_{A1,0} + \mathbf{a}^c_{A2,1} + \cdots + \mathbf{a}^c_{A0,n}\right]$$
$$+ [\mathbf{OA}_1 \times (\alpha_{1,0} + \omega_1 \times \omega_{1,0}) + \cdots + \mathbf{OA}_0 \times (\alpha_{0,n} + \omega_0 \times \omega_{0,n})]$$
$$+ \omega_1 \times (\omega_1 \times \mathbf{A}_1\mathbf{A}_2) + \omega_2 \times (\omega_2 \times \mathbf{A}_2\mathbf{A}_3) + \cdots$$
$$+ \omega_0 \times (\omega_0 \times \mathbf{A}_0\mathbf{A}_1) = \mathbf{0}. \tag{4.23}$$

Equation (4.23) can be rewritten as

$$\sum_{(i)} \mathbf{a}^r_{Ai,i-1} + \sum_{(i)} \mathbf{a}^c_{Ai,i-1} + \sum_{(i)} \mathbf{OA}_i \times (\alpha_{i,i-1} + \omega_i \times \omega_{i,i-1})$$
$$+ \sum_{(i)} \omega_i \times (\omega_i \times \mathbf{A}_i\mathbf{A}_{i+1}) = \mathbf{0}. \tag{4.24}$$

Summing all the expressions in Eq. (4.22), one can obtain

$$\alpha_{2,1} + \alpha_{3,2} + \cdots + \alpha_{1,0} + \omega_2 \times \omega_{2,1} + \cdots + \omega_1 \times \omega_{1,0} = \mathbf{0}. \tag{4.25}$$

Equation (4.25) can be rewritten as

$$\sum_{(i)} \alpha_{i,i-1} + \sum_{(i)} \omega_i \times \omega_{i,i-1} = \mathbf{0}. \tag{4.26}$$

Equations (4.24) and (4.26) are the acceleration equations for the case of a simple closed kinematic chain.

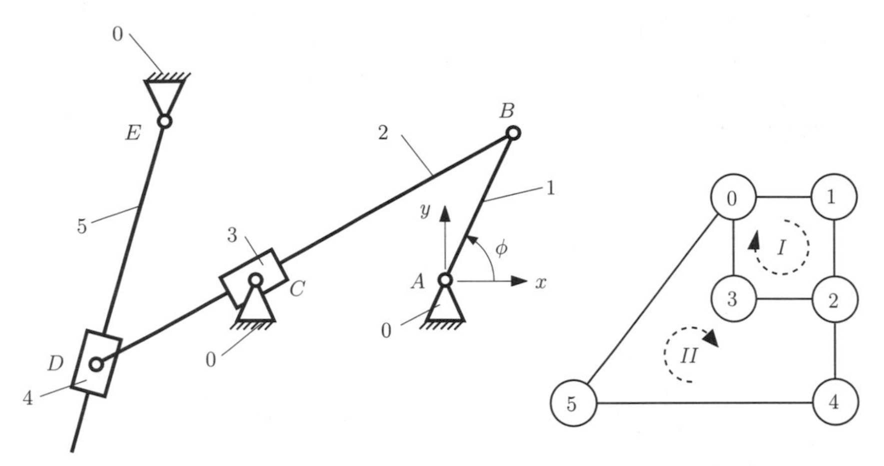

Figure 4.3. Planar mechanism and the diagram that represents the mechanism.

Remarks

1. For a planar kinematic chain, simplified relations are obtained because

$$\omega_i \times (\omega_i \times \mathbf{A}_i\mathbf{A}_{i+1}) = -\omega_i^2 \mathbf{A}_i\mathbf{A}_{i+1}, \qquad \omega_i \times \omega_{i,i-1} = \mathbf{0}. \tag{4.27}$$

2. The Coriolis acceleration, given by the expression

$$\mathbf{a}^c_{Ai,i-1} = 2\omega_{i-1} \times \mathbf{v}^r_{Ai,i-1}, \tag{4.28}$$

vanishes when $\omega_{i-1} = \mathbf{0}$, or $\mathbf{v}_{Ai,i-1} = \mathbf{0}$, or when ω_{i-1} is parallel to $\mathbf{v}_{Ai,i-1}$.

4.5 Diagram Representing a Mechanism

A diagram can be used to represent a mechanism in the following way: the numbered links are the nodes of the diagram and are represented by circles, and the joints are represented by lines that connect the nodes.

Figure 4.3 shows a diagram that represents a planar mechanism, while Fig. 4.4 shows a diagram representing a spatial mechanism. The maximum number of independent contours is given by

$$n_c = l - p + 1, \tag{4.29}$$

where l is the number of joints and p is the number of links.

4.6 Independent Contour Equation Method for Mechanisms

The equations for velocities and accelerations can be written for any closed contour of the mechanism. However, it is best to write the contour equations

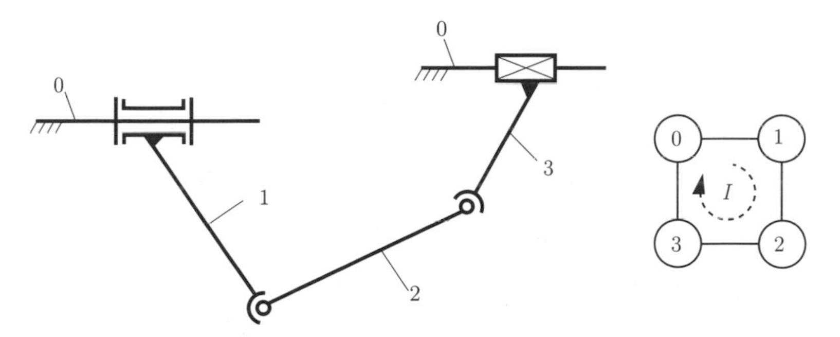

Figure 4.4. Spatial mechanism and the diagram that represents the mechanism.

only for the independent loops of the diagram representing the mechanism.

Step 1. Determine the position analysis of the planar or spatial mechanism.
Step 2. Draw a diagram representing the mechanism and select the independent contours. Determine a path for each contour.
Step 3. For each closed loop, write the contour velocity relations, Eqs. (4.13) and (4.15), and the contour acceleration relations, Eqs. (4.24) and (4.26).
Step 4. Project, on a Cartesian reference system, the velocity and acceleration equations. One can obtain linear algebraic equations where the unknowns are

- the components of the relative angular velocities $\boldsymbol{\omega}_{j,j-1}$;
- the components of the relative angular accelerations $\boldsymbol{\alpha}_{j,j-1}$;
- the components of the relative linear velocities $\mathbf{v}^r_{Aj,j-1}$; and
- the components of the relative linear accelerations $\mathbf{a}^r_{Aj,j-1}$.

Solve the algebraic system of equations and determine the unknown kinematic parameters.
Step 5. Determine the absolute angular velocities $\boldsymbol{\omega}_j$ and the absolute angular accelerations $\boldsymbol{\alpha}_j$. Compute the velocities and accelerations of the characteristic points and joints.

In the following examples, the contour method is applied to determine the distribution of velocities and accelerations for several planar mechanisms.

4.7 Notation

The following notation will be used.

- $\boldsymbol{\omega}_{ij}$ is the relative angular velocity vector of link i with respect to link j. When link j is the ground (denoted as link 0), then $\boldsymbol{\omega}_i = \boldsymbol{\omega}_{i0}$ also denotes the absolute angular velocity vector of link i. The magnitude of $\boldsymbol{\omega}_{ij}$ is ω_{ij}; that is, $|\boldsymbol{\omega}_{ij}| = \omega_{ij}$

- $\mathbf{v}_{ij} = \mathbf{v}^r_{Aij}$ is the relative linear velocity of point A_i on link i with respect to point A_j on link j. Point A_i belonging to link j and Point A_j belonging to link j are coincident at the instant of motion under consideration. When link j is the ground, $\mathbf{v}_i = \mathbf{v}_{i0}$ denotes the absolute linear velocity vector of point A on link i.
- $\boldsymbol{\alpha}_{ij}$ is the relative angular acceleration vector of link i with respect to rigid body j. When link j is the ground, then $\boldsymbol{\alpha}_i = \boldsymbol{\alpha}_{i0}$ also denotes the absolute angular acceleration vector of rigid body i.
- $\mathbf{a}_{ij} = \mathbf{a}^r_{Aij}$ is the relative linear acceleration vector of A_i on link i with respect to A_j on link j. When link j is the ground, then $\mathbf{a}_i = \mathbf{a}_{i0}$.
- $\mathbf{a}^c_{ij}$ is the Coriolis acceleration of A_i with respect to A_j.
- **BC** denotes a vector from joint B to joint C.
- If B is a joint, then x_B, y_B, z_B denote the coordinates of the point B with respect to the fixed reference frame.
- If B is a joint, then $\mathbf{v}_B$ denotes the linear velocity vector of B with respect to the fixed reference frame, and $\mathbf{a}_B$ denotes the linear acceleration vector of joint B with respect to the fixed reference frame.

4.8 R–RRR–RRT Mechanism

The R–RRR–RRT planar mechanism considered is depicted in Fig. 4.5(a). The positions of the mechanism were analyzed in Section 2.3. The mechanism considered has six links and seven full joints. With the use of Eq. (4.29), the number of independent loops is

$$n_c = l - p + 1 = 7 - 6 + 1 = 2.$$

This mechanism has two independent contours. The first contour I contains the links 0, 1, 2, and 3, while the second contour II contains the links 0, 3, 4, and 5. A diagram of the mechanism is represented in Fig. 4.5(b). Clockwise paths are chosen for each closed loop I and II.

First Contour Analysis

According to Fig. 4.6(a), which depicts the first closed-loop I, one can write

- rotational joint R between links 0 and 1 (joint A);
- rotational joint R between links 1 and 2 (joint B);
- rotational joint R between links 2 and 3 (joint C); and
- rotational joint R between links 3 and 0 (joint D).

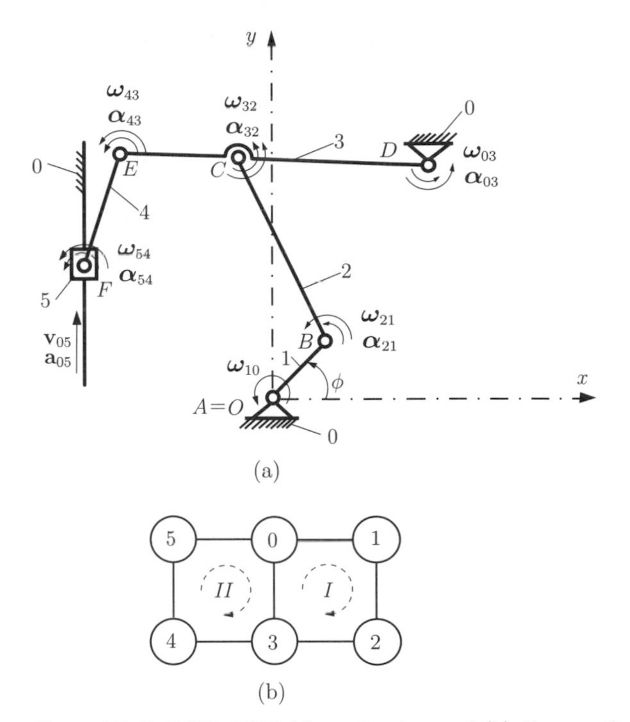

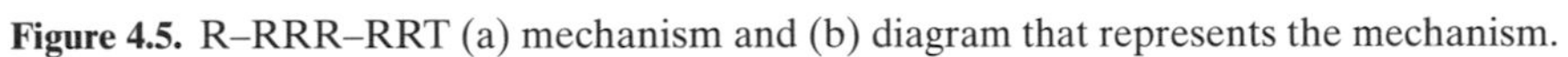

Figure 4.5. R–RRR–RRT (a) mechanism and (b) diagram that represents the mechanism.

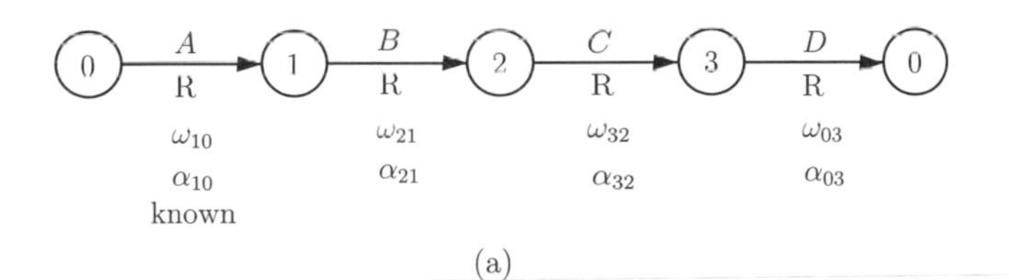

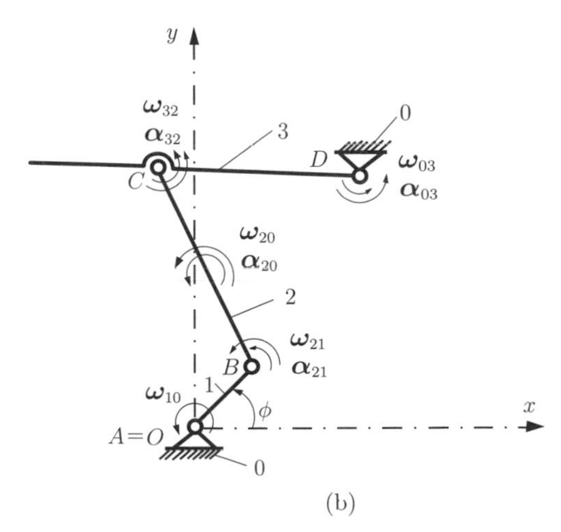

Figure 4.6. First contour RRRR: (a) diagram and (b) mechanism.

The angular velocity of the driver link, ω_{10}, is known:

$$\omega_{10} = \omega_1 = \omega = 100\frac{\pi}{30}\ \text{rad/s} = 10.47\ \text{rad/s}.$$

The following vectorial equations can be written for the *velocity analysis*:

$$\begin{aligned} &\boldsymbol{\omega}_{10} + \boldsymbol{\omega}_{21} + \boldsymbol{\omega}_{32} + \boldsymbol{\omega}_{03} = \mathbf{0}, \\ &\mathbf{AB} \times \boldsymbol{\omega}_{21} + \mathbf{AC} \times \boldsymbol{\omega}_{32} + \mathbf{AD} \times \boldsymbol{\omega}_{03} = \mathbf{0}, \end{aligned} \tag{4.30}$$

where

$$\begin{aligned} \boldsymbol{\omega}_{10} &= \omega_{10}\,\mathbf{k} = [0, 0, \omega_{10}], \\ \boldsymbol{\omega}_{21} &= \omega_{21}\,\mathbf{k} = [0, 0, \omega_{21}], \\ \boldsymbol{\omega}_{32} &= \omega_{32}\,\mathbf{k} = [0, 0, \omega_{32}], \\ \boldsymbol{\omega}_{03} &= \omega_{03}\,\mathbf{k} = [0, 0, \omega_{03}]. \end{aligned}$$

The sign of the relative angular velocities is selected as positive, as shown in Figs. 4.5(a) and 4.6(b). Then the numerical computation will give the correct orientation of the unknown vectors.

The vectors **AB**, **AC**, and **AD** are defined as follows:

$$\begin{aligned} \mathbf{AB} &= x_B\,\mathbf{\imath} + y_B\,\mathbf{\jmath} = [x_B, y_B, 0], \\ \mathbf{AC} &= x_C\,\mathbf{\imath} + y_C\,\mathbf{\jmath} = [x_C, y_C, 0], \\ \mathbf{AD} &= x_D\,\mathbf{\imath} + y_D\,\mathbf{\jmath} = [x_D, y_D, 0]. \end{aligned}$$

The components of the vectors **AB**, **AC**, and **AD** are already known from the position analysis of the mechanism. With these notations, Eq. (4.30) becomes

$$\begin{aligned} &\omega_{10}\,\mathbf{k} + \omega_{21}\,\mathbf{k} + \omega_{32}\,\mathbf{k} + \omega_{03}\,\mathbf{k} = \mathbf{0}, \\ &\begin{vmatrix} \mathbf{\imath} & \mathbf{\jmath} & \mathbf{k} \\ x_B & y_B & 0 \\ 0 & 0 & \omega_{21} \end{vmatrix} + \begin{vmatrix} \mathbf{\imath} & \mathbf{\jmath} & \mathbf{k} \\ x_C & y_C & 0 \\ 0 & 0 & \omega_{32} \end{vmatrix} + \begin{vmatrix} \mathbf{\imath} & \mathbf{\jmath} & \mathbf{k} \\ x_D & y_D & 0 \\ 0 & 0 & \omega_{03} \end{vmatrix} = \mathbf{0}. \end{aligned} \tag{4.31}$$

Equation (4.31) can be projected onto the fixed reference frame $Oxyz$:

$$\begin{aligned} \omega_{10} + \omega_{21} + \omega_{32} + \omega_{03} &= 0, \\ y_B\omega_{21} + y_C\omega_{32} + y_D\omega_{03} &= 0, \\ x_B\omega_{21} + x_C\omega_{32} + x_D\omega_{03} &= 0. \end{aligned} \tag{4.32}$$

Equation (4.32) represents a system of three equations with three unknowns: ω_{21}, ω_{32}, and ω_{03}. When the algebraic equations are solved, the following numerical values are obtained:

$$\omega_{21} = -13.758 \text{ rad/s},$$
$$\omega_{32} = -1.280 \text{ rad/s},$$
$$\omega_{03} = 4.567 \text{ rad/s}.$$

The absolute angular velocities of links BC and DC can be computed from the expressions

$$\boldsymbol{\omega}_{20} = \boldsymbol{\omega}_{10} + \boldsymbol{\omega}_{21} = -3.286\mathbf{k} \text{ rad/s},$$
$$\boldsymbol{\omega}_{30} = \boldsymbol{\omega}_{20} + \boldsymbol{\omega}_{32} = -\boldsymbol{\omega}_{03} = -4.567\mathbf{k} \text{ rad/s}.$$

The absolute linear velocities of joints B and C can be computed with the following expressions:

$$\mathbf{v}_B = \mathbf{v}_A + \boldsymbol{\omega}_{10} \times \mathbf{AB} = -1.110\mathbf{\imath} + 1.110\mathbf{\jmath} \text{ m/s},$$
$$\mathbf{v}_C = \mathbf{v}_B + \boldsymbol{\omega}_{20} \times \mathbf{BC} = \mathbf{v}_D + \boldsymbol{\omega}_{30} \times \mathbf{DC} = 0.070\mathbf{\imath} + 1.688\mathbf{\imath} \text{ m/s},$$

where $\mathbf{v}_A = \mathbf{0}$ and $\mathbf{v}_D = \mathbf{0}$, because joints A and D are grounded and

$$\mathbf{BC} = \mathbf{AC} - \mathbf{AB}, \qquad \mathbf{DC} = \mathbf{AC} - \mathbf{AD}.$$

The following equations can be used for the *acceleration analysis*:

$$\boldsymbol{\alpha}_{10} + \boldsymbol{\alpha}_{21} + \boldsymbol{\alpha}_{32} + \boldsymbol{\alpha}_{03} = \mathbf{0},$$
$$\mathbf{AB} \times \boldsymbol{\alpha}_{21} + \mathbf{AC} \times \boldsymbol{\alpha}_{32} + \mathbf{AD} \times \boldsymbol{\alpha}_{03} - \omega_{10}^2\mathbf{AB} - \omega_{20}^2\mathbf{BC} - \omega_{03}^2\mathbf{CD} = \mathbf{0}. \tag{4.33}$$

Because the driver link has a constant angular velocity, then $\boldsymbol{\alpha}_{10} = \dot{\boldsymbol{\omega}}_{10} = \mathbf{0}$. The components of the angular velocity vectors can be obtained:

$$\boldsymbol{\alpha}_{21} = [0, 0, \alpha_{21}], \qquad \boldsymbol{\alpha}_{32} = [0, 0, \alpha_{32}], \qquad \boldsymbol{\alpha}_{03} = [0, 0, \alpha_{03}].$$

With this notation, Eq. (4.33) can be written as

$$\alpha_{21}\,\mathbf{k} + \alpha_{32}\,\mathbf{k} + \alpha_{03}\,\mathbf{k} = \mathbf{0},$$
$$\begin{vmatrix} \mathbf{\imath} & \mathbf{\jmath} & \mathbf{k} \\ x_B & y_B & 0 \\ 0 & 0 & \alpha_{21} \end{vmatrix} + \begin{vmatrix} \mathbf{\imath} & \mathbf{\jmath} & \mathbf{k} \\ x_C & y_C & 0 \\ 0 & 0 & \alpha_{32} \end{vmatrix} + \begin{vmatrix} \mathbf{\imath} & \mathbf{\jmath} & \mathbf{k} \\ x_D & y_D & 0 \\ 0 & 0 & \alpha_{03} \end{vmatrix} - \omega_{10}^2[x_B\,\mathbf{\imath} + y_B\,\mathbf{\jmath}]$$
$$- \omega_{20}^2[(x_C - x_B)\,\mathbf{\imath} + (y_C - y_B)\,\mathbf{\jmath}] - \omega_{03}^2[(x_D - x_C)\mathbf{\imath} + (y_D - y_C)\mathbf{\jmath}] = \mathbf{0}. \tag{4.34}$$

In the second vectorial relation of Eq. (4.34), only the first two components (for $\mathbf{\imath}$ and $\mathbf{\jmath}$) are not identical to zero. Equation (4.34) can be rewritten as

$$\begin{aligned}
&\alpha_{21} + \alpha_{32} + \alpha_{03} = 0,\\
&y_B\alpha_{21} + y_C\alpha_{32} + y_D\alpha_{03} - \omega_{10}^2 x_B - \omega_{20}^2(x_C - x_B) - \omega_{03}^2(x_D - x_C) = 0,\\
&x_B\alpha_{21} + x_C\alpha_{32} + x_D\alpha_{03} - \omega_{10}^2 y_B - \omega_{20}^2(y_C - y_B) - \omega_{03}^2(y_D - y_C) = 0.
\end{aligned} \tag{4.35}$$

The unknowns in Eq. (4.35) are α_{21}, α_{32}, and α_{03}. When the system is solved, the following numerical values can be obtained:

$$\begin{aligned}
\alpha_{21} &= -47.758 \text{ rad/s}^2,\\
\alpha_{32} &= 66.149 \text{ rad/s}^2,\\
\alpha_{03} &= -18.391 \text{ rad/s}^2.
\end{aligned}$$

The absolute angular accelerations of links BC and DE can be obtained by using the relations

$$\begin{aligned}
\boldsymbol{\alpha}_{20} &= \boldsymbol{\alpha}_{10} + \boldsymbol{\alpha}_{21} = -47.758\mathbf{k} \text{ rad/s}^2,\\
\boldsymbol{\alpha}_{30} &= \boldsymbol{\alpha}_{20} + \boldsymbol{\alpha}_{32} = -\boldsymbol{\alpha}_{03} = 18.391\mathbf{k} \text{ rad/s}^2.
\end{aligned}$$

The absolute linear accelerations of joints B and C can be computed with the expressions

$$\begin{aligned}
\mathbf{a}_B &= \mathbf{a}_A + \boldsymbol{\alpha}_{10} \times \mathbf{AB} - \omega_{10}^2\mathbf{AB} = -11.631\mathbf{\imath} - 11.631\mathbf{\jmath} \text{ m/s}^2,\\
\mathbf{a}_C &= \mathbf{a}_B + \boldsymbol{\alpha}_{20} \times \mathbf{BC} - \omega_{20}^2\mathbf{BC} = \mathbf{a}_D + \boldsymbol{\alpha}_{30} \times \mathbf{DC} - \omega_{30}^2\mathbf{DC}\\
&= 7.427\mathbf{\imath} - 7.119\mathbf{\jmath} \text{ m/s}^2,
\end{aligned} \tag{4.36}$$

where $\mathbf{a}_A = \mathbf{0}$ and $\mathbf{a}_D = \mathbf{0}$, because joints A and D are grounded.

Second Contour Analysis

According to Fig. 4.7(a), one can write

- rotational joint R between links 0 and 3 (joint D);
- rotational joint R between links 3 and 4 (joint E);
- rotational joint R between links 4 and 5 (joint F_R); and
- translational joint T between links 5 and 0 (joint F_T).

The following equations can be written for the *velocity analysis*:

$$\begin{aligned}
&\boldsymbol{\omega}_{30} + \boldsymbol{\omega}_{43} + \boldsymbol{\omega}_{54} = \mathbf{0},\\
&\mathbf{AD} \times \boldsymbol{\omega}_{30} + \mathbf{AE} \times \boldsymbol{\omega}_{43} + \mathbf{AF} \times \boldsymbol{\omega}_{54} + \mathbf{v}_{05} = \mathbf{0}.
\end{aligned} \tag{4.37}$$

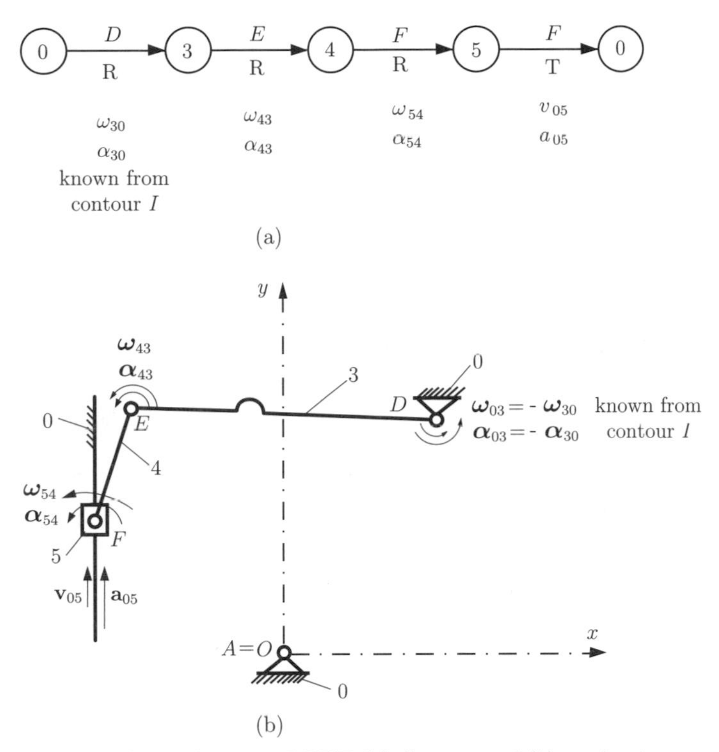

Figure 4.7. Second contour RRRT: (a) diagram and (b) mechanism.

The relative linear velocity vector $\mathbf{v}_{F05} = \mathbf{v}_{05}$ is oriented in the positive sense of y, as shown in Fig. 4.7(b). Since only the y coordinate of this vector is nonzero, one can write

$$\mathbf{v}_{05} = v_{05}\mathbf{J} = [0,\ v_{05},\ 0]. \tag{4.38}$$

With the previous relation, Eq. (4.37) becomes

$$\omega_{30}\mathbf{k} + \omega_{43}\mathbf{k} + \omega_{54}\mathbf{k} = \mathbf{0},$$

$$\begin{vmatrix} \mathbf{\imath} & \mathbf{\jmath} & \mathbf{k} \\ x_D & y_D & 0 \\ 0 & 0 & \omega_{30} \end{vmatrix} + \begin{vmatrix} \mathbf{\imath} & \mathbf{\jmath} & \mathbf{k} \\ x_E & y_E & 0 \\ 0 & 0 & \omega_{43} \end{vmatrix} + \begin{vmatrix} \mathbf{\imath} & \mathbf{\jmath} & \mathbf{k} \\ x_F & y_F & 0 \\ 0 & 0 & \omega_{54} \end{vmatrix} + v_{05}\,\mathbf{\jmath} = \mathbf{0}. \tag{4.39}$$

Equation (4.39) can be rewritten as

$$\begin{aligned} &\omega_{30} + \omega_{43} + \omega_{54} = 0, \\ &y_B\omega_{21} + y_C\omega_{32} + y_E\omega_{43} + y_F\omega_{54} = 0, \\ &x_B\omega_{21} + x_C\omega_{32} + x_E\omega_{43} + x_F\omega_{54} + v_{05} = 0. \end{aligned} \tag{4.40}$$

The unknowns in Eq. (4.40) are ω_{43}, ω_{54}, and v_{05}. When this system is solved, the numerical values can be obtained:

$$\omega_{43} = 4.046 \text{ rad/s},$$
$$\omega_{54} = 0.521 \text{ rad/s},$$
$$v_{05} = -2.7746 \text{ rad/s}.$$

The absolute angular velocity of link EF can be computed as

$$\boldsymbol{\omega}_{40} = \boldsymbol{\omega}_{30} + \boldsymbol{\omega}_{43} = -\boldsymbol{\omega}_{54} = -0.520\mathbf{k} \text{ rad/s}. \tag{4.41}$$

The absolute linear velocities of joints E and F can be calculated by using the expressions

$$\mathbf{v}_E = \mathbf{v}_D + \boldsymbol{\omega}_{30} \times \mathbf{DE} = 0.113\mathbf{ı} + 2.737\mathbf{ȷ} \text{ rad/s},$$
$$\mathbf{v}_F = \mathbf{v}_E + \boldsymbol{\omega}_{40} \times \mathbf{EF} = -\mathbf{v}_{05} = 2.774\mathbf{ȷ} \text{ rad/s}.$$

The following equations can be used for the *acceleration analysis*:

$$\boldsymbol{\alpha}_{30} + \boldsymbol{\alpha}_{43} + \boldsymbol{\alpha}_{54} = \mathbf{0},$$
$$\mathbf{AD} \times \boldsymbol{\alpha}_{30} + \mathbf{AE} \times \boldsymbol{\alpha}_{43} + \mathbf{AF} \times \boldsymbol{\alpha}_{54} + \mathbf{a}_{05}^c + \mathbf{a}_{05} - \omega_{30}^2\mathbf{DE} - \omega_{40}^2\mathbf{EF} = \mathbf{0}. \tag{4.42}$$

In Eq. (4.42), $\mathbf{a}_{05}^c$ is the Coriolis acceleration and $\mathbf{a}_{05}$ is the relative acceleration at F. The relative linear acceleration $\mathbf{a}_{05}$ has the direction shown in Fig. 4.7(b) and only the y component is nonzero. The Coriolis acceleration $\mathbf{a}_{05}^c$ is

$$\mathbf{a}_{05}^c = 2\boldsymbol{\omega}_{50} \times \mathbf{v}_{05} = \mathbf{0}, \tag{4.43}$$

since slider 5 performs only a translation movement ($\boldsymbol{\omega}_{50} = \mathbf{0}$). With the previous remark, Eq. (4.42) becomes

$$\alpha_{30}\mathbf{k} + \alpha_{43}\mathbf{k} + \alpha_{54}\mathbf{k} = \mathbf{0},$$
$$\begin{vmatrix} \mathbf{ı} & \mathbf{ȷ} & \mathbf{k} \\ x_D & y_D & 0 \\ 0 & 0 & \alpha_{30} \end{vmatrix} + \begin{vmatrix} \mathbf{ı} & \mathbf{ȷ} & \mathbf{k} \\ x_E & y_E & 0 \\ 0 & 0 & \alpha_{43} \end{vmatrix} + \begin{vmatrix} \mathbf{ı} & \mathbf{ȷ} & \mathbf{k} \\ x_F & y_F & 0 \\ 0 & 0 & \alpha_{54} \end{vmatrix} + a_{05}\mathbf{ȷ}$$
$$- \omega_{30}^2[(x_E - x_D)\,\mathbf{ı} + (y_E - y_D)\,\mathbf{ȷ}] - \omega_{40}^2[(x_F - x_E)\,\mathbf{ı} + (y_F - y_E)\mathbf{ȷ}] = \mathbf{0}. \tag{4.44}$$

The unknowns in Eq. (4.44) are α_{43}, α_{54}, and a_{05}. The numerical values can be obtained as

$$\alpha_{43} = -73.498 \text{ rad/s}^2,$$
$$\alpha_{54} = 55.107 \text{ rad/s}^2,$$
$$a_{05} = 7.600 \text{ rad/s}^2.$$

The absolute angular acceleration of the link EF is

$$\alpha_{40} = \alpha_{30} + \alpha_{43} = -\alpha_{54} = -55.107\mathbf{k} \text{ rad/s}^2.$$

The absolute linear accelerations of joints E and F can be computed with the following expressions:

$$\mathbf{a}_E = \mathbf{a}_D + \alpha_{30} \times \mathbf{DE} - \omega_{30}^2 \mathbf{DE} = 12.045\mathbf{\imath} - 11.545\mathbf{\jmath} \text{ rad/s}^2,$$
$$\mathbf{a}_F = \mathbf{a}_E + \alpha_{40} \times \mathbf{EF} - \omega_{40}^2 \mathbf{EF} = -\mathbf{a}_{05} = -7.600\mathbf{\jmath} \text{ rad/s}^2.$$

A *Mathematica* program for velocity and acceleration analysis using the contour method is given in Appendix 6.

4.9 R–TRR–RRT Mechanism

The R–TRR–RRT planar mechanism considered in this example is depicted in Fig. 4.8(a). The positions of the mechanism were analyzed in Section 2.4. The mechanism has six links and seven full joints. With the use of Eq. (4.29), the number of independent loops is given by

$$n_c = l - p + 1 = 7 - 6 + 1 = 2.$$

This mechanism has two independent contours. The first contour I contains links 0, 1, 2, and 3, while the second contour II contains links 0, 3, 4, and 5. The

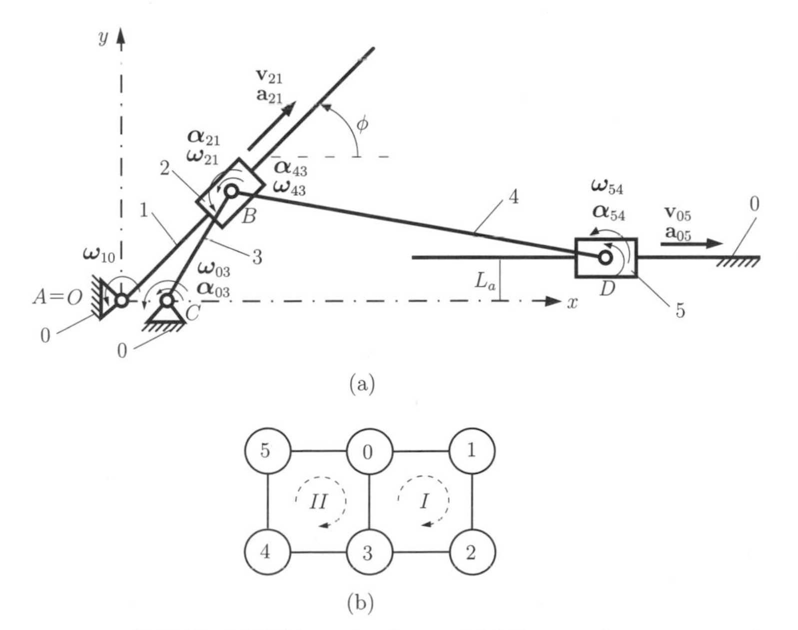

Figure 4.8. R–TRR–RRT (a) mechanism and (b) diagram that represents the mechanism.

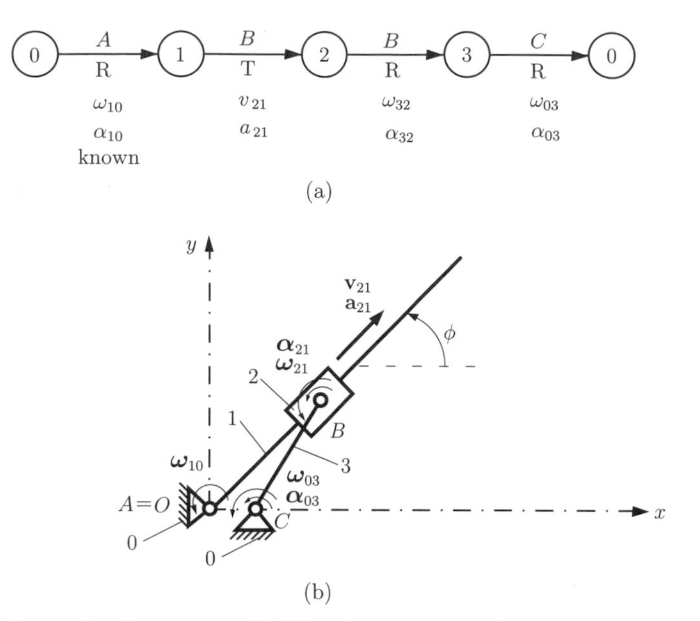

Figure 4.9. First contour RTRR: (a) diagram and (b) mechanism.

diagram representing the mechanism is given in Fig. 4.8(b). Clockwise paths are chosen for each closed-loop *I* and *II*. The angular velocity of the driver link is

$$\omega_{10} = \omega_1 = \omega = 100\frac{\pi}{30} \text{ rad/s} = 10.47 \text{ rad/s}.$$

First Contour Analysis

According to Fig. 4.9, the first contour has

- rotational joint R between links 0 and 1 (joint A);
- translational joint T between links 1 and 2 (joint B_{T});
- rotational joint R between links 2 and 3 (joint B_{R}); and
- rotational joint R between links 3 and 0 (joint C).

The following equations can be written for the *velocity analysis*:

$$\begin{aligned} &\boldsymbol{\omega}_{10} + \boldsymbol{\omega}_{32} + \boldsymbol{\omega}_{03} = \mathbf{0}, \\ &\mathbf{AB} \times \boldsymbol{\omega}_{32} + \mathbf{AC} \times \boldsymbol{\omega}_{03} + \mathbf{v}_{21} = \mathbf{0}. \end{aligned} \tag{4.45}$$

The relative velocity of B_2 on link 2 with respect to B_1 on link 1, $\mathbf{v}_{B21} = \mathbf{v}_{21}$, has $\imath$ and $\jmath$ components:

$$\mathbf{v}_{21} = v_{21x}\boldsymbol{\imath} + v_{21y}\boldsymbol{\jmath} = [v_{21x},\ v_{21y},\ 0].$$

The two components v_{21x} and v_{21y} are not independent,

$$v_{21x} = v_{21}\cos\phi, \qquad v_{21y} = v_{21}\sin\phi, \tag{4.46}$$

where v_{21} is the magnitude of vector $\mathbf{v}_{21}$; that is, $|\mathbf{v}_{21}| = v_{21}$. The unknowns in Eqs. (4.45) and (4.46) are v_{21}, ω_{32}, and ω_{03}. The following numerical results can be obtained:

$$\begin{aligned} v_{21} &= -0.92 \text{ m/s}, \\ \omega_{32} &= 2.54 \text{ rad/s}, \\ \omega_{03} &= -13.01 \text{ rad/s}. \end{aligned}$$

The absolute angular velocity of link 3 is

$$\boldsymbol{\omega}_{30} = -\boldsymbol{\omega}_{03} = 13.01\mathbf{k} \text{ rad/s}. \tag{4.47}$$

The velocity of joint B can be computed with the expression

$$\mathbf{v}_B = \mathbf{v}_C + \boldsymbol{\omega}_{30} \times \mathbf{CB} = -3.333\mathbf{\imath} + 2.031\mathbf{\jmath} \text{ m/s}, \tag{4.48}$$

where $\mathbf{v}_C = \mathbf{0}$ because joint C is grounded.
Link 2 and the driver link have the samc angular velocity:

$$\boldsymbol{\omega}_{10} = \boldsymbol{\omega}_{20} = \boldsymbol{\omega}_{30} + \boldsymbol{\omega}_{23}.$$

One can write for the *acceleration analysis*:

$$\begin{aligned} &\boldsymbol{\alpha}_{32} + \boldsymbol{\alpha}_{03} = \mathbf{0}, \\ &\mathbf{a}_{21} + \mathbf{a}_{21}^c + \mathbf{AB} \times \boldsymbol{\alpha}_{32} + \mathbf{AC} \times \boldsymbol{\alpha}_{03} - \omega_{10}^2\mathbf{AB} - \omega_{03}^2\mathbf{BC} = \mathbf{0}. \end{aligned} \tag{4.49}$$

The expression for the Coriolis acceleration is

$$\mathbf{a}_{21}^c = 2\boldsymbol{\omega}_{10} \times \mathbf{v}_{21}. \tag{4.50}$$

In Eq. (4.49), the x and y components of the relative linear acceleration $\mathbf{a}_{21}$ are not independent. They must satisfy the conditions

$$a_{21x} = a_{21}\cos\phi, \qquad a_{21y} = a_{21}\sin\phi, \tag{4.51}$$

where a_{21} is the magnitude of vector $\mathbf{a}_{21}$. The unknowns in Eq. (4.49) are a_{21}, α_{32}, and α_{03}. The following numerical values are obtained:

$$\begin{aligned} a_{21} &= -7.865 \text{ m/s}^2, \\ \alpha_{03} &= 25.032 \text{ rad/s}^2, \\ \alpha_{32} &= -25.032 \text{ rad/s}^2. \end{aligned}$$

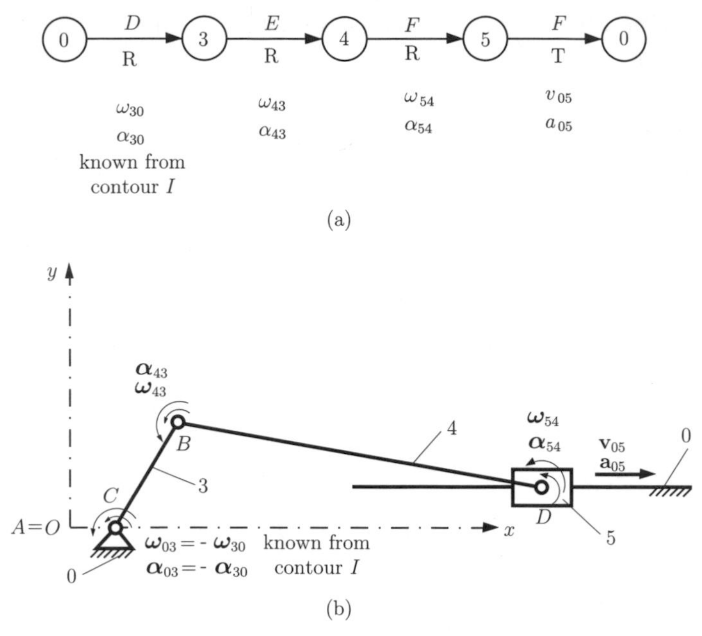

Figure 4.10. Second contour RRRT: (a) diagram and (b) mechanism.

The angular acceleration of link 3 is

$$\boldsymbol{\alpha}_{30} = -\boldsymbol{\alpha}_{03} = \boldsymbol{\alpha}_{32} = -25.032\mathbf{k} \text{ rad/s}^2.$$

The absolute linear acceleration of joint B is computed as follows:

$$\mathbf{a}_B = \mathbf{a}_C + \boldsymbol{\alpha}_{30} \times \mathbf{CB} - \omega_{30}^2 \mathbf{CB} = -20.026\mathbf{\imath} - 47.277\mathbf{\jmath} \text{ m/s}^2.$$

Second Contour Analysis

According to Fig. 4.10, the second contour is described as

- rotational joint R between links 0 and 3 (joint C);
- rotational joint R between links 3 and 4 (joint B);
- rotational joint R between links 4 and 5 (joint D_R); and
- translational joint T between links 5 and 0 (joint D_T).

The following equations can be written for the *velocity analysis*:

$$\begin{aligned} &\boldsymbol{\omega}_{30} + \boldsymbol{\omega}_{43} + \boldsymbol{\omega}_{54} = \mathbf{0}, \\ &\mathbf{AC} \times \boldsymbol{\omega}_{30} + \mathbf{AB} \times \boldsymbol{\omega}_{43} + \mathbf{AD} \times \boldsymbol{\omega}_{54} + \mathbf{v}_{05} = \mathbf{0}. \end{aligned} \tag{4.52}$$

The relative linear velocity $\mathbf{v}_{D05} = \mathbf{v}_{05}$ has only one component, along the x axis:

$$\mathbf{v}_{05} = [v_{05}, 0, 0].$$

The unknown parameters in Eq. (4.52) are ω_{43}, ω_{54}, and v_{05}. The following numerical values are obtained:

$$\omega_{43} = -15.304 \text{ rad/s},$$
$$\omega_{54} = 2.292 \text{ rad/s},$$
$$v_{05} = 3.691 \text{ m/s}.$$

The angular velocity of link BD is

$$\boldsymbol{\omega}_{40} = \boldsymbol{\omega}_{30} + \boldsymbol{\omega}_{43} = -\boldsymbol{\omega}_{54} = -2.292\mathbf{k} \text{ rad/s}.$$

The absolute linear velocity of joint D can be computed as follows:

$$\mathbf{v}_D = \mathbf{v}_B + \boldsymbol{\omega}_{40} \times \mathbf{BD} = -\mathbf{v}_{05} = -3.691\mathbf{ı} \text{ m/s}.$$

The following equations exist for the *acceleration analysis*:

$$\begin{aligned} &\boldsymbol{\alpha}_{30} + \boldsymbol{\alpha}_{43} + \boldsymbol{\alpha}_{54} = \mathbf{0}, \\ &\mathbf{a}_{05} + \mathbf{a}_{05}^c + \mathbf{AC} \times \boldsymbol{\alpha}_{30} + \mathbf{AB} \times \boldsymbol{\alpha}_{43} + \mathbf{AD} \times \boldsymbol{\alpha}_{54} \\ &\quad - \omega_{30}^2 \mathbf{CB} - \omega_{40}^2 \mathbf{BD} = \mathbf{0}. \end{aligned} \tag{4.53}$$

Because slider 5 does not rotate ($\boldsymbol{\omega}_{50} = \mathbf{0}$), the Coriolis acceleration is

$$\mathbf{a}_{05}^c = 2\boldsymbol{\omega}_{50} \times \mathbf{v}_{05} = \mathbf{0}.$$

The unknowns in Eq. (4.53) are α_{43}, α_{54}, and a_{05}. The following numerical results can be obtained:

$$\alpha_{43} = 77.446 \text{ rad/s}^2,$$
$$\alpha_{54} = -52.414 \text{ rad/s}^2,$$
$$a_{05} = 16.499 \text{ m/s}^2.$$

The absolute angular acceleration of link BD is

$$\boldsymbol{\alpha}_{40} = \boldsymbol{\alpha}_{30} + \boldsymbol{\alpha}_{43} = \boldsymbol{\alpha}_{45} = 52.414\mathbf{k} \text{ rad/s}^2,$$

and the linear acceleration of joint D is

$$\mathbf{a}_D = \mathbf{a}_B + \boldsymbol{\alpha}_{40} \times \mathbf{BD} - \omega_{40}^2 \mathbf{BD} = -16.499\mathbf{ı} \text{ m/s}^2.$$

A *Mathematica* program for velocity and acceleration analysis using the contour method for this mechanism is given in Appendix 7.

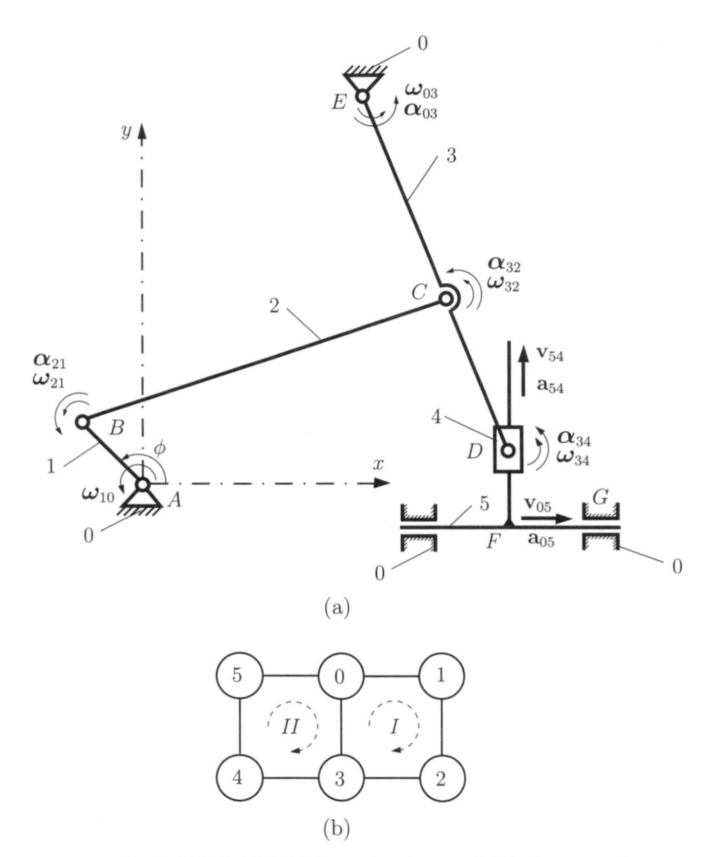

Figure 4.11. R–RRR–RTT (a) mechanism and (b) diagram that represents the mechanism.

4.10 R–RRR–RTT Mechanism

The R–RRR–RTT planar mechanism considered in this example is depicted in Fig. 4.11(a). The positions of the mechanism were analyzed in Section 2.5. The mechanism has six links and seven full joints. With the use of Eq. (4.29), the number of independent loops is given by

$$n_c = l - p + 1 = 7 - 6 + 1 = 2.$$

This mechanism has two independent contours. The first contour I contains links 0, 1, 2, and 3, while the second contour II contains links 0, 3, 4, and 5. The diagram of the mechanism is represented in Fig. 4.11(b). Clockwise paths are chosen for each closed-loop I and II. The angular velocity of the driver link is

$$\omega_{10} = \omega_1 = \omega = 100\frac{\pi}{30} \text{ rad/s} = 10.472 \text{ rad/s}.$$

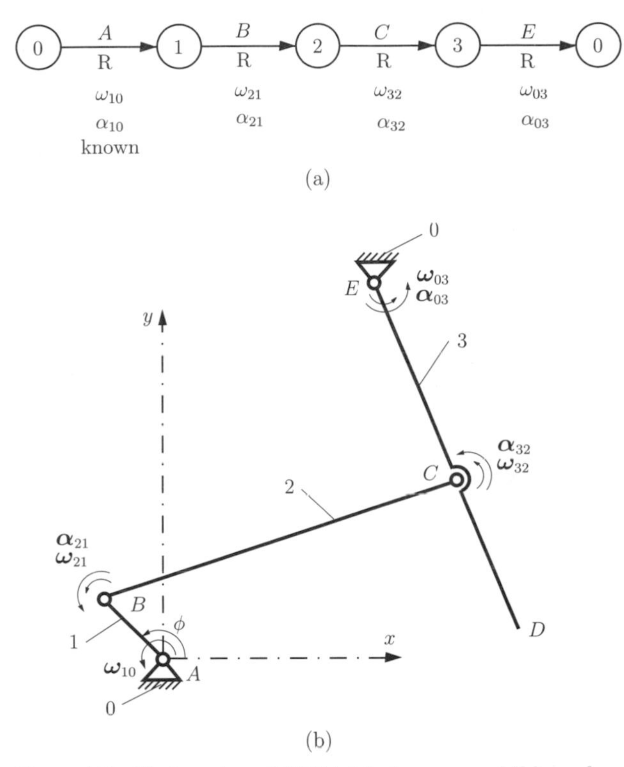

Figure 4.12. First contour RRRR: (a) diagram and (b) mechanism.

First Contour Analysis

The first contour (Fig. 4.12) has

- rotational joint R between links 0 and 1 (joint A);
- rotational joint R between links 1 and 2 (joint B);
- rotational joint R between links 2 and 3 (joint C);
- rotational joint R between links 3 and 0 (joint D).

The following equations can be written for the *velocity analysis*:

$$\boldsymbol{\omega}_{10} + \boldsymbol{\omega}_{21} + \boldsymbol{\omega}_{32} + \boldsymbol{\omega}_{03} = \mathbf{0},$$
$$\mathbf{AB} \times \boldsymbol{\omega}_{21} + \mathbf{AC} \times \boldsymbol{\omega}_{32} + \mathbf{AE} \times \boldsymbol{\omega}_{03} = \mathbf{0}. \tag{4.54}$$

The unknowns in Eq. (4.54) are ω_{21}, ω_{32}, and ω_{03}. When the algebraic system of equations is solved, the following numerical values are found:

$$\omega_{21} = -9.549 \text{ rad/s},$$
$$\omega_{32} = -4.667 \text{ rad/s},$$
$$\omega_{03} = 3.744 \text{ rad/s}.$$

The angular velocities of links 2 and 3 are

$$\omega_{20} = \omega_{10} + \omega_{21} = 0.922\mathbf{k} \text{ rad/s},$$
$$\omega_{30} = \omega_{20} + \omega_{32} = -\omega_{03} = -3.744\mathbf{k} \text{ rad/s}.$$

The absolute velocities of joints B and C are

$$\mathbf{v}_B = \mathbf{v}_A + \omega_{10} \times \mathbf{AB} = -0.592\mathbf{\imath} - 0.592\mathbf{\imath} \text{ m/s},$$
$$\mathbf{v}_C = \mathbf{v}_B + \omega_{20} \times \mathbf{BC} = \mathbf{v}_E + \omega_{30} \times \mathbf{EC} = -0.692\mathbf{\imath} - 0.285\mathbf{\jmath} \text{ m/s},$$

where $\mathbf{v}_A = \mathbf{v}_E = \mathbf{0}$, because joints A and E are grounded.

One can write for the *acceleration analysis*:

$$\boldsymbol{\alpha}_{21} + \boldsymbol{\alpha}_{32} + \boldsymbol{\alpha}_{03} = \mathbf{0},$$
$$\mathbf{AB} \times \boldsymbol{\alpha}_{21} + \mathbf{AC} \times \boldsymbol{\alpha}_{32} + \mathbf{AE} \times \boldsymbol{\alpha}_{03} - \omega_{10}^2\mathbf{AB} - \omega_{20}^2\mathbf{BC} - \omega_{30}^2\mathbf{CE} = \mathbf{0}. \tag{4.55}$$

The unknowns in Eq. (4.55) are α_{21}, α_{32}, and α_{03}. The numerical values, obtained after the algebraic system of equations is solved, are

$$\alpha_{21} = 31.178 \text{ rad/s}^2,$$
$$\alpha_{32} = -11.681 \text{ rad/s}^2,$$
$$\alpha_{03} = -19.496 \text{ rad/s}^2.$$

The accelerations of joints B, C, and D are

$$\mathbf{a}_B = \mathbf{a}_A + \boldsymbol{\alpha}_{10} \times \mathbf{AB} - \omega_{10}^2\mathbf{AB} = 6.203\mathbf{\imath} - 6.203\mathbf{\jmath} \text{ m/s}^2,$$
$$\mathbf{a}_C = \mathbf{a}_B + \boldsymbol{\alpha}_{20} \times \mathbf{BC} - \omega_{20}^2\mathbf{BC}$$
$$= \mathbf{a}_E + \boldsymbol{\alpha}_{30} \times \mathbf{EC} - \omega_{30}^2\mathbf{EC} = 2.536\mathbf{\imath} + 4.078\mathbf{\jmath} \text{ m/s}^2,$$
$$\mathbf{a}_D = \mathbf{a}_E + \boldsymbol{\alpha}_{30} \times \mathbf{ED} - \omega_{30_z}^2\mathbf{ED} = 4.439\mathbf{\imath} + 7.137\mathbf{\jmath} \text{ m/s}^2,$$

where $\mathbf{a}_A = \mathbf{a}_E = \mathbf{0}$ (joints A and E are grounded) and $\boldsymbol{\alpha}_{10} = \mathbf{0}$ (the driver link has a constant angular velocity).

Second Contour Analysis

According to Fig. 4.13 the second contour can be described by

- rotational joint R between links 0 and 3 (joint E);
- rotational joint R between links 3 and 4 (joint D);
- translational joint T between links 4 and 5 (joint D); and
- translational joint T between links 5 and 0 (joint G).

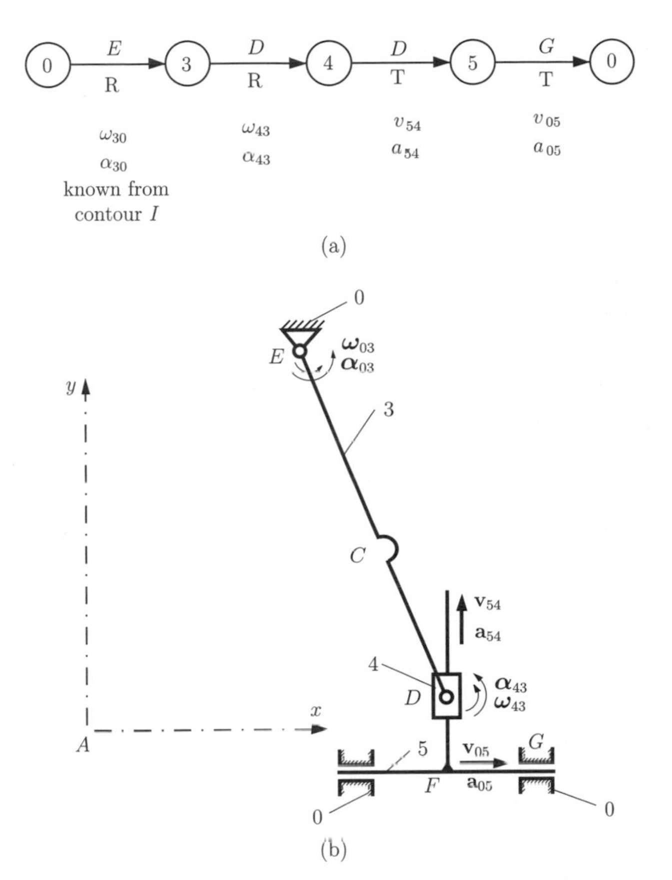

Figure 4.13. Second contour RRTT: (a) diagram and (b) mechanism.

The following equations can be written for the *velocity analysis*:

$$\boldsymbol{\omega}_{30} + \boldsymbol{\omega}_{43} = \mathbf{0},$$

$$\mathbf{AE} \times \boldsymbol{\omega}_{30} + \mathbf{AD} \times \boldsymbol{\omega}_{43} + \mathbf{v}_{54} + \mathbf{v}_{05} = \mathbf{0}. \tag{4.56}$$

The relative linear velocity $\mathbf{v}_{54}$ has a nonzero component only in the y direction,

$$\mathbf{v}_{54} = v_{54y}\mathbf{J} = v_{54}\mathbf{J},$$

while $\mathbf{v}_{05}$ has a nonzero component only in the x direction,

$$\mathbf{v}_{05} = v_{05x}\mathbf{I} = v_{05}\mathbf{I}.$$

The unknown parameters of Eq. (4.56) are ω_{43}, v_{54}, and v_{05}. The following numerical values are obtained:

$$\omega_{43} = 3.744 \text{ rad/s},$$

$$v_{54} = 0.499 \text{ m/s},$$

$$v_{05} = 1.211 \text{ m/s}.$$

The following equations are used for the *acceleration analysis*:

$$\boldsymbol{\alpha}_{30} + \boldsymbol{\alpha}_{43} = \mathbf{0},$$
$$\mathbf{AE} \times \boldsymbol{\alpha}_{30} + \mathbf{AD} \times \boldsymbol{\alpha}_{43} + \mathbf{a}_{54} + \mathbf{a}_{54}^{c} + \mathbf{a}_{05} - \omega_{30}^{2}\mathbf{ED} = \mathbf{0}. \tag{4.57}$$

The Coriolis acceleration $\mathbf{a}_{54}^{c} = 2\boldsymbol{\omega}_{40} \times \mathbf{v}_{54}$ is zero because link 4 has only a translation motion ($\boldsymbol{\omega}_{40} = \mathbf{0}$). Relative acceleration $\mathbf{a}_{54}$ has a nonzero component only in the y direction (a_{54}), and acceleration $\mathbf{a}_{05}$ has a nonzero component only in the x direction (a_{05}). The unknowns in Eq. (4.57) are α_{43}, a_{54}, and a_{05}. The numerical values obtained are

$$\alpha_{43} = -19.496 \text{ rad/s}^2,$$
$$a_{54} = -7.137 \text{ m/s}^2,$$
$$a_{05} = -4.439 \text{ m/s}^2.$$

5 Dynamic Force Analysis

For a kinematic chain it is important to know how forces and torques are transmitted from the input to the output of the system, so that the links can be properly designated. The friction effects are assumed to be negligible in the force analysis presented here. The first part of this chapter is a review of general force analysis principles and conventions.

5.1 Force, Moment, Couple

A *force vector* **F** has a magnitude, an orientation, and a sense. The magnitude of a vector is specified by a positive number and a unit having appropriate dimensions. The orientation of a vector is specified by the relationship between the vector and given reference lines and/or planes. The sense of a vector is specified by the order of two points on a line parallel to the vector. Orientation and sense together determine the direction of a vector. The line of action of a vector is a hypothetical infinite straight line collinear with the vector.

The force vector **F** can be expressed in terms of a Cartesian reference frame, with the unit vectors **ı**, **ȷ**, and **k**, as shown in Fig. 5.1:

$$\mathbf{F} = F_x\boldsymbol{\imath} + F_y\boldsymbol{\jmath} + F_z\mathbf{k}. \tag{5.1}$$

The components of the force in the x, y, and z directions are F_x, F_y, and F_z, respectively. The resultant of two forces $\mathbf{F}_1 = F_{1x}\boldsymbol{\imath} + F_{1y}\boldsymbol{\jmath} + F_{1z}\mathbf{k}$ and $\mathbf{F}_2 = F_{2x}\boldsymbol{\imath} + F_{2y}\boldsymbol{\jmath} + F_{2z}\mathbf{k}$ is the vector sum of those forces:

$$\mathbf{R} = \mathbf{F}_1 + \mathbf{F}_2 = (F_{1x} + F_{2x})\boldsymbol{\imath} + (F_{1y} + F_{2y})\boldsymbol{\jmath} + (F_{1z} + F_{2z})\mathbf{k}. \tag{5.2}$$

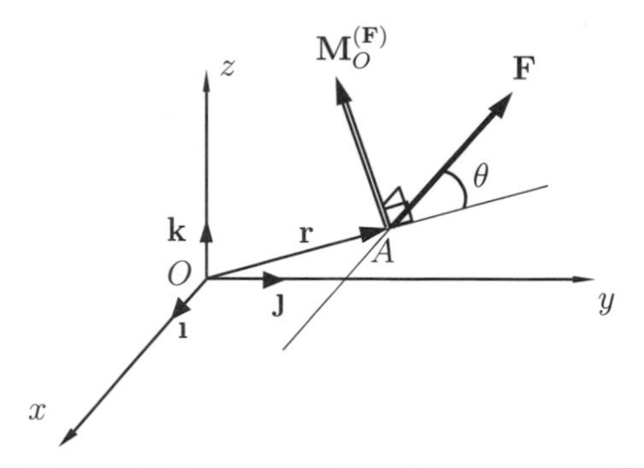

Figure 5.1. Force vector **F** and the moment of **F** about point O.

A *moment* (*torque*) is defined as the moment of a force about (with respect to) a point. The moment of force **F** about point O is the cross-product vector

$$\begin{aligned}
\mathbf{M}_O^{(\mathbf{F})} &= \mathbf{r} \times \mathbf{F} \\
&= \begin{vmatrix} \imath & \jmath & \mathbf{k} \\ r_x & r_y & r_z \\ F_x & F_y & F_z \end{vmatrix} \\
&= (r_y F_z - r_z F_y)\imath + (r_z F_x - r_x F_z)\jmath + (r_x F_y - r_y F_x)\mathbf{k},
\end{aligned} \tag{5.3}$$

where $\mathbf{r} = \mathbf{OA} = r_x \imath + r_y \jmath + r_z \mathbf{k}$ is a position vector directed from the point about which the moment is taken (O in this case) to any point on the line of action of the force; see Fig. 5.1.

The magnitude of $\mathbf{M}_O^{(\mathbf{F})}$ is

$$\left|\mathbf{M}_O^{(\mathbf{F})}\right| = M_O^{(\mathbf{F})} = rF|\sin\theta|,$$

where $\theta = \angle(\mathbf{r}, \mathbf{F})$ is the angle between vectors **r** and **F**, and $r = |\mathbf{r}|$ and $F = |\mathbf{F}|$ are the magnitudes of the vectors.

The line of action of $\mathbf{M}_O^{(\mathbf{F})}$ is perpendicular to the plane containing **r** and **F**, $\mathbf{M}_O^{(\mathbf{F})} \perp \mathbf{r}$ and $\mathbf{M}_O^{(\mathbf{F})} \perp \mathbf{F}$, and the sense is given by the right-hand rule.

Two forces, $\mathbf{F}_1$ and $\mathbf{F}_2$, that have equal magnitudes $|\mathbf{F}_1| = |\mathbf{F}_2|$, opposite senses $\mathbf{F}_1 = -\mathbf{F}_2$, and parallel directions ($\mathbf{F}_1 || \mathbf{F}_2$) are called a *couple*. The resultant force **R** of a couple is zero, $\mathbf{R} = \mathbf{F}_1 + \mathbf{F}_2 = \mathbf{0}$. The resultant moment $\mathbf{M} \neq \mathbf{0}$ about an arbitrary point is

$$\mathbf{M} = \mathbf{r}_1 \times \mathbf{F}_1 + \mathbf{r}_2 \times \mathbf{F}_2,$$

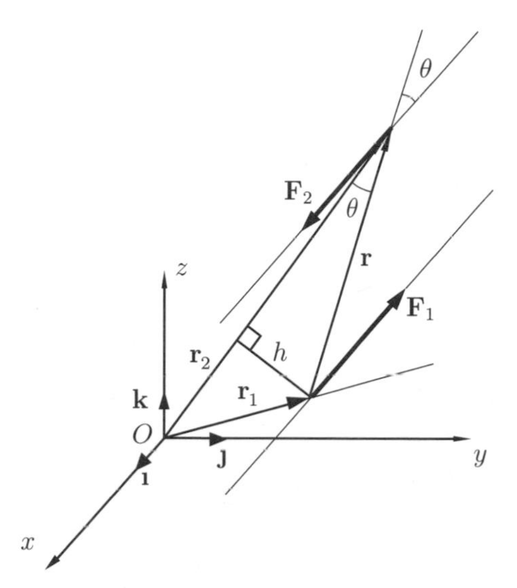

Figure 5.2. Couple.

or

$$\mathbf{M} = \mathbf{r}_1 \times (-\mathbf{F}_2) + \mathbf{r}_2 \times \mathbf{F}_2 = (\mathbf{r}_2 - \mathbf{r}_1) \times \mathbf{F}_2 = \mathbf{r} \times \mathbf{F}_2, \tag{5.4}$$

where $\mathbf{r} = \mathbf{r}_2 - \mathbf{r}_1$ is a vector from any point on the line of action of $\mathbf{F}_1$ to any point of the line of action of $\mathbf{F}_2$. The direction of the torque is perpendicular to the plane of the couple, and the magnitude is given by (see Fig. 5.2)

$$|\mathbf{M}| = M = rF_2|\sin\theta| = hF_2, \tag{5.5}$$

where $h = r|\sin\theta|$ is the perpendicular distance between the lines of action. The resultant moment of a couple is independent of the point with respect to which moments are taken.

5.2 Newton's Second Law of Motion

Newton's second law of motion states that *a particle acted on by forces whose resultant is not zero will move in such a way that the time rate of change of its momentum will at any instant be proportional to the resultant force.* In the case of a particle with constant mass (m = constant), Newton's second law can be expressed as

$$\mathbf{F} = m\mathbf{a}, \tag{5.6}$$

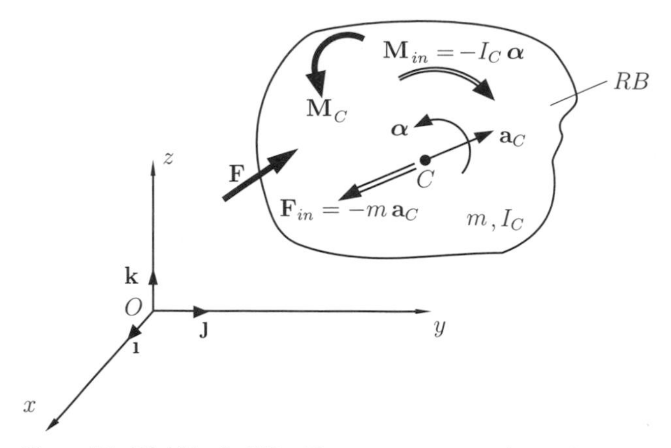

Figure 5.3. Rigid body RB with constant mass m in motion.

where the vector **F** is the resultant of all the external forces on the particle and **a** is the acceleration vector of the particle.

In the case of a rigid body, RB, with constant mass m, Newton's second law is (see Fig. 5.3)

$$\mathbf{F} = m\mathbf{a}_C, \tag{5.7}$$

$$\mathbf{M}_C = I_C\boldsymbol{\alpha}, \tag{5.8}$$

where

- **F** is the resultant of the external forces on the rigid body,
- $\mathbf{a}_C$ is the acceleration of the center of mass, C, of the rigid body,
- $\mathbf{M}_C$ is the resultant external moment on the rigid body about the center of mass, C,
- I_C is the mass moment of inertia of the rigid body with respect to an axis passing through the center of mass, C, and perpendicular to the plane of rotation of the rigid body, and
- $\boldsymbol{\alpha}$ is the angular acceleration of the rigid body.

Equations (5.7) and (5.8) can be interpreted in two ways.

1. The forces and moments are known and the equations can be solved for the motion of the rigid body (direct dynamics).
2. The motion of the RB is known and the equations can be solved for the force and moments (inverse dynamics).

The dynamic force analysis in this chapter is based on the known motion of the mechanism.

5.3 D'Alembert's Principle

D'Alembert's principle is derived from Newton's second law and is expressed as

$$\mathbf{F} + (-m\mathbf{a}_C) = \mathbf{0}, \tag{5.9}$$

$$\mathbf{M}_C + (-I_C\boldsymbol{\alpha}) = \mathbf{0}. \tag{5.10}$$

The terms in parentheses in Eqs. (5.9) and (5.10) are called the *inertia force* and the *inertia moment*, respectively. The inertia force $\mathbf{F}_{\text{in}}$ is

$$\mathbf{F}_{\text{in}} = -m\mathbf{a}_C, \tag{5.11}$$

and the inertia moment is

$$\mathbf{M}_{\text{in}} = -I_C\boldsymbol{\alpha}. \tag{5.12}$$

The dynamic force analysis can be expressed in a form similar to that of static force analysis:

$$\sum \mathbf{R} = \sum \mathbf{F} + \mathbf{F}_{\text{in}} = \mathbf{0}, \tag{5.13}$$

$$\sum \mathbf{T}_C = \sum \mathbf{M}_C + \mathbf{M}_{\text{in}} = \mathbf{0}, \tag{5.14}$$

where $\sum \mathbf{F}$ is the vector sum of all external forces (resultant of external forces), and $\sum \mathbf{M}_C$ is the sum of all external moments about the center of mass C (resultant external moment).

For a rigid body in plane motion in the xy plane,

$$\mathbf{a}_C = \ddot{x}_C\mathbf{\imath} + \ddot{y}_C\mathbf{\jmath}, \qquad \boldsymbol{\alpha} = \alpha\mathbf{k}.$$

With all external forces in the xy plane, Eqs. (5.13) and (5.14) become

$$\sum R_x = \sum F_x + F_{\text{in}\,x} = \sum F_x + (-m\ddot{x}_C) = 0, \tag{5.15}$$

$$\sum R_y = \sum F_y + F_{\text{in}\,y} = \sum F_y + (-m\ddot{y}_C) = 0, \tag{5.16}$$

$$\sum T_C = \sum M_C + M_{\text{in}} = \sum M_C + (-I_C\alpha) = 0. \tag{5.17}$$

With the use of d'Alembert's principle, the moment summation can be about any arbitrary point P,

$$\sum \mathbf{T}_P = \sum \mathbf{M}_P + \mathbf{M}_{\text{in}} + \mathbf{r}_{PC} \times \mathbf{F}_{\text{in}} = \mathbf{0}, \tag{5.18}$$

where

- $\sum \mathbf{M}_P$ is the sum of all external moments about P,
- $\mathbf{M}_{\text{in}} = -I_C\boldsymbol{\alpha}$ is the inertia moment,
- $\mathbf{F}_{\text{in}} = -m\mathbf{a}_C$ is the inertia force, and
- $\mathbf{r}_{PC} = \mathbf{PC}$ is a vector from P to C.

The dynamic analysis problem is reduced to a static force and moment balance problem in which the inertia forces and moments are treated in the same way as external forces and torques.

5.4 Free-Body Diagrams

A free-body diagram is a drawing of a part of a complete system, which is isolated in order to determine the forces acting on that rigid body.

The following force convention is defined: $\mathbf{F}_{ij}$ represents the force exerted by link i on link j.

Figure 5.4 shows various free-body diagrams that can be considered in the analysis of a crank slider mechanism, as shown in Fig. 5.4(a). In Fig. 5.4(b), the free body consists of the three moving links isolated from the frame 0. The

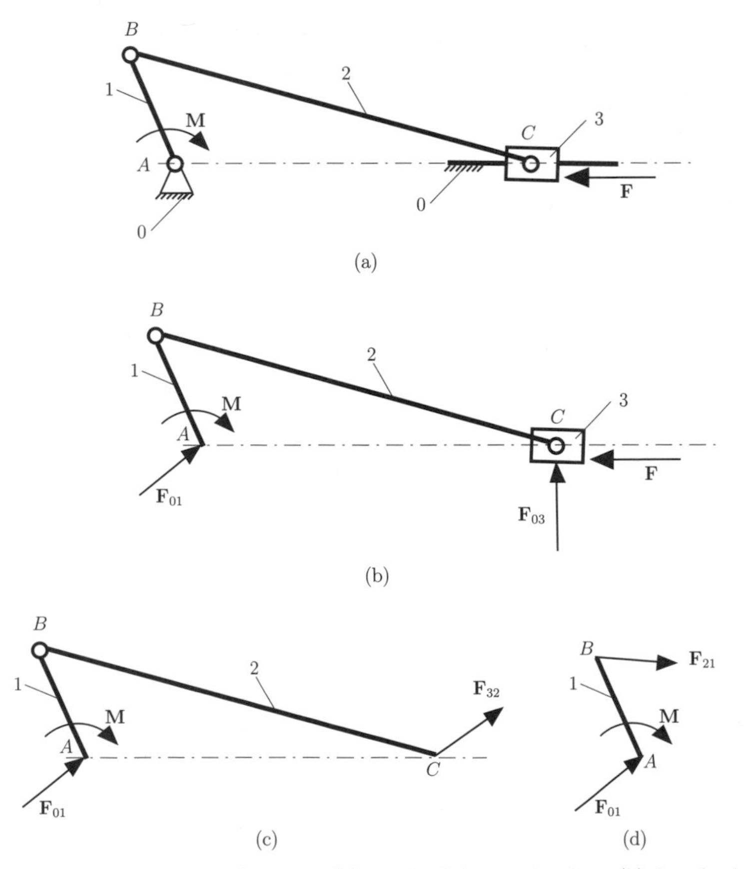

Figure 5.4. Free-body diagrams: (a) crank slider mechanism; (b) free-body diagram of the three links 1, 2, and 3; (c) free-body diagram of the two links 1 and 2; and (d) free-body diagram of link 1.

forces acting on the system include a driving torque **M**, an external driven force **F**, and the forces transmitted from the frame at joint A, $\mathbf{F}_{01}$, and at joint C, $\mathbf{F}_{03}$. Figure 5.4(c) is a free-body diagram of the two links 1 and 2. Figure 5.4(d) is a free-body diagram of a single link.

The force analysis can be accomplished by examining individual links or a subsystem of links. In this way, the joint forces between links, as well as the required input force or moment for a given output load, are computed.

5.5 Computation of Joint Forces by Using Individual Links

Figure 5.5(a) is a schematic diagram of a slider-crank mechanism composed of a crank 1, a connecting rod 2, and a slider 3. The center of mass of link 1 is C_1, the center of mass of link 2 is C_2, and the center of mass of slider 3 is C. The mass of the crank is m_1, the mass of the connecting rod is m_2, and the mass of the slider is m_3. The moment of inertia of link i is I_{Ci}, $i = 1, 2, 3$. The gravitational force is $\mathbf{G}_i = -m_i g\mathbf{J}$, $i = 1, 2, 3$, where $g = 9.81$ m/s^2 is the acceleration of gravity.

For a given value of the crank angle ϕ and a known driven force $\mathbf{F}_{\text{ext}}$, the joint reactions and the drive moment **M** on the crank can be computed by using free-body diagrams of the individual links.

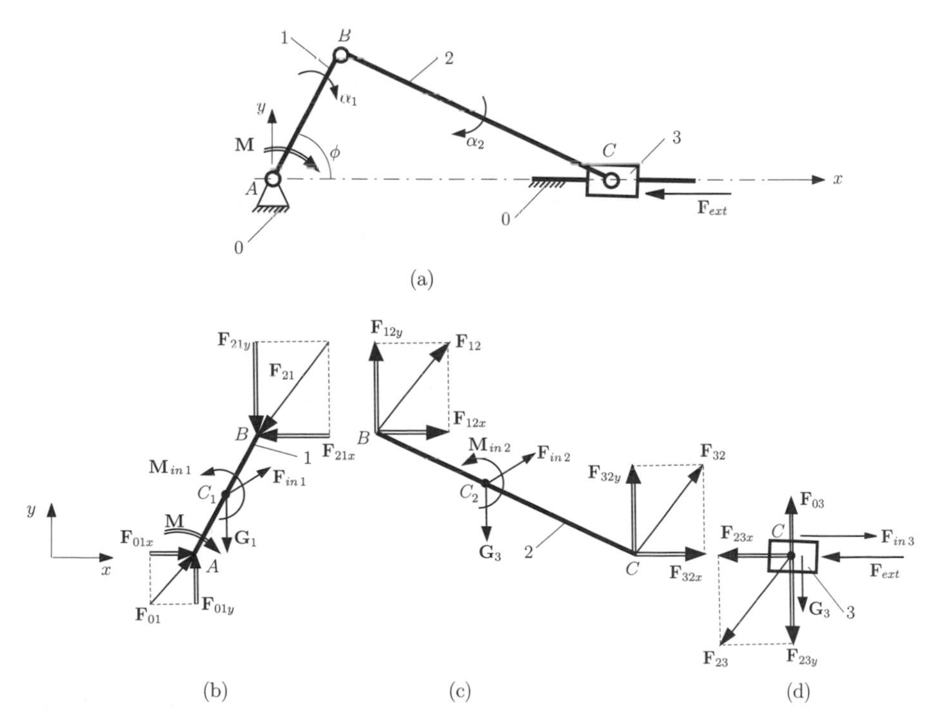

Figure 5.5. (a) Schematic diagram of a slider-crank mechanism; free-body diagrams of (b) crank 1, (c) rod 2, and (d) slider 3.

Figures 5.5(b), (c), and (d) show free-body diagrams of crank 1, connecting rod 2, and slider 3. For each moving link the dynamic equilibrium equations are applied.

For slider 3, the vector sum of all the forces (external forces $\mathbf{F}_{\text{ext}}$, gravitational force $\mathbf{G}_3$, inertia forces $\mathbf{F}_{\text{in}3}$, and joint forces $\mathbf{F}_{23}$, $\mathbf{F}_{03}$) is zero, as shown in Fig. 5.5(d):

$$\sum \mathbf{F}^{(3)} = \mathbf{F}_{23} + \mathbf{F}_{\text{in}3} + \mathbf{G}_3 + \mathbf{F}_{\text{ext}} + \mathbf{F}_{03} = \mathbf{0}.$$

Projecting this total force onto x and y axes gives

$$\sum \mathbf{F}^{(3)} \cdot \mathbf{\imath} = F_{23x} + (-m_3 \ddot{x}_C) + F_{\text{ext}} = 0, \tag{5.19}$$

$$\sum \mathbf{F}^{(3)} \cdot \mathbf{\jmath} = F_{23y} - m_3 g + F_{03y} = 0. \tag{5.20}$$

For connecting rod 2, shown in Fig. 5.5(c), two vectorial equations can be written:

$$\sum \mathbf{F}^{(2)} = \mathbf{F}_{32} + \mathbf{F}_{\text{in}2} + \mathbf{G}_2 + \mathbf{F}_{12} = \mathbf{0},$$

$$\sum \mathbf{M}_B^{(2)} = (\mathbf{r}_C - \mathbf{r}_B) \times \mathbf{F}_{32} + (\mathbf{r}_{C2} - \mathbf{r}_B) \times (\mathbf{F}_{\text{in}2} + \mathbf{G}_2) + \mathbf{M}_{\text{in}2} = \mathbf{0},$$

or

$$\sum \mathbf{F}^{(2)} \cdot \mathbf{\imath} = F_{32x} + (-m_2 \ddot{x}_{C2}) + F_{12x} = 0, \tag{5.21}$$

$$\sum \mathbf{F}^{(2)} \cdot \mathbf{\jmath} = F_{32y} + (-m_2 \ddot{y}_{C2}) - m_2 g + F_{12y} = 0, \tag{5.22}$$

$$\begin{vmatrix} \mathbf{\imath} & \mathbf{\jmath} & \mathbf{k} \\ x_C - x_B & y_C - y_B & 0 \\ F_{32x} & F_{32y} & 0 \end{vmatrix} + \begin{vmatrix} \mathbf{\imath} & \mathbf{\jmath} & \mathbf{k} \\ x_{C2} - x_B & y_{C2} - y_B & 0 \\ -m_2 \ddot{x}_{C2} & -m_2 \ddot{y}_{C2} - m_2 g & 0 \end{vmatrix} - I_{C2} \alpha_2 \mathbf{k} = \mathbf{0}. \tag{5.23}$$

For crank 1, shown in Fig. 5.5(b), there are two vectorial equations:

$$\sum \mathbf{F}^{(1)} = \mathbf{F}_{21} + \mathbf{F}_{\text{in}1} + \mathbf{G}_1 + \mathbf{F}_{01} = \mathbf{0},$$

$$\sum \mathbf{M}_A^{(1)} = \mathbf{r}_B \times \mathbf{F}_{21} + \mathbf{r}_{C1} \times (\mathbf{F}_{\text{in}1} + \mathbf{G}_1) + \mathbf{M}_{\text{in}1} + \mathbf{M} = \mathbf{0},$$

or

$$\sum \mathbf{F}^{(1)} \cdot \mathbf{\imath} = F_{21x} + (-m_1 \ddot{x}_{C1}) + F_{01x} = 0, \tag{5.24}$$

$$\sum \mathbf{F}^{(1)} \cdot \mathbf{\jmath} = F_{21y} + (-m_1 \ddot{y}_{C1}) - m_1 g + F_{01y} = 0, \tag{5.25}$$

$$\begin{vmatrix} \mathbf{\imath} & \mathbf{\jmath} & \mathbf{k} \\ x_B & y_B & 0 \\ F_{21x} & F_{21y} & 0 \end{vmatrix} + \begin{vmatrix} \mathbf{\imath} & \mathbf{\jmath} & \mathbf{k} \\ x_{C1} & y_{C1} & 0 \\ -m_1 \ddot{x}_{C1} & -m_1 \ddot{y}_{C1} - m_1 g & 0 \end{vmatrix} - I_{C1} \alpha_1 \mathbf{k} + M\mathbf{k} = \mathbf{0}, \tag{5.26}$$

where $M = |\mathbf{M}|$ is the magnitude of the input torque on the crank.

The eight scalar unknowns F_{03y}, $F_{23x} = -F_{32x}$, $F_{23y} = -F_{32y}$, $F_{12x} = -F_{21x}$, $F_{12y} = -F_{21y}$, F_{01x}, F_{01y}, and M are computed from the set of eight equations, Eqs. (5.20), (5.20), (5.21), (5.22), (5.23), (5.24), (5.25), and (5.26).

5.6 Computation of Joint Forces by Using the Contour Method

An analytical method to compute joint forces that can be applied for both planar and spatial mechanisms is presented. The method is based on the decoupling of a closed kinematic chain and the writing of the dynamic equilibrium equations. The kinematic links are loaded with external forces and inertia forces and moments.

A general monocontour closed kinematic chain is considered in Fig. 5.6. The joint force between links $i-1$ and i (joint A_i) will be determined. When these two links $i-1$ and i are separated, as in Fig. 5.6(b), the joint forces $\mathbf{F}_{i-1,i}$ and $\mathbf{F}_{i,i-1}$ are introduced and

$$\mathbf{F}_{i-1,i} + \mathbf{F}_{i,i-1} = \mathbf{0}. \tag{5.27}$$

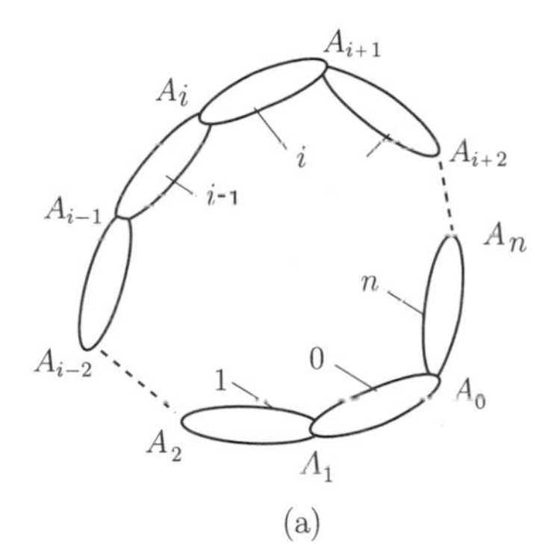

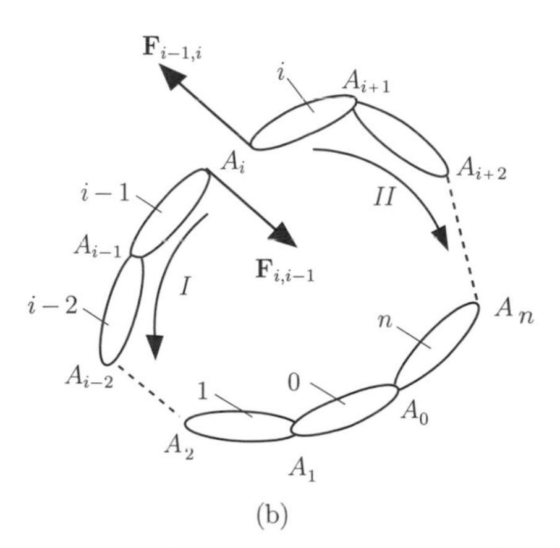

Figure 5.6. (a) General monocontour closed kinematic chain, and (b) joint at A_i replaced by joint forces $\mathbf{F}_{i-1,i}$ and $\mathbf{F}_{i,i-1}$: $\mathbf{F}_{i-1,i} + \mathbf{F}_{i,i-1} = \mathbf{0}$.

Table 5.1. Joint Forces

Type of Kinematic Pair	Joint Force or Moment	Unknowns	Equilibrium Condition
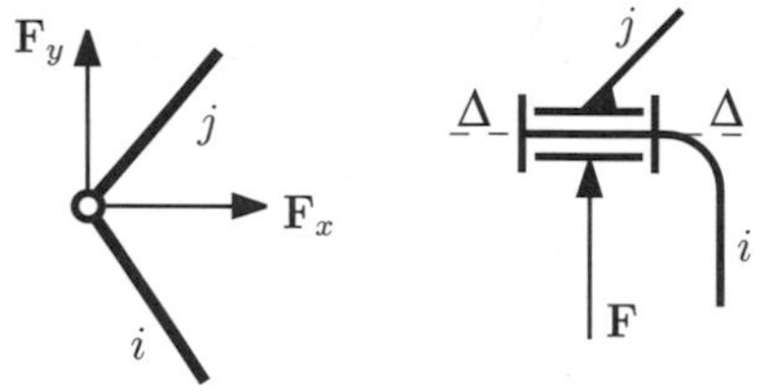 rotational joint	$\mathbf{F}_x + \mathbf{F}_y = \mathbf{F}$ $\mathbf{F} \perp \Delta\Delta$	$\lvert\mathbf{F}_x\rvert = F_x$ $\lvert\mathbf{F}_y\rvert = F_y$	$\mathbf{M}_\Delta = 0$
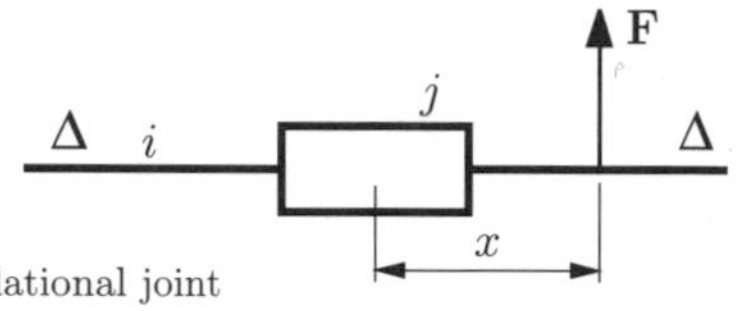 translational joint	$\mathbf{F} \perp \Delta\Delta$	$\lvert\mathbf{F}\rvert = F$ x	$\mathbf{F}_\Delta = 0$
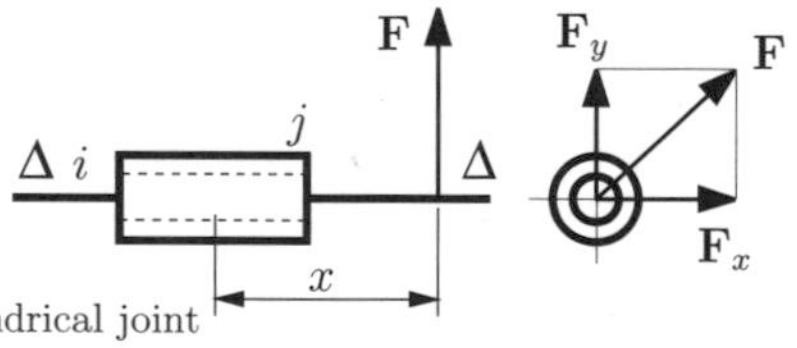 cylindrical joint	$\mathbf{F}_x + \mathbf{F}_y = \mathbf{F}$ $\mathbf{F} \perp \Delta\Delta$	$\lvert\mathbf{F}\rvert = F_x$ $\lvert\mathbf{F}_y\rvert = F_y$ x	$\mathbf{F}_\Delta = 0$ $\mathbf{M}_\Delta = 0$
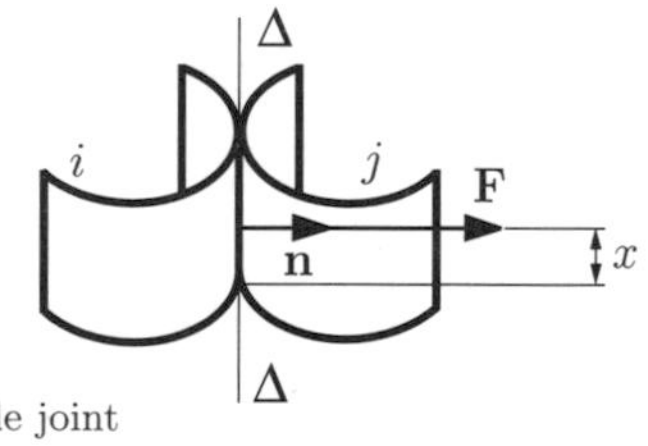 roll-slide joint	$\mathbf{F} \perp \Delta\Delta$ $\mathbf{F} \parallel n$	$\lvert\mathbf{F}\rvert = F$ x	$\mathbf{F}_\Delta = 0$ $\mathbf{M}_\Delta = 0$
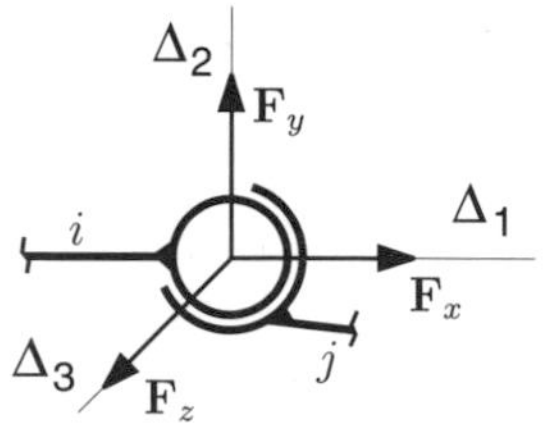 sphere joint	$\mathbf{F}_x + \mathbf{F}_y + \mathbf{F}_z = \mathbf{F}$	$\lvert\mathbf{F}_x\rvert = F_x$ $\lvert\mathbf{F}_y\rvert = F_y$ $\lvert\mathbf{F}_z\rvert = F_z$	$\mathbf{M}_{\Delta_1} = 0$ $\mathbf{M}_{\Delta_2} = 0$ $\mathbf{M}_{\Delta_3} = 0$

Table 5.1 shows the joint forces for several joints. The following notations have been used: $\mathbf{M}_\Delta$ is the moment with respect to axis Δ, and F_Δ is the projection of force vector $\mathbf{F}$ onto axis Δ.

It is helpful to "mentally disconnect" the two links $(i-1)$ and i, which create the joint A_i, from the rest of the mechanism. The joint at A_i will be replaced by joint forces $\mathbf{F}_{i-1,i}$ and $\mathbf{F}_{i,i-1}$. The closed kinematic chain has been transformed into two open kinematic chains, and two paths I and II can be discerned. The two paths start from A_i.

For the path I (counterclockwise), starting at A_i and following I, the first joint encountered is A_{i-1}. For the link $i-1$ left behind, dynamic equilibrium equations can be written according to the type of the joint at A_{i-1}. Following the same path I, the next joint encountered is A_{i-2}. For the subsystem ($i-1$ and $i-2$), equilibrium conditions corresponding to the type of the joint at A_{i-2} can be specified, and so on. A similar analysis can be performed for the path II of the open kinematic chain. The number of equilibrium equations written is equal to the number of unknown scalars introduced by joint A_i (joint forces at this joint). For a joint, the number of equilibrium conditions is equal to the number of relative mobilities of the joint.

The five-link ($j = 1, 2, 3, 4, 5$) mechanism shown in Fig. 5.7(a) has the center of mass locations designated by $C_j(x_{Cj}, y_{Cj}, 0)$. The following analysis will consider the relationships of inertia forces $\mathbf{F}_{\text{in}\,j}$, inertia moments $\mathbf{M}_{\text{in}\,j}$, gravitational force $\mathbf{G}_{\text{in}\,j}$, and driven force, $\mathbf{F}_{\text{ext}}$, to joint reactions $\mathbf{F}_{ij}$ and driver torque $\mathbf{M}$ on crank 1.

To simplify the notation, the total vector force at C_j is written as $\mathbf{F}_j = \mathbf{F}_{\text{in}\,j} + \mathbf{G}_j$ and the inertia torque of link j is written as $\mathbf{M}_j = \mathbf{M}_{\text{in}\,j}$.

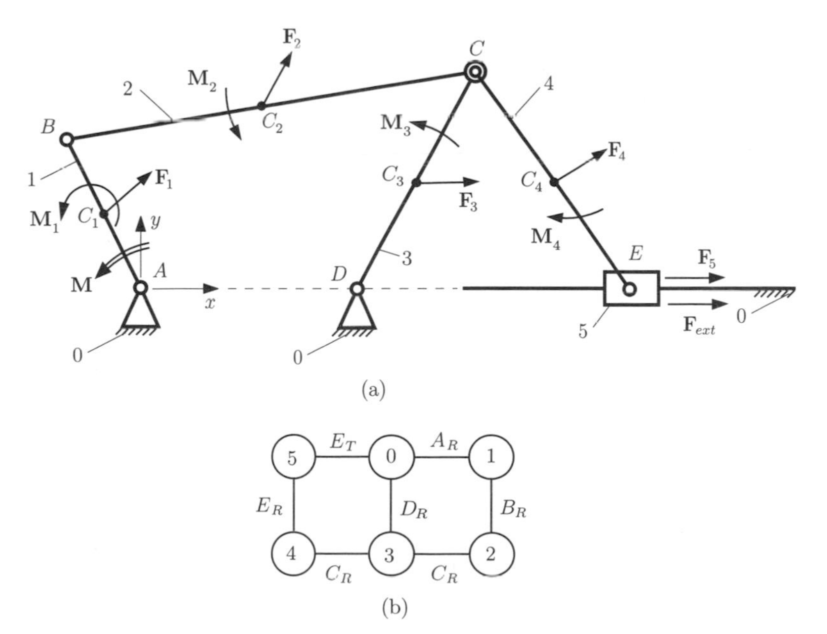

Figure 5.7. (a) Five-link ($j = 1, 2, 3, 4, 5$) mechanism and (b) diagram representing the mechanism.

A diagram representing the mechanism is depicted in Fig. 5.7(b) and has two contours, 0-1-2-3-0 and 0-3-4-5-0.

Remark

The joint at C represents a ramification point for the mechanism and the diagram, and the dynamic force analysis will start with this joint. The force computation starts with contour 0-3-4-5-0 because driven load $\mathbf{F}_{\text{ext}}$ on link 5 is given.

I. Contour 0-3-4-5-0

REACTION $\mathbf{F}_{34}$

The rotation joint at C (or C_R, where the subscript R means rotation), between 3 and 4, is replaced with the unknown reaction (Fig. 5.8):

$$\mathbf{F}_{34} = -\mathbf{F}_{43} = F_{34x}\mathbf{\imath} + F_{34y}\mathbf{\jmath}.$$

If path I is followed, as shown in Fig. 5.8(a), for the rotation joint at $E(E_R)$, a moment equation can be written as

$$\sum \mathbf{M}_E^{(4)} = (\mathbf{r}_C - \mathbf{r}_E) \times \mathbf{F}_{32} + (\mathbf{r}_{C4} - \mathbf{r}_D) \times \mathbf{F}_4 + \mathbf{M}_4 = \mathbf{0},$$

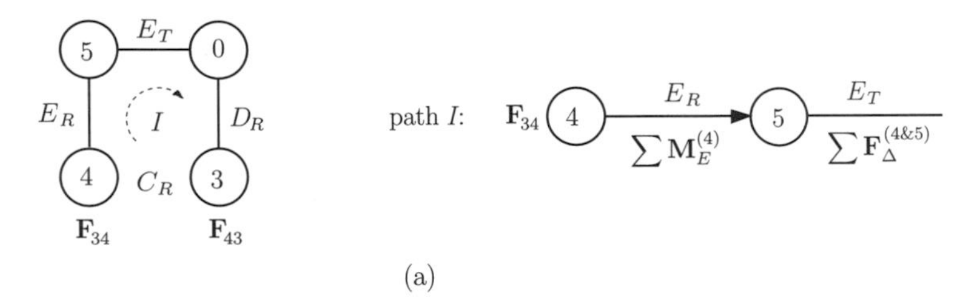

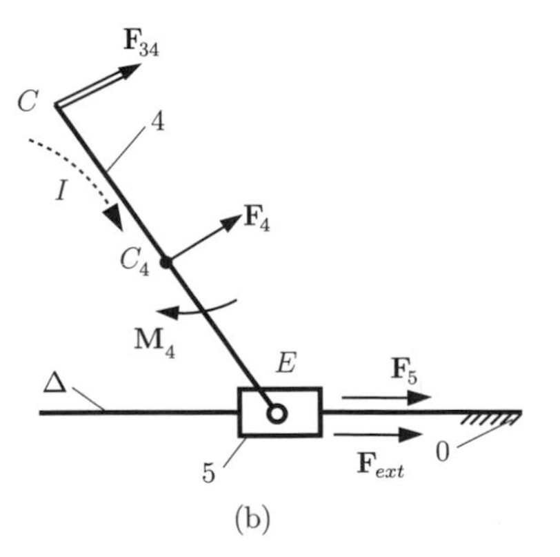

Figure 5.8. Reaction force $\mathbf{F}_{34}$: (a) diagram calculation and (b) free-body diagram.

or

$$\begin{vmatrix} \mathbf{\imath} & \mathbf{\jmath} & \mathbf{k} \\ x_C - x_E & y_C - y_E & 0 \\ F_{34x} & F_{34y} & 0 \end{vmatrix} + \begin{vmatrix} \mathbf{\imath} & \mathbf{\jmath} & \mathbf{k} \\ x_{C4} - x_E & y_{C4} - y_E & 0 \\ F_{4x} & F_{4y} & 0 \end{vmatrix} + M_4\mathbf{k} = \mathbf{0}. \quad (5.28)$$

Continuing on path I, the next joint is the translational joint at $D(D_T)$. The projection of all the forces that act on 4 and 5 onto the sliding direction Δ (x axis) should be zero.

$$\sum \mathbf{F}_{\Delta}^{(4\&5)} = \sum \mathbf{F}^{(4\&5)} \cdot \mathbf{\imath} = (\mathbf{F}_{34} + \mathbf{F}_4 + \mathbf{F}_5 + \mathbf{F}_{\text{ext}}) \cdot \mathbf{\imath}$$
$$= F_{34x} + F_{4x} + F_{5x} + F_{\text{ext}} = 0. \quad (5.29)$$

After the system of equations, Eqs. (5.28) and (5.48), is solved, the two unknowns F_{34x} and F_{34y} are obtained.

REACTION $\mathbf{F}_{45}$

The rotation joint at E (E_R), between 4 and 5, is replaced with the unknown reaction (Fig. 5.9):

$$\mathbf{F}_{45} = -\mathbf{F}_{54} = F_{45x}\mathbf{\imath} + F_{45y}\mathbf{\jmath}.$$

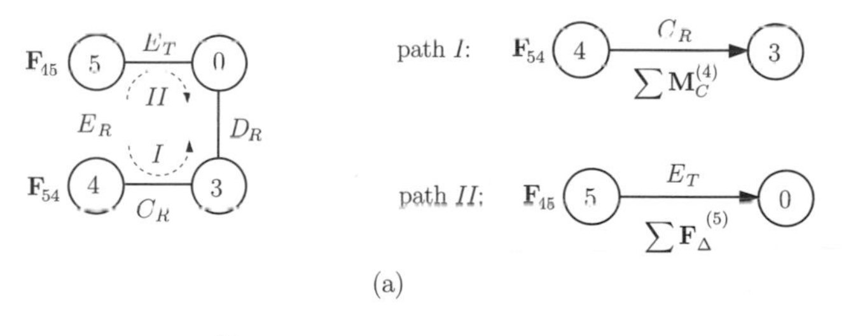

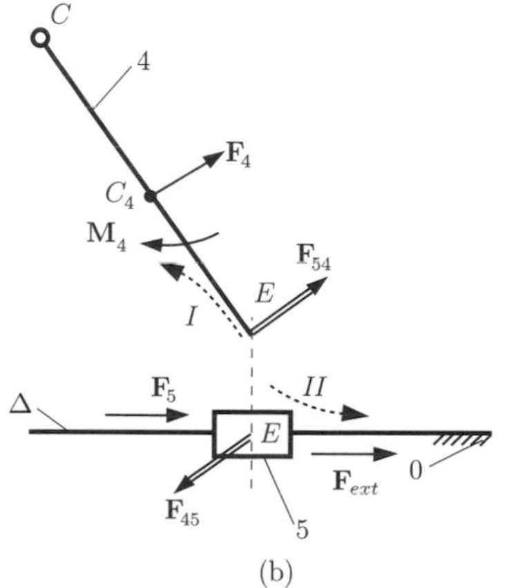

Figure 5.9. Reaction force $\mathbf{F}_{45}$: (a) diagram calculation and (b) free-body diagram.

If path I is traced, as shown in Fig. 5.9(a), for the pin joint at C (C_R) a moment equation can be written as

$$\sum \mathbf{M}_C^{(4)} = (\mathbf{r}_E - \mathbf{r}_C) \times \mathbf{F}_{54} + (\mathbf{r}_{C4} - \mathbf{r}_C) \times \mathbf{F}_4 + \mathbf{M}_4 = \mathbf{0},$$

or

$$\begin{vmatrix} \mathbf{\imath} & \mathbf{\jmath} & \mathbf{k} \\ x_E - x_C & y_E - y_C & 0 \\ -F_{45x} & -F_{45y} & 0 \end{vmatrix} + \begin{vmatrix} \mathbf{\imath} & \mathbf{\jmath} & \mathbf{k} \\ x_{C4} - x_C & y_{C4} - y_C & 0 \\ F_{4x} & F_{4y} & 0 \end{vmatrix} + M_4\mathbf{k} = \mathbf{0}. \quad (5.30)$$

For path II, the slider joint at E (E_T) is encountered. The projection of all the forces that act on 5 onto the sliding direction Δ (x axis) should be zero.

$$\sum \mathbf{F}_\Delta^{(5)} = \sum \mathbf{F}^{(5)} \cdot \mathbf{\imath} = (\mathbf{F}_{45} + \mathbf{F}_5 + \mathbf{F}_{\text{ext}}) \cdot \mathbf{\imath} = F_{45x} + F_{5x} + F_{\text{ext}} = 0. \quad (5.31)$$

Unknown force components F_{45x} and F_{45y} can be calculated from Eqs. (5.30) and (5.31).

REACTION $\mathbf{F}_{05}$

The slider joint at $E(E_T)$, between 0 and 5, is replaced with the unknown reaction (Fig. 5.10):

$$\mathbf{F}_{05} = F_{05y}\mathbf{\jmath}.$$

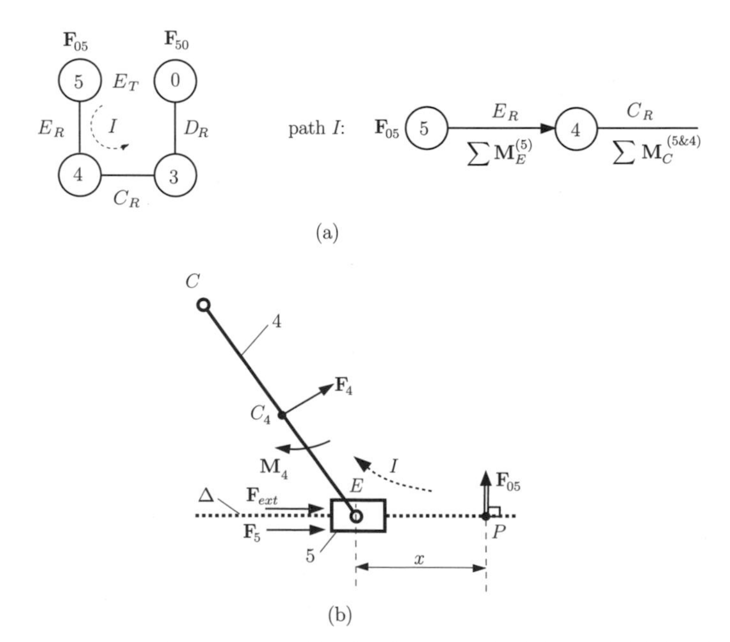

Figure 5.10. Reaction force $\mathbf{F}_{05}$: (a) diagram calculation and (b) free-body diagram.

The reaction joint introduced by the translational joint is perpendicular to the sliding direction $\mathbf{F}_{05} \perp \Delta$. The application point P of the force $\mathbf{F}_{05}$ is unknown.

If the path I is followed, as in Fig. 5.10(a), for the pin joint at $E(E_R)$, a moment equation can be written for link 5 as

$$\sum \mathbf{M}_E^{(5)} = (\mathbf{r}_P - \mathbf{r}_E) \times \mathbf{F}_{05} = \mathbf{0},$$

or

$$x F_{05y} = 0 \quad \Rightarrow \quad x = 0. \tag{5.32}$$

The application point is at $E(P \equiv E)$.

Continuing on path I, the next joint is the pin joint $C(C_R)$.

$$\sum \mathbf{M}_C^{(4\&5)} = (\mathbf{r}_E - \mathbf{r}_C) \times (\mathbf{F}_{05} + \mathbf{F}_5 + \mathbf{F}_{\text{ext}}) + (\mathbf{r}_{C4} - \mathbf{r}_C) \times \mathbf{F}_4 + \mathbf{M}_4 = \mathbf{0},$$

or

$$\begin{vmatrix} \mathbf{\imath} & \mathbf{\jmath} & \mathbf{k} \\ x_E - x_C & y_E - y_C & 0 \\ F_{5x} + F_{\text{ext}} & F_{05y} & 0 \end{vmatrix} + \begin{vmatrix} \mathbf{\imath} & \mathbf{\jmath} & \mathbf{k} \\ x_{C4} - x_C & y_{C4} - y_C & 0 \\ F_{4x} & F_{4y} & 0 \end{vmatrix} + M_4 \mathbf{k} = \mathbf{0}. \tag{5.33}$$

Joint reaction force F_{05y} can be computed from Eq. (5.33) above.

II. Contour 0-1-2-3-0

For this contour, joint force $\mathbf{F}_{43} = -\mathbf{F}_{34}$ at ramification point C is considered as a known external force.

REACTION $\mathbf{F}_{03}$

The pin joint D_R, between 0 and 3, is replaced with the unknown reaction force (Fig. 5.11):

$$\mathbf{F}_{03} = F_{03x}\mathbf{\imath} + F_{03y}\mathbf{\jmath}.$$

If path I is followed, as shown in Fig. 5.11(a), a moment equation can be written for pin joint C_R for link 3 as

$$\sum \mathbf{M}_C^{(3)} = (\mathbf{r}_D - \mathbf{r}_C) \times \mathbf{F}_{03} + (\mathbf{r}_{C3} - \mathbf{r}_C) \times \mathbf{F}_3 + \mathbf{M}_3 = \mathbf{0},$$

or

$$\begin{vmatrix} \mathbf{\imath} & \mathbf{\jmath} & \mathbf{k} \\ x_D - x_C & y_D - y_C & 0 \\ F_{03x} & F_{03y} & 0 \end{vmatrix} + \begin{vmatrix} \mathbf{\imath} & \mathbf{\jmath} & \mathbf{k} \\ x_{C3} - x_C & y_{C3} - y_C & 0 \\ F_{3x} & F_{3y} & 0 \end{vmatrix} + M_3 \mathbf{k} = \mathbf{0}. \tag{5.34}$$

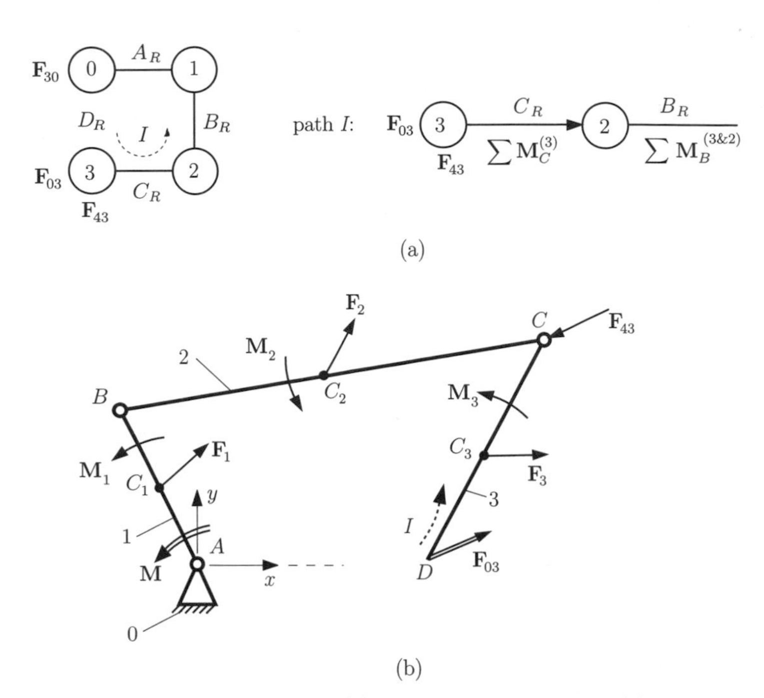

Figure 5.11. Reaction force $\mathbf{F}_{03}$: (a) diagram calculation and (b) free-body diagram.

Continuing on path I, the next joint is the pin joint B_R, and a moment equation can be written for links 3 and 2 as

$$\sum \mathbf{M}_B^{(3\&2)} = (\mathbf{r}_D - \mathbf{r}_B) \times \mathbf{F}_{03} + (\mathbf{r}_{C3} - \mathbf{r}_B) \times \mathbf{F}_3 + \mathbf{M}_3 + (\mathbf{r}_C - \mathbf{r}_B) \times \mathbf{F}_{43} + (\mathbf{r}_{C2} - \mathbf{r}_B) \times \mathbf{F}_2 + \mathbf{M}_2 = \mathbf{0},$$

or

$$\begin{vmatrix} \mathbf{i} & \mathbf{j} & \mathbf{k} \\ x_D - x_B & y_D - y_B & 0 \\ F_{03x} & F_{03y} & 0 \end{vmatrix} + \begin{vmatrix} \mathbf{i} & \mathbf{j} & \mathbf{k} \\ x_{C3} - x_B & y_{C3} - y_B & 0 \\ F_{3x} & F_{3y} & 0 \end{vmatrix} + M_3\mathbf{k} + \begin{vmatrix} \mathbf{i} & \mathbf{j} & \mathbf{k} \\ x_C - x_B & y_C - y_B & 0 \\ F_{43x} & F_{43y} & 0 \end{vmatrix} + \begin{vmatrix} \mathbf{i} & \mathbf{j} & \mathbf{k} \\ x_{C2} - x_B & y_{C2} - y_B & 0 \\ F_{2x} & F_{2y} & 0 \end{vmatrix} + M_2\mathbf{k} = \mathbf{0}. \quad (5.35)$$

The two components, F_{03x} and F_{03y}, of the joint force are obtained from Eqs. (5.34) and (5.36).

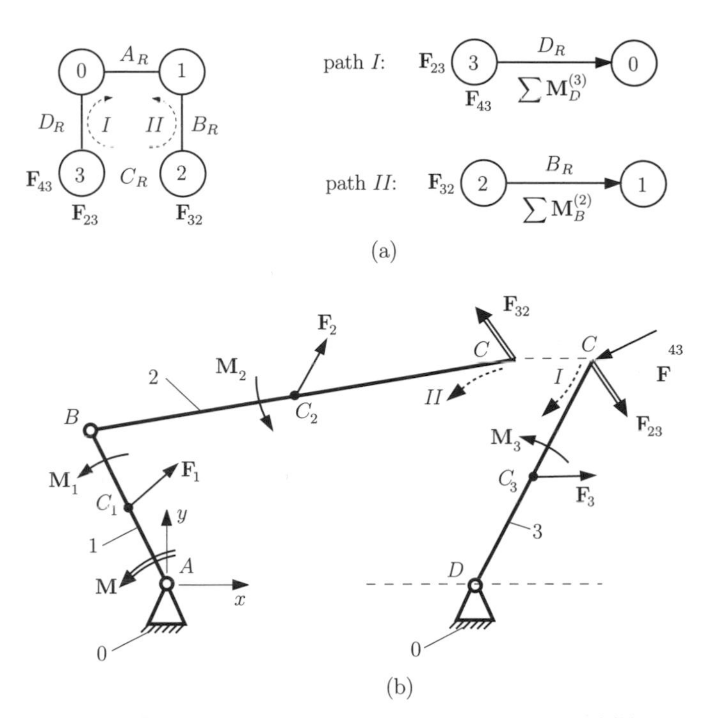

Figure 5.12. Reaction force $\mathbf{F}_{23}$: (a) diagram calculation and (b) free-body diagram.

REACTION $\mathbf{F}_{23}$

The pin joint C_R, between 2 and 3, is replaced with the unknown reaction force (Fig. 5.12):

$$\mathbf{F}_{23} = F_{23x}\mathbf{\imath} + F_{23y}\mathbf{\jmath}.$$

If path *I* is followed, as in Fig. 5.12(a), a moment equation can be written for pin joint D_R for link 3 as

$$\sum \mathbf{M}_D^{(3)} = (\mathbf{r}_C - \mathbf{r}_D) \times (\mathbf{F}_{23} + \mathbf{F}_{43})(\mathbf{r}_{C3} - \mathbf{r}_D) \times \mathbf{F}_3 + \mathbf{M}_3 = \mathbf{0},$$

or

$$\begin{vmatrix} \mathbf{\imath} & \mathbf{\jmath} & \mathbf{k} \\ x_C - x_D & y_C - y_D & 0 \\ F_{23x} + F_{43x} & F_{23y} + F_{43y} & 0 \end{vmatrix} + \begin{vmatrix} \mathbf{\imath} & \mathbf{\jmath} & \mathbf{k} \\ x_{C3} - x_D & y_{C3} - y_D & 0 \\ F_{3x} & F_{3y} & 0 \end{vmatrix} + M_3\mathbf{k} = \mathbf{0}. \tag{5.36}$$

For path *II* the first joint encountered is pin joint B_R, and a moment equation can be written for link 2 as

$$\sum \mathbf{M}_B^{(2)} = (\mathbf{r}_C - \mathbf{r}_B) \times (-\mathbf{F}_{23}) + (\mathbf{r}_{C2} - \mathbf{r}_B) \times \mathbf{F}_2 + \mathbf{M}_2 = \mathbf{0},$$

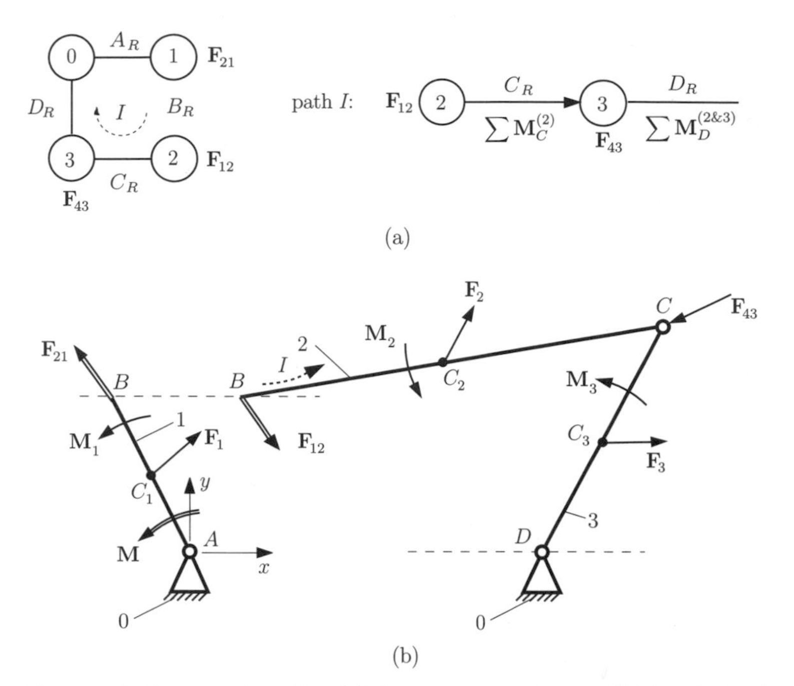

Figure 5.13. Reaction force $\mathbf{F}_{12}$: (a) diagram calculation and (b) free-body diagram.

or

$$\begin{vmatrix} \mathbf{\imath} & \mathbf{\jmath} & \mathbf{k} \\ x_C - x_B & y_C - y_B & 0 \\ -F_{23x} & -F_{23y} & 0 \end{vmatrix} + \begin{vmatrix} \mathbf{\imath} & \mathbf{\jmath} & \mathbf{k} \\ x_{C2} - x_B & y_{C2} - y_B & 0 \\ F_{2x} & F_{2y} & 0 \end{vmatrix} + M_2\mathbf{k} = \mathbf{0}. \quad (5.37)$$

The two force components, F_{23x} and F_{23y}, of the joint force are obtained from Eqs. (5.36) and (5.37).

REACTION $\mathbf{F}_{12}$

The pin joint B_R, between 1 and 2, is replaced with the unknown reaction force (Fig. 5.13):

$$\mathbf{F}_{12} = F_{12x}\mathbf{\imath} + F_{12y}\mathbf{\jmath}.$$

If path I is followed, as in Fig. 5.13(a), a moment equation can be written for pin joint C_R for link 2 as

$$\sum \mathbf{M}_C^{(2)} = (\mathbf{r}_B - \mathbf{r}_C) \times \mathbf{F}_{12} + (\mathbf{r}_{C2} - \mathbf{r}_C) \times \mathbf{F}_2 + \mathbf{M}_2 = \mathbf{0},$$

or

$$\begin{vmatrix} \mathbf{\imath} & \mathbf{\jmath} & \mathbf{k} \\ x_B - x_C & y_B - y_C & 0 \\ F_{12x} & F_{12y} & 0 \end{vmatrix} + \begin{vmatrix} \mathbf{\imath} & \mathbf{\jmath} & \mathbf{k} \\ x_{C2} - x_C & y_{C2} - y_C & 0 \\ F_{2x} & F_{2y} & 0 \end{vmatrix} + M_2\mathbf{k} = \mathbf{0}. \quad (5.38)$$

Continuing on path I, the next joint encountered is pin joint D_R, and a moment equation can be written for links 2 and 3 as

$$\sum \mathbf{M}_D^{(2\&3)} = (\mathbf{r}_B - \mathbf{r}_D) \times \mathbf{F}_{12} + (\mathbf{r}_{C2} - \mathbf{r}_D) \times \mathbf{F}_2 + \mathbf{M}_2 + (\mathbf{r}_C - \mathbf{r}_D) \times \mathbf{F}_{43} + (\mathbf{r}_{C3} - \mathbf{r}_D) \times \mathbf{F}_3 + \mathbf{M}_3 = \mathbf{0},$$

or

$$\begin{vmatrix} \mathbf{i} & \mathbf{j} & \mathbf{k} \\ x_B - x_D & y_B - y_D & 0 \\ F_{12x} & F_{12y} & 0 \end{vmatrix} + \begin{vmatrix} \mathbf{i} & \mathbf{j} & \mathbf{k} \\ x_{C2} - x_D & y_{C2} - y_D & 0 \\ F_{2x} & F_{2y} & 0 \end{vmatrix} + M_2\mathbf{k} + M_3\mathbf{k} = \mathbf{0}.$$

$$\begin{vmatrix} \mathbf{i} & \mathbf{j} & \mathbf{k} \\ x_C - x_D & y_C - y_D & 0 \\ F_{43x} & F_{43y} & 0 \end{vmatrix} + \begin{vmatrix} \mathbf{i} & \mathbf{j} & \mathbf{k} \\ x_{C3} - x_D & y_{C3} - y_D & 0 \\ F_{3x} & F_{3y} & 0 \end{vmatrix} \tag{5.39}$$

The two components F_{12x} and F_{12y} of the joint force are computed from Eqs. (5.38) and (5.39).

REACTION $\mathbf{F}_{01}$ AND DRIVER TORQUE M

The pin joint A_R, between 0 and 1, is replaced with the unknown reaction force (Fig. 5.14):

$$\mathbf{F}_{01} = F_{01x}\mathbf{i} + F_{01y}\mathbf{j}.$$

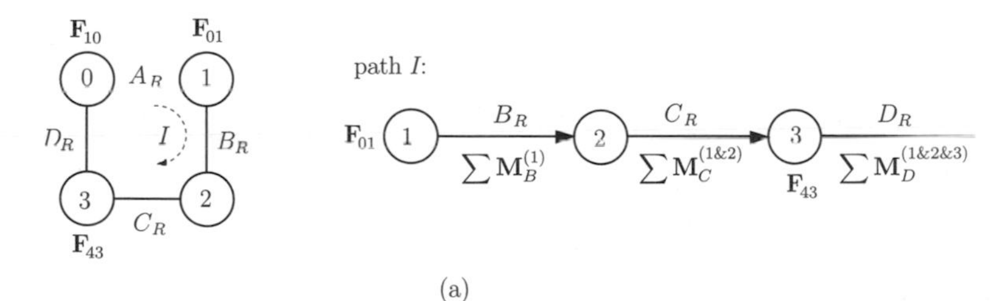

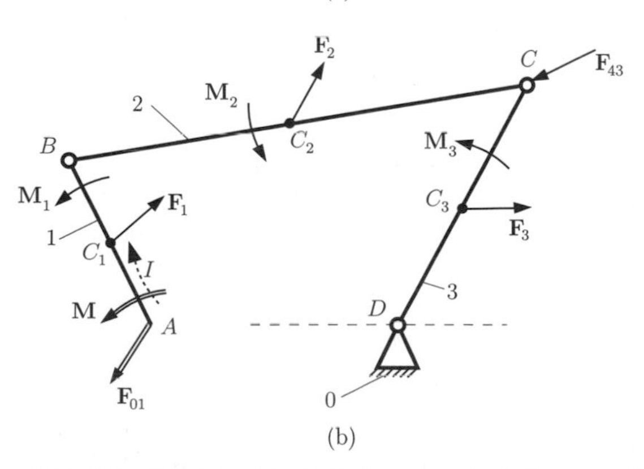

Figure 5.14. Reaction force $\mathbf{F}_{01}$: (a) diagram calculation and (b) free-body diagram.

The unknown driver torque is $\mathbf{M} = M\mathbf{k}$. If path I is followed, as shown in Fig. 5.14(a), a moment equation can be written for pin joint B_R for link 1 as

$$\sum \mathbf{M}_B^{(1)} = (\mathbf{r}_A - \mathbf{r}_B) \times \mathbf{F}_{01} + (\mathbf{r}_{C1} - \mathbf{r}_B) \times \mathbf{F}_1 + \mathbf{M}_1 + \mathbf{M} = \mathbf{0},$$

or

$$\begin{vmatrix} \mathbf{1} & \mathbf{J} & \mathbf{k} \\ x_A - x_B & y_A - y_B & 0 \\ F_{01x} & F_{01y} & 0 \end{vmatrix} + \begin{vmatrix} \mathbf{1} & \mathbf{J} & \mathbf{k} \\ x_{C1} - x_B & y_{C1} - y_B & 0 \\ F_{1x} & F_{1y} & 0 \end{vmatrix} + M_1\mathbf{k} + M\mathbf{k} = \mathbf{0}. \tag{5.40}$$

Continuing on path I, the next joint encountered is pin joint C_R, and a moment equation can be written for links 1 and 2 as

$$\sum \mathbf{M}_C^{(1\&2)} = (\mathbf{r}_A - \mathbf{r}_C) \times \mathbf{F}_{01} + (\mathbf{r}_{C1} - \mathbf{r}_C) \times \mathbf{F}_1 + \mathbf{M}_1 + \mathbf{M} + (\mathbf{r}_{C2} - \mathbf{r}_C) \times \mathbf{F}_2 + \mathbf{M}_2 = \mathbf{0}. \tag{5.41}$$

Equation (5.58) is the vector sum of the moments about D_R of all forces and torques that act on links 1, 2, and 3.

$$\sum \mathbf{M}_D^{(1\&2\&3)} = (\mathbf{r}_A - \mathbf{r}_D) \times \mathbf{F}_{01} + (\mathbf{r}_{C1} - \mathbf{r}_D) \times \mathbf{F}_1 + \mathbf{M}_1 + \mathbf{M} + (\mathbf{r}_{C2} - \mathbf{r}_D) \times \mathbf{F}_2 + \mathbf{M}_2 + (\mathbf{r}_C - \mathbf{r}_D) \times \mathbf{F}_{43} + (\mathbf{r}_{C3} - \mathbf{r}_D) \times \mathbf{F}_3 + \mathbf{M}_3 = \mathbf{0}. \tag{5.42}$$

The components F_{01x}, F_{01y}, and M are computed from Eqs. (5.40), (5.58), and (5.42).

■ EXAMPLE: R-TRR-RRT MECHANISM

Calculate the torque $\mathbf{M}$ required for dynamic equilibrium of the five-link mechanism (Sections 2.4 and 4.9) in the position when the crank angle $\phi = \pi/4$ rad, as shown in Fig. 5.15. The dimensions are $AC = 0.10$ m, $BC = 0.30$ m, $BD = 0.90$ m, and $L_a = 0.10$ m, and the external force on slider 5 is $F_{\text{ext}} = 100$ N. The angular speed of the crank 1 is $n_1 = 100$ rpm, or $\omega_1 = 100\pi/30$ rad/s. The center of mass locations of links $j = 1, 2, \ldots, 5$ (with the masses m_j) are designated by $C_j(x_{Cj}, y_{Cj}, 0)$. The position vectors of the joints and the centers of mass are

$$\mathbf{r}_A = 0.000\mathbf{1} + 0.000\mathbf{J}\ \text{m},$$
$$\mathbf{r}_{C1} = 0.212\mathbf{1} + 0.212\mathbf{J}\ \text{m},$$
$$\mathbf{r}_B = \mathbf{r}_{C2} = 0.256\mathbf{1} + 0.256\mathbf{J}\ \text{m},$$

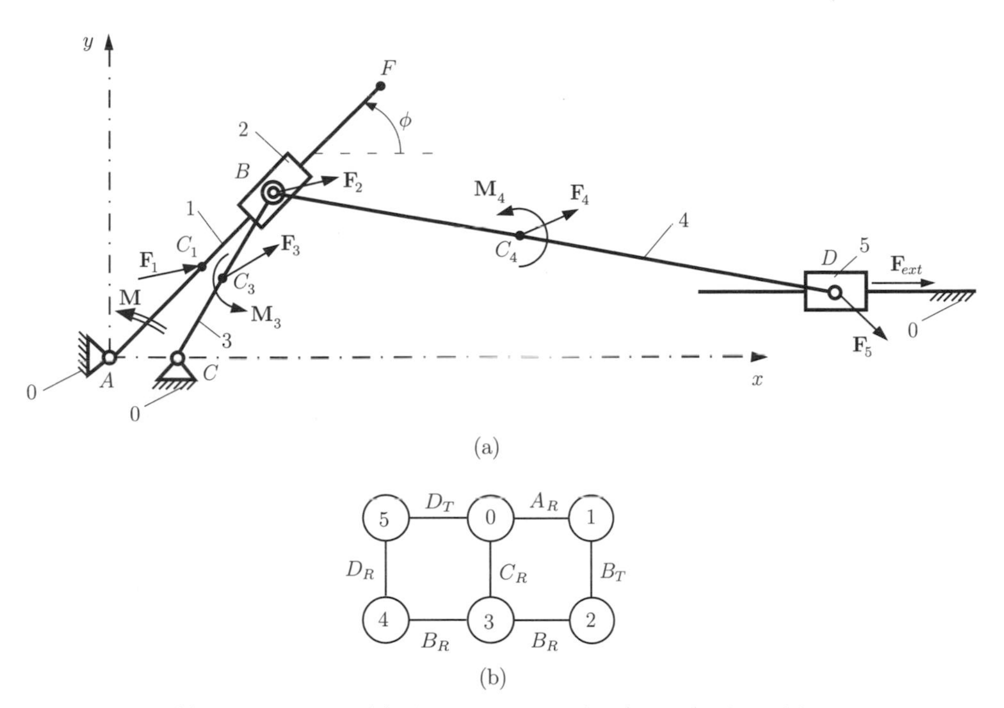

Figure 5.15. (a) Mechanism and (b) diagram representing the mechanism with two contours.

$$\begin{aligned}
\mathbf{r}_{C3} &= 0.178\imath + 0.128\jmath \text{ m},\\
\mathbf{r}_{C} &= 0.100\imath + 0.000\jmath \text{ m},\\
\mathbf{r}_{C4} &= 0.699\imath + 0.178\jmath \text{ m},\\
\mathbf{r}_{D} &= \mathbf{r}_{C5} = 1.142\imath + 0.100\jmath \text{ m}.
\end{aligned} \tag{5.43}$$

The total forces and moments at C_j are $\mathbf{F}_j = \mathbf{F}_{\text{in}\,j} + \mathbf{G}_j$ and $\mathbf{M}_j = \mathbf{M}_{\text{in}\,j}$, where $\mathbf{F}_{\text{in}\,j}$ is the inertia force and $\mathbf{M}_j$ is the inertia moment and $\mathbf{G}_j = -m_j g\jmath$ is the gravity force with the gravity acceleration $g = 9.81$ m/s^2.

$$\begin{aligned}
\mathbf{F}_1 &= 5.514\imath + 3.189\jmath \text{ N},\\
\mathbf{F}_2 &= 0.781\imath + 1.843\jmath \text{ N},\\
\mathbf{F}_3 &= 1.202\imath + 1.660\jmath \text{ N},\\
\mathbf{F}_4 &= 6.466\imath + 4.896\jmath \text{ N},\\
\mathbf{F}_5 &= 0.643\imath - 0.382\jmath \text{ N},\\
\mathbf{M}_1 &= \mathbf{M}_2 = \mathbf{M}_5 = 0.000\mathbf{k} \text{ N m},\\
\mathbf{M}_3 &= 0.023\mathbf{k} \text{ N m},\\
\mathbf{M}_4 &= -1.274\mathbf{k} \text{ N m}.
\end{aligned} \tag{5.44}$$

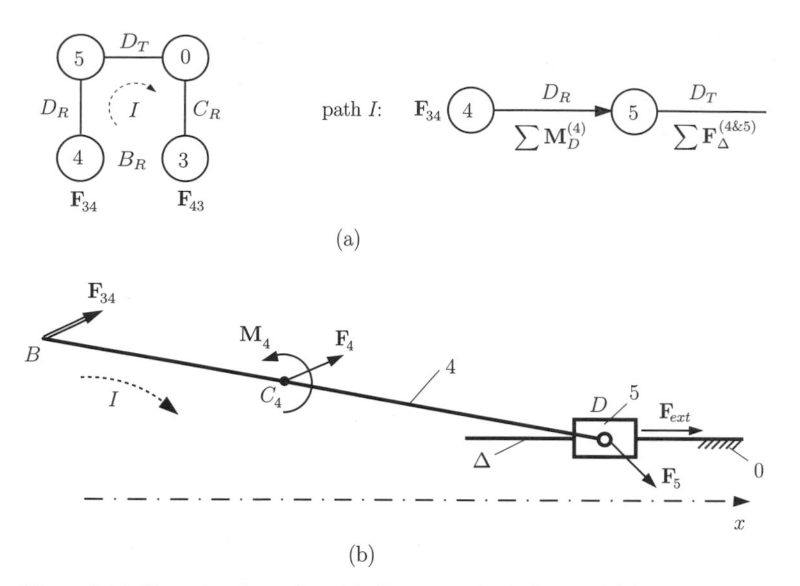

Figure 5.16. Reaction force $\mathbf{F}_{34}$: (a) diagram calculation and (b) free-body diagram.

A diagram representing the mechanism is shown in Fig. 5.15(b) and has two contours, 0-1-2-3-0 and 0-3-4-5-0.

I. Contour 0-3-4-5-0

The joint at B represents a ramification point, and the dynamic force analysis will start with this joint.

REACTION $\mathbf{F}_{34}$

The rotation joint at B_R between links 3 and 4 is replaced with the unknown reaction (Fig. 5.16):

$$\mathbf{F}_{34} = -\mathbf{F}_{43} = F_{34x}\mathbf{\imath} + F_{34y}\mathbf{\jmath}.$$

If path I is followed, as shown in Fig. 5.16(a), a moment equation can be written for rotation joint D_R as

$$\sum \mathbf{M}_D^{(4)} = (\mathbf{r}_B - \mathbf{r}_D) \times \mathbf{F}_{32} + (\mathbf{r}_{C4} - \mathbf{r}_D) \times \mathbf{F}_4 + \mathbf{M}_4 = \mathbf{0}. \tag{5.45}$$

Continuing on path I, the next joint is slider joint D_T, and a force equation can be written. The projection of all the forces that act on links 4 and 5 onto the sliding direction x is zero.

$$\sum \mathbf{F}^{(4\&5)} \cdot \mathbf{\imath} = (\mathbf{F}_{34} + \mathbf{F}_4 + \mathbf{F}_5 + \mathbf{F}_{\text{ext}}) \cdot \mathbf{\imath} = F_{34x} + F_{4x} + F_{5x} + F_{\text{ext}} = 0. \tag{5.46}$$

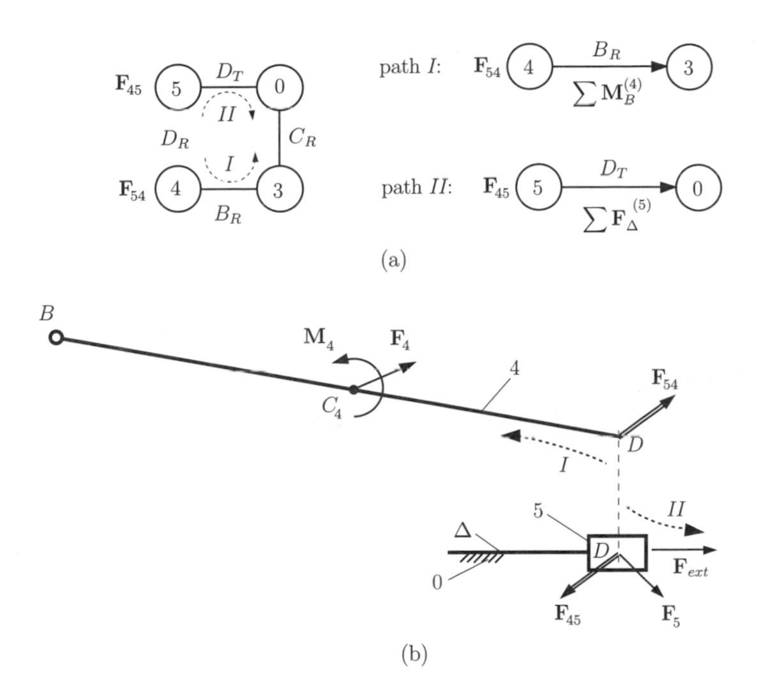

Figure 5.17. Reaction force $\mathbf{F}_{45}$: (a) diagram calculation and (b) free-body diagram.

When the system of equations, Eqs. (5.45) and (5.46), is solved,

$$F_{34x} = -107.110 \text{ N}, \qquad F_{34y} = 14.415 \text{ N}.$$

REACTION $\mathbf{F}_{45}$

The pin joint at D_R between 4 and 7 is replaced with the reaction force (Fig. 5.17):

$$\mathbf{F}_{45} = -\mathbf{F}_{54} = F_{45x}\mathbf{\imath} + F_{45y}\mathbf{\jmath}.$$

For path I, shown Fig. 5.17(a), a moment equation about B_R can be written for link 4 as

$$\sum \mathbf{M}_B^{(4)} = (\mathbf{r}_D - \mathbf{r}_B) \times \mathbf{F}_{54} + (\mathbf{r}_{C4} - \mathbf{r}_B) \times \mathbf{F}_4 + \mathbf{M}_4 = \mathbf{0}. \tag{5.47}$$

For path II, an equation for the forces projected onto the sliding direction of joint D_T can be written for link 5 as

$$\sum \mathbf{F}^{(5)} \cdot \mathbf{\imath} = (\mathbf{F}_{45} + \mathbf{F}_5 + \mathbf{F}_{\text{ext}}) \cdot \mathbf{\imath} = F_{45x} + F_{5x} + F_{\text{ext}} = 0. \tag{5.48}$$

Joint force $\mathbf{F}_{45}$ is obtained from the system of Eqs. (5.47) and (5.48):

$$F_{45x} = -100.643 \text{ N}, \qquad F_{45y} = 19.310 \text{ N}.$$

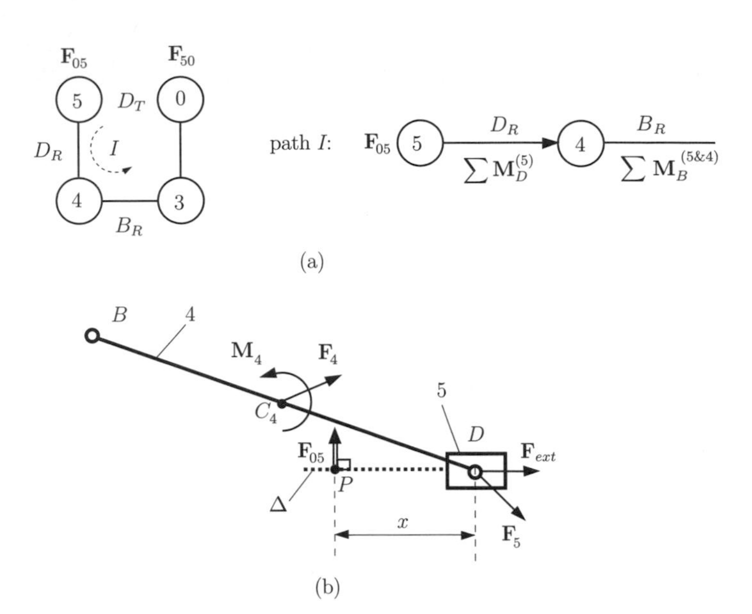

Figure 5.18. Reaction force $\mathbf{F}_{05}$: (a) diagram calculation and (b) free-body diagram.

REACTION $\mathbf{F}_{05}$

The reaction force $\mathbf{F}_{05}$ is perpendicular to the sliding direction of joint D_T (Fig. 5.18):

$$\mathbf{F}_{05} = F_{05y}\mathbf{J}.$$

The application point of unknown reaction force $\mathbf{F}_{05}$ can be computed from a moment equation about D_R, for link 5 (path I), as shown in Fig. 5.18(a):

$$\sum \mathbf{M}_D^{(5)} = (\mathbf{r}_P - \mathbf{r}_D) \times \mathbf{F}_{05} = \mathbf{0}, \tag{5.49}$$

or

$$x F_{05y} = 0 \quad \Rightarrow \quad x = 0. \tag{5.50}$$

The application point of reaction force $\mathbf{F}_{05}$ is at D ($P \equiv D$). The magnitude of reaction force F_{05y} is obtained from a moment equation about B_R, for links 5 and 4 (path I), as

$$\sum \mathbf{M}_B^{(5\&4)} = (\mathbf{r}_D - \mathbf{r}_B) \times (\mathbf{F}_{05} + \mathbf{F}_5 + \mathbf{F}_{\text{ext}}) + (\mathbf{r}_{C4} - \mathbf{r}_B) \times \mathbf{F}_4 + \mathbf{M}_4 = \mathbf{0}. \tag{5.51}$$

Solving the above equation gives

$$F_{05y} = -18.928 \text{ N}.$$

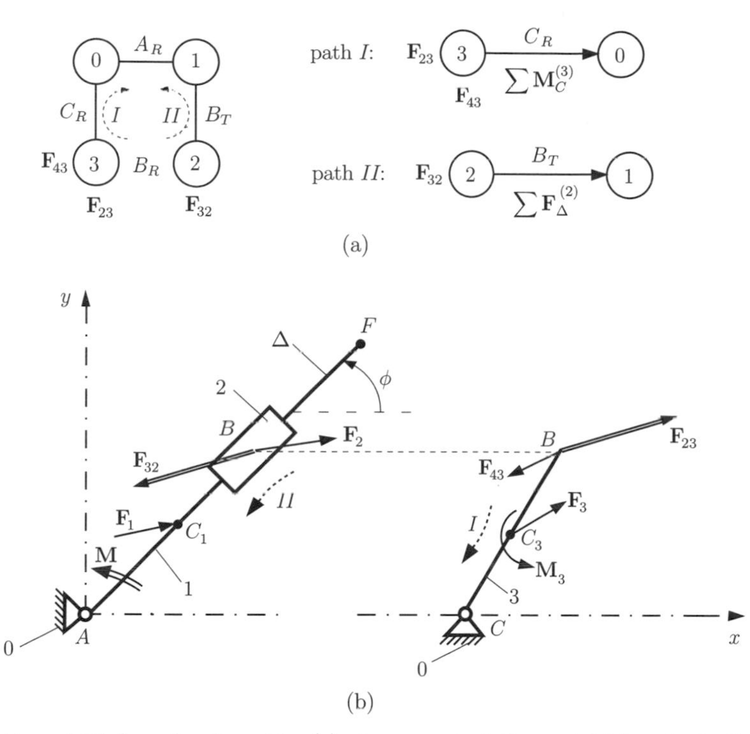

Figure 5.19. Reaction force $\mathbf{F}_{23}$: (a) diagram calculation and (b) free-body diagram.

II. Contour 0-1-2-3-0

The reaction force $\mathbf{F}_{43} = 107.110\mathbf{ı} - 14.415\mathbf{ȷ}$ N is considered as an external force for this contour at B.

REACTION $\mathbf{F}_{23}$

The rotation joint at B_R between 2 and 3 is replaced with the unknown reaction force (Fig. 5.19):

$$\mathbf{F}_{23} = -\mathbf{F}_{32} = F_{23x}\mathbf{ı} + F_{23y}\mathbf{ȷ}.$$

If path I is followed, as in Fig. 5.19(a), a moment equation can be written for pin joint C_R for link 3 as

$$\sum \mathbf{M}_C^{(3)} = (\mathbf{r}_B - \mathbf{r}_C) \times (\mathbf{F}_{23} + \mathbf{F}_{43}) + (\mathbf{r}_{C3} - \mathbf{r}_C) \times \mathbf{F}_3 + \mathbf{M}_3 = \mathbf{0}. \quad (5.52)$$

For path II an equation for the forces projected in direction Δ, the sliding direction of joint B_T, can be written for link 2 as

$$\sum \mathbf{F}^{(2)} \cdot \Delta = (\mathbf{F}_{32} + \mathbf{F}_2) \cdot (\cos\phi\mathbf{ı} + \sin\phi\mathbf{ȷ}) = 0. \quad (5.53)$$

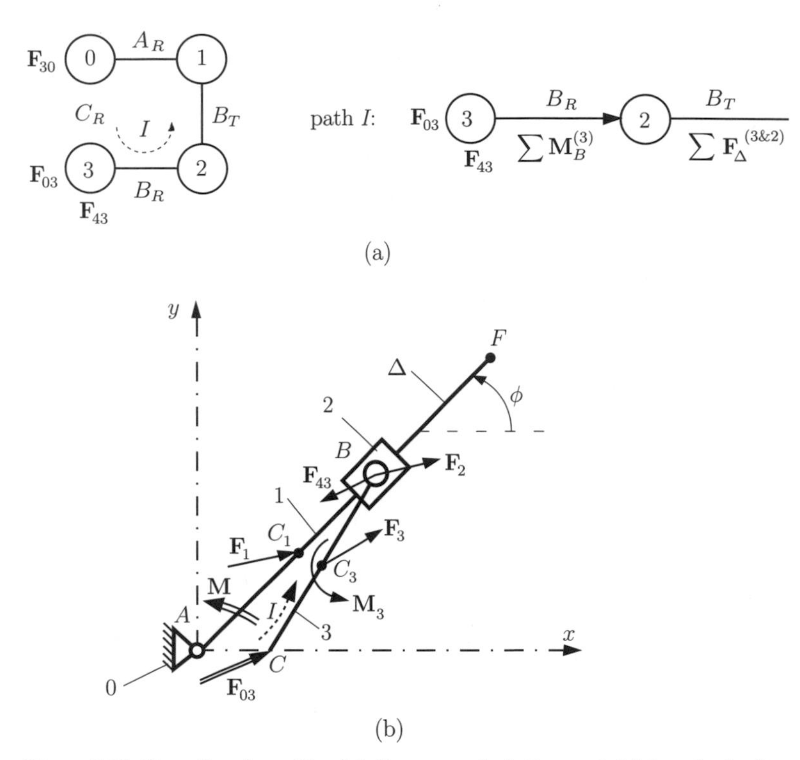

Figure 5.20. Reaction force $\mathbf{F}_{03}$: (a) diagram calculation and (b) free-body diagram.

Joint force $\mathbf{F}_{23}$ is calculated from Eqs. (5.52) and (5.53):

$$F_{23x} = -71.155 \text{ N}, \qquad F_{23y} = 73.397 \text{ N}.$$

REACTION $\mathbf{F}_{03}$

For joint reaction force $\mathbf{F}_{03}$ at C_R, there is only path I. For pin joint B_R, one moment equation can be written for link 3 (Fig. 5.20):

$$\sum \mathbf{M}_B^{(3)} = (\mathbf{r}_C - \mathbf{r}_B) \times \mathbf{F}_{03} + (\mathbf{r}_{C3} - \mathbf{r}_B) \times \mathbf{F}_3 + \mathbf{M}_3 = \mathbf{0}. \qquad (5.54)$$

A force equation can be written for links 3 and 2 for slider joint B_T as

$$\sum \mathbf{F}^{(3\&2)} \cdot \Delta = (\mathbf{F}_{03} + \mathbf{F}_3 + \mathbf{F}_{43} + \mathbf{F}_2) \cdot (\cos\phi\,\mathbf{\imath} + \sin\phi\,\mathbf{\jmath}) = 0. \qquad (5.55)$$

The components of the unknown force are obtained by solving the system of Eqs. (5.54) and (5.55):

$$F_{03x} = -37.156 \text{ N}, \qquad F_{03y} = -60.643 \text{ N}.$$

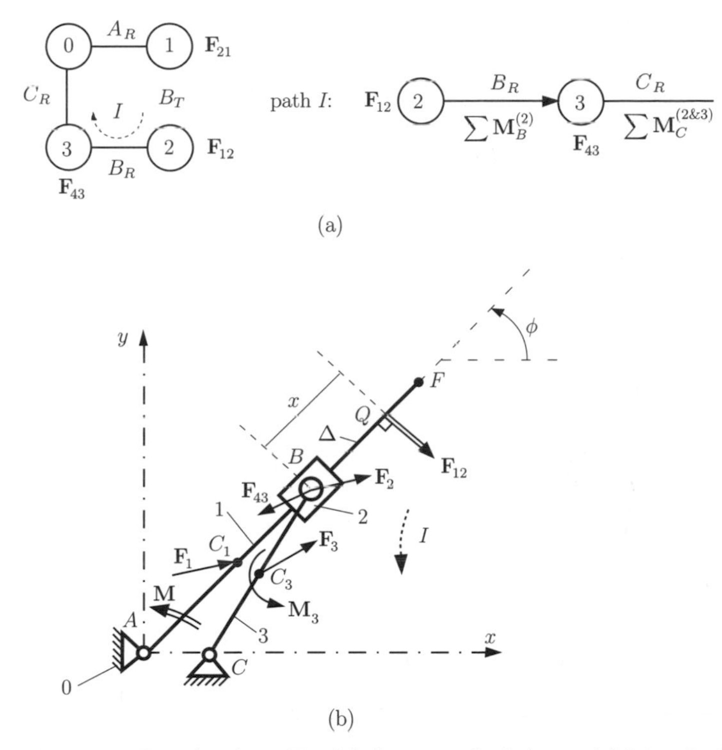

Figure 5.21. Reaction force $\mathbf{F}_{12}$: (a) diagram calculation and (b) free-body diagram.

REACTION $\mathbf{F}_{12}$

The slider joint at B_T between 1 and 2 is replaced with the reaction force (Fig. 5.21)

$$\mathbf{F}_{12} = -\mathbf{F}_{21} = F_{12x}\mathbf{\imath} + F_{12y}\mathbf{\jmath}.$$

Reaction force $\mathbf{F}_{12}$ is perpendicular to sliding direction Δ:

$$\mathbf{F}_{12} \cdot \Delta = (F_{12x}\mathbf{\imath} + F_{12y}\mathbf{\jmath}) \cdot (\cos\phi\mathbf{\imath} + \sin\phi\mathbf{\jmath}) = F_{12x}\cos\phi + F_{12y}\sin\phi = 0. \quad (5.56)$$

The point of application of force $\mathbf{F}_{12}$ is determined from the equation (path I)

$$\sum \mathbf{M}_B^{(2)} = (\mathbf{r}_Q - \mathbf{r}_B) \times \mathbf{F}_{12} = \mathbf{0}, \quad (5.57)$$

or

$$xF_{12} = 0 \Rightarrow x = 0, \quad (5.58)$$

and force $\mathbf{F}_{12}$ acts at B.

Continuing on path I, a moment equation can be written for links 2 and 3 with respect to pin joint C_R as

$$\sum \mathbf{M}_C^{(2\&3)} = (\mathbf{r}_B - \mathbf{r}_C) \times (\mathbf{F}_{12} + \mathbf{F}_2 + \mathbf{F}_{43}) + (\mathbf{r}_{C3} - \mathbf{r}_C) \times \mathbf{F}_3 + \mathbf{M}_3 = \mathbf{0}. \tag{5.59}$$

The two components of joint force $\mathbf{F}_{12}$ are computed from Eqs. (5.56) and (5.59):

$$F_{12x} = -71.936 \text{ N}, \qquad F_{12y} = 71.936 \text{ N}.$$

REACTION $\mathbf{F}_{01}$ AND EQUILIBRIUM TORQUE M

The pin joint A_R, between 0 and 1, is replaced with the unknown reaction (Fig. 5.22):

$$\mathbf{F}_{01} = F_{01x}\mathbf{\imath} + F_{01y}\mathbf{\jmath}.$$

The unknown equilibrium torque is $\mathbf{M} = M\mathbf{k}$. If path I is followed, as in Fig. 5.22(a) for slider joint B_T, a force equation can be written for link 1 as

$$\sum \mathbf{F}^{(1)} \cdot \Delta = (\mathbf{F}_{01} + \mathbf{F}_1) \cdot (\cos\phi\mathbf{\imath} + \sin\phi\mathbf{\jmath}) = 0. \tag{5.60}$$

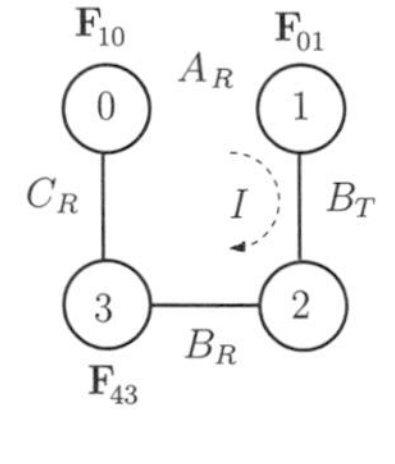

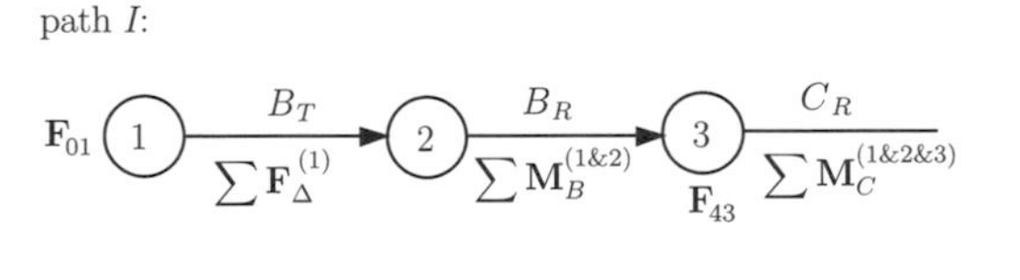

(a)

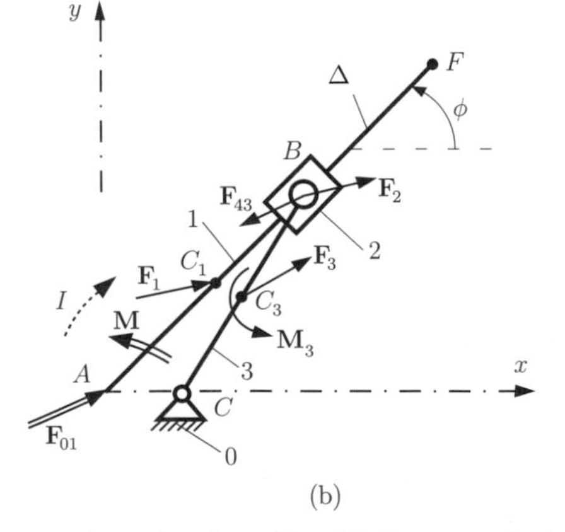

(b)

Figure 5.22. Reaction force $\mathbf{F}_{01}$: (a) diagram calculation and (b) free-body diagram.

Continuing on path I, the next joint encountered is pin joint B_R, and a moment equation can be written for links 1 and 2 as

$$\sum \mathbf{M}_B^{(1\&2)} = -\mathbf{r}_B \times \mathbf{F}_{01} + (\mathbf{r}_{C1} - \mathbf{r}_B) \times \mathbf{F}_1 + \mathbf{M} = \mathbf{0}. \quad (5.61)$$

Equation (5.61) is the vector sum of the moments about C_R of all forces and torques that act on links 1, 2, and 3.

$$\sum \mathbf{M}_C^{(1\&2\&3)} = -\mathbf{r}_C \times \mathbf{F}_{01} + (\mathbf{r}_{C1} - \mathbf{r}_C) \times \mathbf{F}_1 + \mathbf{M} + (\mathbf{r}_B - \mathbf{r}_C) \times (\mathbf{F}_2 + \mathbf{F}_{43}) + \mathbf{M}_3 + (\mathbf{r}_{C3} - \mathbf{r}_C) \times \mathbf{F}_3 = \mathbf{0}. \quad (5.62)$$

From Eqs. (5.60), (5.61), and (5.62), the components F_{01x}, F_{01y}, and M are computed:

$$F_{01x} = -77.451 \text{ N}, \qquad F_{01y} = 68.747 \text{ N}, \qquad M = 37.347 \text{ Nm}.$$

A *Mathematica* program for the calculation of the joint reactions is presented in Appendix 8.

5.7 Computation of Joint Force for Dyads

RRR Dyad

Figure 5.23 shows an RRR dyad with two links 2 and 3, and three pin joints B, C, and D. The unknowns are the joint reaction forces:

$$\mathbf{F}_{12} = F_{12x}\mathbf{\imath} + F_{12y}\mathbf{\jmath},$$
$$\mathbf{F}_{43} = F_{43x}\mathbf{\imath} + F_{43y}\mathbf{\jmath},$$
$$\mathbf{F}_{23} = -\mathbf{F}_{32} = F_{23x}\mathbf{\imath} + F_{23y}\mathbf{\jmath}. \quad (5.63)$$

The inertia forces and external forces $\mathbf{F}_j = F_j\mathbf{\imath} + F_j\mathbf{\jmath}$, inertia torques and external torques $\mathbf{M}_j = M_j\mathbf{k}$, $(j = 2, 3)$ are given.

For $\mathbf{F}_{12}$ and $\mathbf{F}_{43}$ to be determined, the following equations can be written.

- The sum of all forces on links 2 and 3 is zero:

$$\sum \mathbf{F}^{(2\&3)} = \mathbf{F}_{12} + \mathbf{F}_2 + \mathbf{F}_3 + \mathbf{F}_{43} = \mathbf{0},$$

or

$$\sum \mathbf{F}^{(2\&3)} \cdot \mathbf{\imath} = F_{12x} + F_{2x} + F_{3x} + F_{43x} = 0,$$
$$\sum \mathbf{F}^{(2\&3)} \cdot \mathbf{\jmath} = F_{12y} + F_{2y} + F_{3y} + F_{43y} = 0. \quad (5.64)$$

- The sum of moments of all forces and torques on link 2 about C is zero:

$$\sum \mathbf{M}_C^{(2)} = (\mathbf{r}_B - \mathbf{r}_C) \times \mathbf{F}_{12} + (\mathbf{r}_{C2} - \mathbf{r}_C) \times \mathbf{F}_2 + \mathbf{M}_2 = \mathbf{0}. \quad (5.65)$$

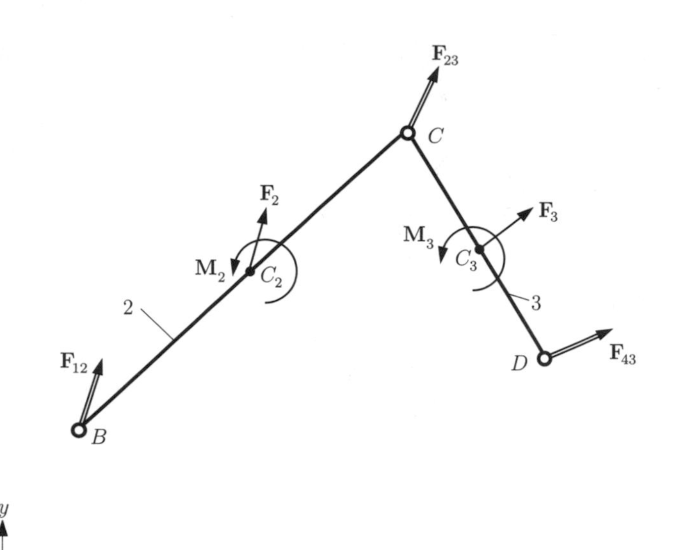

Figure 5.23. RRR dyad.

- The sum of moments of all forces and torques on link 3 about C is zero:

$$\sum \mathbf{M}_C^{(3)} = (\mathbf{r}_D - \mathbf{r}_C) \times \mathbf{F}_{43} + (\mathbf{r}_{C3} - \mathbf{r}_C) \times \mathbf{F}_3 + \mathbf{M}_3 = \mathbf{0}. \tag{5.66}$$

The components $F_{12x}, F_{12y}, F_{43x}$, and F_{43y} are calculated from Eqs. (5.64), (5.65), and (5.66).

Reaction force $\mathbf{F}_{32} = -\mathbf{F}_{23}$ is computed from the sum of all forces on link 2:

$$\sum \mathbf{F}^{(2)} = \mathbf{F}_{12} + \mathbf{F}_2 + \mathbf{F}_{32} = \mathbf{0},$$

or

$$\sum \mathbf{F}^{(2)} \cdot \mathbf{\imath} = F_{12x} + F_{2x} + F_{32x} = 0,$$
$$\sum \mathbf{F}^{(2)} \cdot \mathbf{\jmath} = F_{12y} + F_{2y} + F_{32y} = 0. \tag{5.67}$$

RRT Dyad

Figure 5.24 shows an RRT dyad with unknown joint reaction forces $\mathbf{F}_{12}$, $\mathbf{F}_{43}$, and $\mathbf{F}_{23} = -\mathbf{F}_{32}$. Joint reaction force $\mathbf{F}_{43}$ is perpendicular to the sliding direction

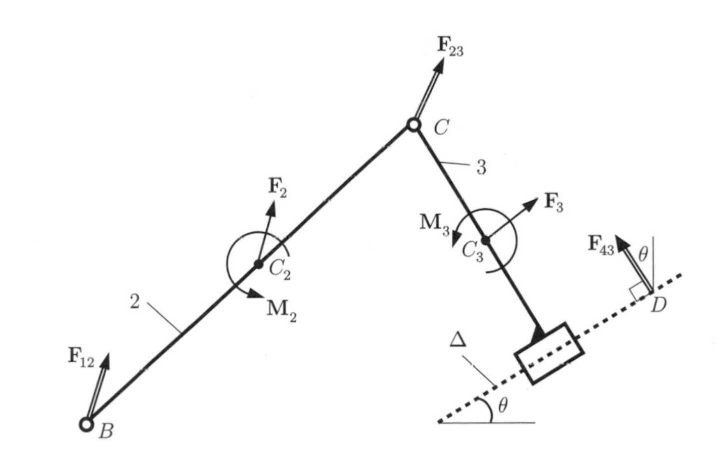

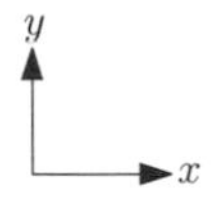

Figure 5.24. RRT dyad.

$\mathbf{F}_{43} \perp \Delta$, or

$$\mathbf{F}_{43} \cdot \Delta = (F_{43x}\mathbf{\imath} + F_{43y}\mathbf{\jmath}) \cdot (\cos\theta\mathbf{\imath} + \sin\theta\mathbf{\jmath}) = 0. \tag{5.68}$$

For $\mathbf{F}_{12}$ and $\mathbf{F}_{43}$ to be determined, the following equations can be written.

- The sum of all the forces on links 2 and 3 is zero:

$$\sum \mathbf{F}^{(2\&3)} = \mathbf{F}_{12} + \mathbf{F}_2 + \mathbf{F}_3 + \mathbf{F}_{43} = \mathbf{0},$$

or

$$\sum \mathbf{F}^{(2\&3)} \cdot \mathbf{\imath} = F_{12x} + F_{2x} + F_{3x} + F_{43x} = 0,$$
$$\sum \mathbf{F}^{(2\&3)} \cdot \mathbf{\jmath} = F_{12y} + F_{2y} + F_{3y} + F_{43y} = 0. \tag{5.69}$$

- The sum of moments of all the forces and the torques on link 2 about C is zero:

$$\sum \mathbf{M}_C^{(2)} = (\mathbf{r}_B - \mathbf{r}_C) \times \mathbf{F}_{12} + (\mathbf{r}_{C2} - \mathbf{r}_C) \times \mathbf{F}_2 + \mathbf{M}_2 = \mathbf{0}. \tag{5.70}$$

Components F_{12x}, F_{12y}, F_{43x}, and F_{43y} are calculated from Eqs. (5.68), (5.69), and (5.70).

Reaction force components F_{32x} and F_{32y} are computed from the sum of all the forces on link 2:

$$\sum \mathbf{F}^{(2)} = \mathbf{F}_{12} + \mathbf{F}_2 + \mathbf{F}_{32} = \mathbf{0},$$

or

$$\sum \mathbf{F}^{(2)} \cdot \mathbf{\imath} = F_{12x} + F_{2x} + F_{32x} = 0,$$
$$\sum \mathbf{F}^{(2)} \cdot \mathbf{\jmath} = F_{12y} + F_{2y} + F_{32y} = 0. \quad (5.71)$$

RTR Dyad

Unknown joint reaction forces $\mathbf{F}_{12}$ and $\mathbf{F}_{43}$ are calculated from the relations shown in Fig. 5.25.

- The sum of all the forces on links 2 and 3 is zero:

$$\sum \mathbf{F}^{(2\&3)} = \mathbf{F}_{12} + \mathbf{F}_2 + \mathbf{F}_3 + \mathbf{F}_{43} = \mathbf{0},$$

or

$$\sum \mathbf{F}^{(2\&3)} \cdot \mathbf{\imath} = F_{12x} + F_{2x} + F_{3x} + F_{43x} = 0,$$
$$\sum \mathbf{F}^{(2\&3)} \cdot \mathbf{\jmath} = F_{12y} + F_{2y} + F_{3y} + F_{43y} = 0. \quad (5.72)$$

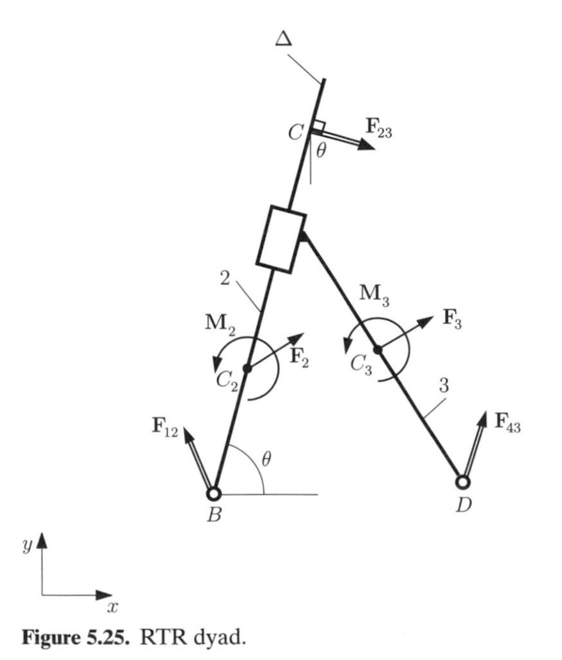

Figure 5.25. RTR dyad.

- The sum of the moments of all the forces and torques on links 2 and 3 about B is zero:

$$\sum \mathbf{M}_B^{(2\&3)} = (\mathbf{r}_D - \mathbf{r}_B) \times \mathbf{F}_{43} + (\mathbf{r}_{C3} - \mathbf{r}_B) \times \mathbf{F}_3 + \mathbf{M}_3 + (\mathbf{r}_{C2} - \mathbf{r}_B) \times \mathbf{F}_2 + \mathbf{M}_2 = \mathbf{0}. \quad (5.73)$$

- The sum of all the forces on link 2 projected onto the sliding direction $\Delta = \cos\theta\mathbf{\imath} + \sin\theta\mathbf{\jmath}$ is zero:

$$\sum \mathbf{F}^{(2)} \cdot \Delta = (\mathbf{F}_{12} + \mathbf{F}_2) \cdot (\cos\theta\mathbf{\imath} + \sin\theta\mathbf{\jmath}) = 0. \quad (5.74)$$

Components F_{12x}, F_{12y}, F_{43x}, and F_{43y} are calculated from Eqs. (5.76), (5.77), and (5.78).

Force components F_{32x} and F_{32y} are computed from the sum of all the forces on link 2:

$$\sum \mathbf{F}^{(2)} = \mathbf{F}_{12} + \mathbf{F}_2 + \mathbf{F}_{32} = \mathbf{0},$$

or

$$\sum \mathbf{F}^{(2)} \cdot \mathbf{\imath} = F_{12x} + F_{2x} + F_{32x} = 0,$$
$$\sum \mathbf{F}^{(2)} \cdot \mathbf{\jmath} = F_{12y} + F_{2y} + F_{32y} = 0. \quad (5.75)$$

EXAMPLE: R-TRR-RRT MECHANISM

The R-TRR-RRT mechanism in Subsection 5.6 will be considered.

$B_R D_R D_T$ Dyad

Figure 5.26(a) shows the last dyad $B_R D_R D_T$ with unknown joint reactions $\mathbf{F}_{34}$, $\mathbf{F}_{05}$, and $\mathbf{F}_{45} = -\mathbf{F}_{54}$. Joint reaction $\mathbf{F}_{05}$ is perpendicular to sliding direction $\mathbf{F}_{05} \perp \Delta = \mathbf{\imath}$ or

$$\mathbf{F}_{05} = F_{05y}\mathbf{\jmath}. \quad (5.76)$$

The following equations can be written to determine $\mathbf{F}_{34}$ and $\mathbf{F}_{05}$.

- The sum of all the forces on links 4 and 5 is zero:

$$\sum \mathbf{F}^{(4\&5)} = \mathbf{F}_{34} + \mathbf{F}_4 + \mathbf{F}_5 + \mathbf{F}_{\text{ext}} + \mathbf{F}_{05} = \mathbf{0},$$

or

$$\sum \mathbf{F}^{(2\&3)} \cdot \mathbf{\imath} = F_{43x} + F_{4x} + F_{5x} + F_{\text{ext}} = 0,$$
$$\sum \mathbf{F}^{(2\&3)} \cdot \mathbf{\jmath} = F_{43y} + F_{4y} + F_{5y} + F_{05y} = 0. \quad (5.77)$$

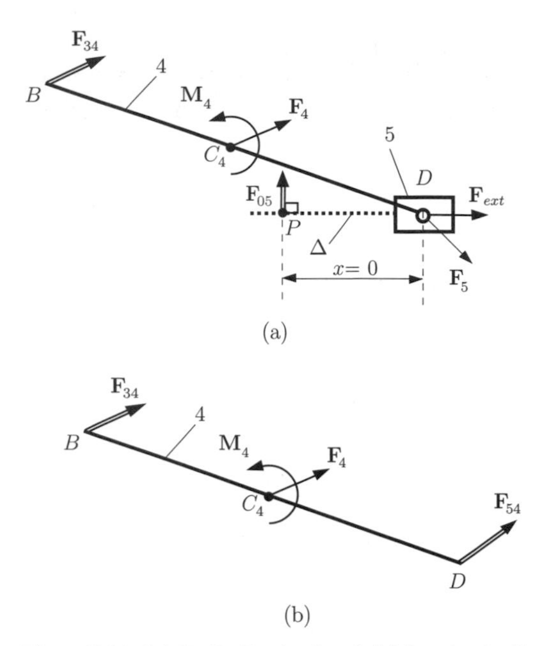

Figure 5.26. (a) $B_R D_R D_T$ dyad and (b) free-body diagram of link 4.

- The sum of moments of all the forces and torques on link 4 about D_R is zero:

$$\sum \mathbf{M}_D^{(4)} = (\mathbf{r}_B - \mathbf{r}_D) \times \mathbf{F}_{43} + (\mathbf{r}_{C3} - \mathbf{r}_D) \times \mathbf{F}_4 + \mathbf{M}_4 = \mathbf{0}. \tag{5.78}$$

The unknown components are calculated from Eqs. (5.77) and (5.78).

$$F_{34x} = -107.110 \text{ N}, \qquad F_{34y} = 14.415 \text{ N}, \qquad F_{05y} = -18.928 \text{ N}.$$

Reaction components F_{54x} and F_{54y} are computed from the sum of all the forces on link 4, as shown in Fig. 5.26(b):

$$\sum \mathbf{F}^{(4)} = \mathbf{F}_{34} + \mathbf{F}_4 + \mathbf{F}_{54} = \mathbf{0},$$

or

$$\sum \mathbf{F}^{(4)} \cdot \mathbf{\imath} = F_{34x} + F_{4x} + F_{54x} = 0,$$
$$\sum \mathbf{F}^{(4)} \cdot \mathbf{\jmath} = F_{34y} + F_{5y} + F_{54y} = 0, \tag{5.79}$$

and

$$F_{54x} = 100.643 \text{ N}, \qquad F_{54y} = -19.310 \text{ N}.$$

$B_T B_R C_R$ Dyad

Figure 5.27(a) shows the first dyad $B_T B_R C_R$ with unknown joint reaction forces $\mathbf{F}_{12}$, $\mathbf{F}_{03}$, and $\mathbf{F}_{23} = -\mathbf{F}_{32}$. Joint reaction force $\mathbf{F}_{12}$ is perpendicular to sliding direction $\mathbf{F}_{12} \perp \Delta$, or

$$\mathbf{F}_{12} \cdot \Delta = (F_{12x}\mathbf{\imath} + F_{12y}\mathbf{\jmath}) \cdot (\cos\phi\mathbf{\imath} + \sin\phi\mathbf{\jmath}) = 0. \tag{5.80}$$

The following equations can be written to determine forces $\mathbf{F}_{12}$ and $\mathbf{F}_{03}$.

- The sum of all forces on links 2 and 3 is zero.

$$\sum \mathbf{F}^{(2\&3)} = \mathbf{F}_{12} + \mathbf{F}_2 + \mathbf{F}_3 + \mathbf{F}_{43} + \mathbf{F}_{03} = \mathbf{0},$$

or

$$\sum \mathbf{F}^{(2\&3)} \cdot \mathbf{\imath} = F_{12x} + F_{2x} + F_{3x} + F_{43x} + F_{03x} = 0,$$

$$\sum \mathbf{F}^{(2\&3)} \cdot \mathbf{\jmath} = F_{12y} + F_{2y} + F_{3y} + F_{43y} + F_{03y} = 0. \tag{5.81}$$

- The sum of moments of all the forces and the torques on link 3 about B_R is zero:

$$\sum \mathbf{M}_B^{(3)} = (\mathbf{r}_C - \mathbf{r}_B) \times \mathbf{F}_{03} + (\mathbf{r}_{C3} - \mathbf{r}_B) \times \mathbf{F}_3 + \mathbf{M}_3 = \mathbf{0}. \tag{5.82}$$

The following components are obtained from Eqs. (5.80), (5.81), and (5.82).

$$F_{12x} = -71.936 \text{ N}, \qquad F_{12y} = 71.936 \text{ N},$$
$$F_{03x} = -37.156 \text{ N}, \qquad F_{03y} = -60.643 \text{ N}.$$

Reaction components F_{23x} and F_{23y} are computed from the sum of all the forces on link 3, as shown in Fig. 5.27(b):

$$\sum \mathbf{F}^{(3)} = \mathbf{F}_{23} + \mathbf{F}_3 + \mathbf{F}_{43} + \mathbf{F}_{03} = \mathbf{0},$$

or

$$\sum \mathbf{F}^{(3)} \cdot \mathbf{\imath} = F_{23x} + F_{3x} + F_{43x} + F_{03x} = 0,$$
$$\sum \mathbf{F}^{(2)} \cdot \mathbf{\jmath} = F_{23y} + F_{3y} + F_{43y} + F_{03y} = 0, \tag{5.83}$$

and solving equations

$$F_{23x} = -71.155 \text{ N}, \qquad F_{23y} = 73.397 \text{ N}.$$

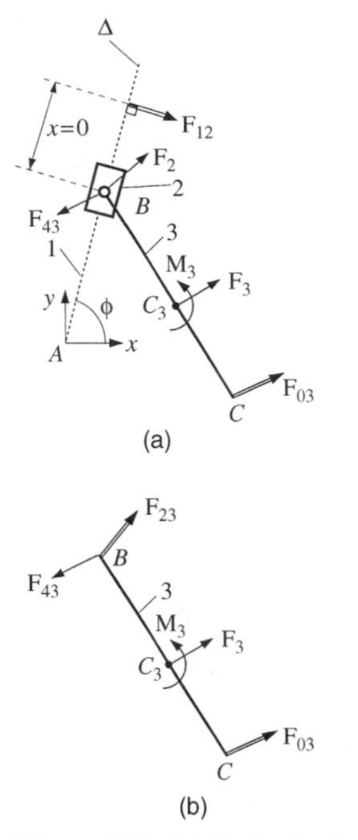

Figure 5.27. (a) $B_T B_R C_R$ dyad and (b) free-body diagram of link 3.

Driver Link

A force equation for the driver can be written to determine the joint reaction $\mathbf{F}_{01}$, as shown in Fig. 5.28:

$$\sum \mathbf{F}^{(1)} = \mathbf{F}_{01} + \mathbf{F}_1 + \mathbf{F}_{21} = \mathbf{0},$$

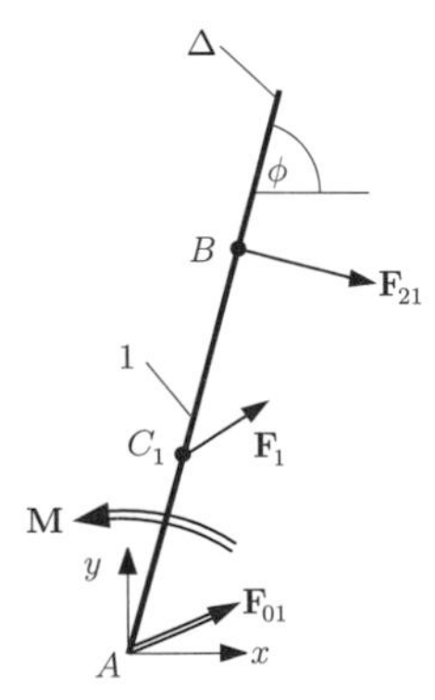

Figure 5.28. Free-body diagram of driver link 1.

or

$$\sum \mathbf{F}^{(1)} \cdot \imath = F_{01x} + F_{1x} + F_{21x} = 0,$$
$$\sum \mathbf{F}^{(1)} \cdot \jmath = F_{01y} + F_{1y} + F_{21y} = 0. \tag{5.84}$$

Solving the above equations gives

$$F_{01x} = -77.451 \text{ N}, \qquad F_{01y} = 68.747 \text{ N}.$$

Summing the moments about A_R for link 1 gives the equilibrium torque:

$$\sum \mathbf{M}_A^{(1)} = \mathbf{r}_B \times \mathbf{F}_{21} + \mathbf{r}_{C1} \times \mathbf{F}_1 + \mathbf{M} = \mathbf{0}, \tag{5.85}$$

and $M = 37.347$ N m.

6 Mechanisms with Gears

6.1 Introduction

Gears are toothed elements that transmit rotary motion from one shaft to another. Gears are generally rugged and durable, and their power transmission efficiency is as high as 98 percent. Gears are usually more costly than chains or belts. The American Gear Manufacturers Association, AGMA, has established standard tolerances for various degrees of gear manufacturing precision. *Spurs gears* are the simplest and most common type of gears. They are used to transfer motion between parallel shafts, and they have teeth that are parallel to the shaft axes.

6.2 Geometry and Nomenclature

The basic requirement determined by the gear-tooth geometry is the condition that the angular velocity ratios are exactly constant; that is; the angular velocity ratio between a 30-tooth and a 90-tooth gear must be precisely 3 in every position. The action of a pair of gear teeth satisfying this criterion is named conjugate gear-tooth action.

Law of Conjugate Gear-Tooth Action

The common normal to the surfaces at the point of contact of two gears in rotation must always intersect the line of centers at the same point P*, called the pitch point.*

The law of conjugate gear-tooth action can be satisfied by various tooth shapes, but the one of current importance is the involute of a circle. An *involute* (of a circle) is the curve generated by any point on a taut thread as it unwinds from a circle, called the base circle, as shown in Fig. 6.1(a). The involute can also be defined as the locus of a point on a taut string that is unwrapped from a

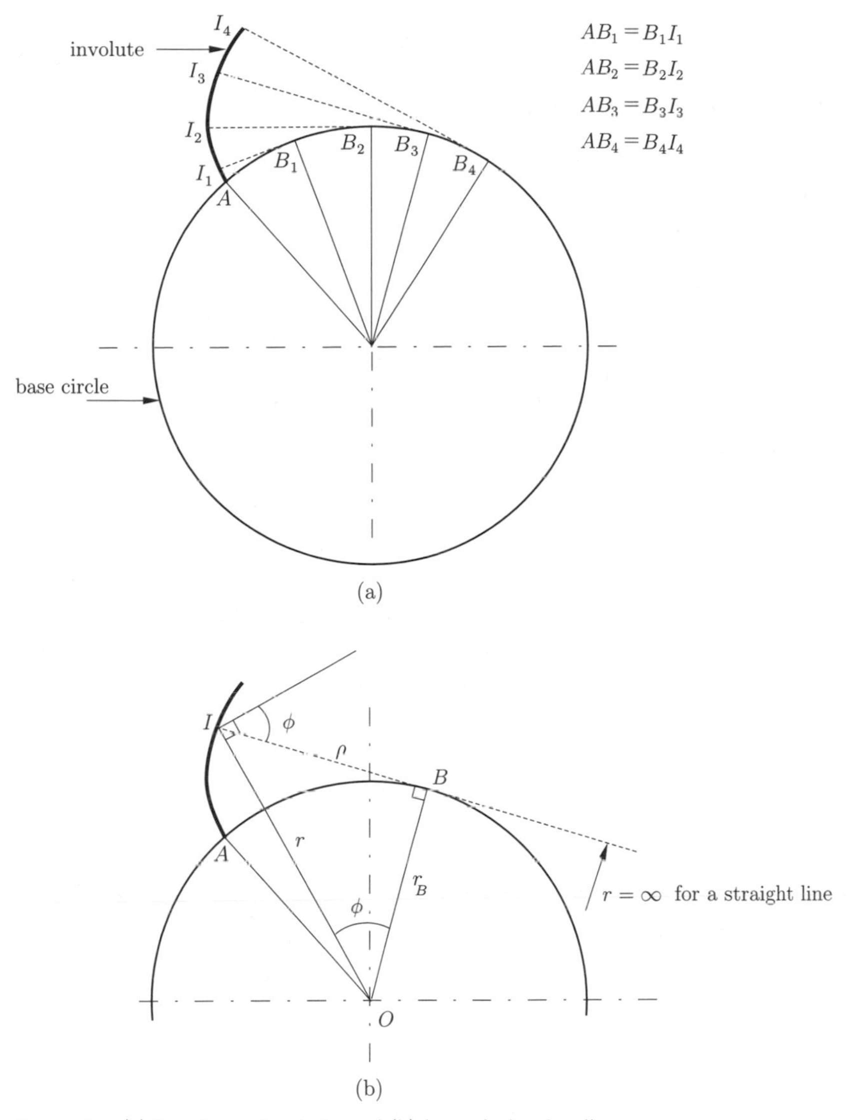

Figure 6.1. (a) Involute of a circle and (b) base circle of radius r_b.

cylinder. The circle that represents the cylinder is the *base circle*. Figure 6.1(b) represents an involute generated from a base circle of radius r_b starting at point A. The radius of curvature of the involute at any point I is given by

$$\rho = \sqrt{r^2 - r_b^2}, \tag{6.1}$$

where $r = OI$. The involute pressure angle at I is defined as the angle between the normal to the involute IB and the normal to OI, $\phi = \angle IOB$.

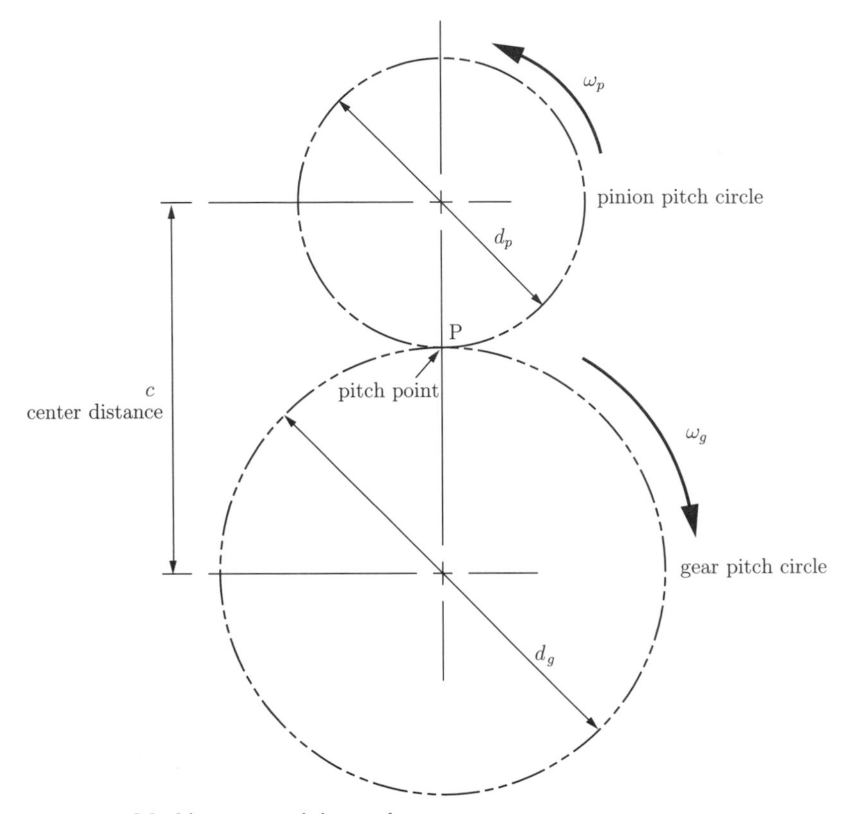

Figure 6.2. Meshing gears: pinion and gear.

In any pair of meshing gears, the smaller of the two is called the *pinion* and the larger one the *gear*. The term *gear* is used in a general sense to indicate either of the members and also in a specific sense to indicate the larger of the two. The angular velocity ratio i between a pinion and a gear is (Fig. 6.2)

$$i = \omega_p/\omega_g = -d_g/d_p, \tag{6.2}$$

where ω is the angular velocity and d is the *pitch diameter*, and the minus sign indicates that the two gears rotate in opposite directions. The *pitch circles* are the two circles, one for each gear, that remain tangent throughout the engagement cycle. The point of tangency is the pitch point. The diameter of the pitch circle is the pitch diameter. If the angular speed is expressed in rpm, then the symbol n is preferred instead of ω. The diameter (without a qualifying adjective) of a gear always refers to its pitch diameter. If other diameters (base, root, outside, etc.) are intended, they are always specified. Similarly, d, without subscripts, refers to pitch diameter. The pitch diameters of a pinion and gear are distinguished

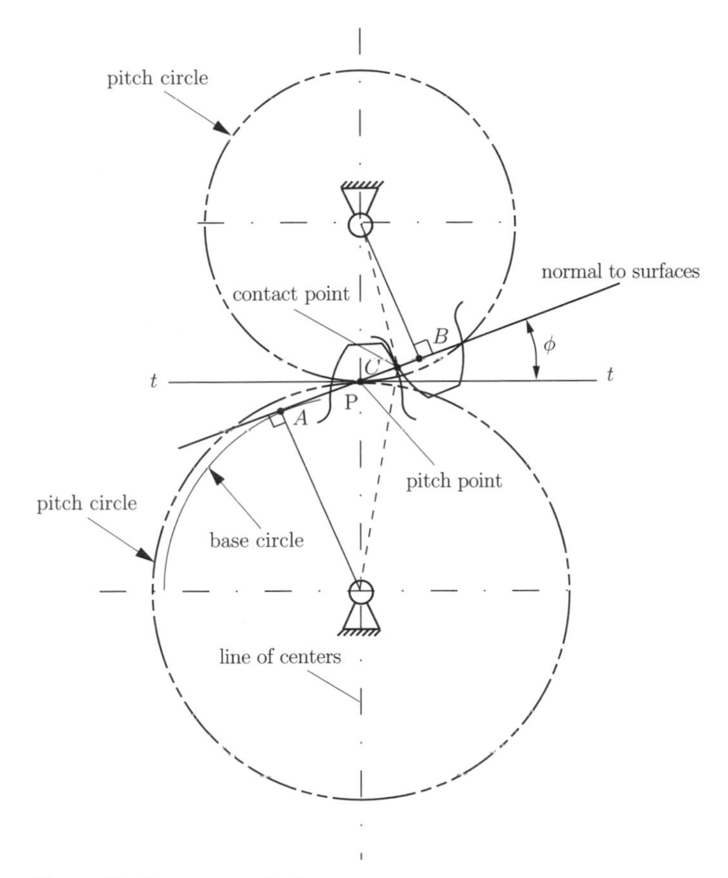

Figure 6.3. Pressure angle ϕ.

by subscripts p and g (d_p and d_g are their symbols), as shown in Fig. 6.2. The *center distance* is

$$c = (d_p + d_g)/2 = r_p + r_g, \tag{6.3}$$

where $r = d/2$ is the *pitch circle radius*.

In Fig. 6.3 the line tt is the common tangent to the pitch circles at the pitch point and AB is the common normal to the surfaces at C, the point of contact of two gears. The inclination of AB with line tt is called the *pressure angle*, ϕ. The most commonly used pressure angle, with both English and SI units, is 20°. In the United States 25° is also standard, and 14.5° was formerly an alternative standard value. The pressure angle affects the force that tends to separate mating gears.

The involute profiles are augmented outward beyond the pitch circle by a distance called the *addendum*, a, as shown in Fig. 6.4. The outer circle is usually

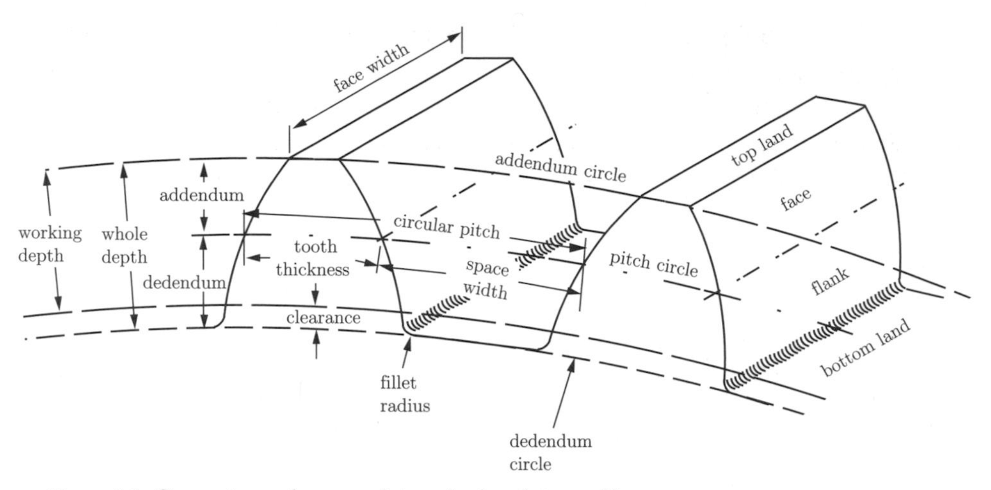

Figure 6.4. Geometry and nomenclature for involute profiles.

termed the *addendum circle*, $r_a = r + a$. Similarly, the tooth profiles are extended inward from the pitch circle a distance called the *dedendum*, b. The involute portion can extend inward only to the base circle. A fillet at the base of the tooth merges the profile into the dedendum circle. The fillet decreases the bending stress concentration. The *clearance* is the amount by which the dedendum in a given gear exceeds the addendum of its meshing gear.

The *circular pitch* is designated as p, and it is measured in inches (English units) or millimeters (SI units). If N is the number of teeth in the gear (or pinion), then

$$p = \pi d/N, \qquad p = \pi d_p/N_p, \qquad p = \pi d_g/N_g. \tag{6.4}$$

The more commonly used indices of gear-tooth size are the *diametral pitch*, P_d (used only with English units), and the *module*, m (used only with SI units). The diametral pitch is defined as the number of teeth per inch of pitch diameter:

$$P_d = N/d, \qquad P_d = N_p/d_p, \qquad P_d = N_g/d_g. \tag{6.5}$$

The module m, which is essentially the complementary of P, is defined as the pitch diameter in millimeters divided by the number of teeth (number of millimeters of pitch diameter per tooth):

$$m = d/N, \qquad m = d_p/N_p, \qquad m = d_g/N_g. \tag{6.6}$$

One can easily verify that

$$pP_d = \pi (p \text{ in inches; } P_d \text{ in teeth per inch}),$$

$$p/m = \pi \ (p \text{ in millimeters; } m \text{ in millimeters per tooth}),$$

$$m = 25.4/P_d.$$

With English units the word "pitch," without a qualifying adjective, denotes diametral pitch (a "12-pitch gear" refers to a gear with $P_d = 12$ teeth per inch of pitch diameter). With SI units "pitch" means the circular pitch (a "gear of pitch = 3.14 mm" refers to a gear having a circular pitch of $p = 3.14$ mm).

Standard diametral pitches P_d (English units) in common use are:

- 1 to 2, in increments of 0.25,
- 2 to 4, in increments of 0.5,
- 4 to 10, in increments of 1,
- 10 to 20, in increments of 2, and
- 20 to 40, in increments of 4.

With SI units, commonly used standard values of module m are:

- 0.2 to 1.0, in increments of 0.1,
- 1.0 to 4.0, in increments of 0.25, and
- 4.0 to 5.0, in increments of 0.5.

The addendum, minimum dedendum, and clearance for standard full-depth involute teeth (the pressure angle is 20°) with English units in common use are:

- addendum $a = 1/P_d$,
- minimum dedendum $b = 1.157/P_d$.

For stub involute teeth with a pressure angle equal to 20° the standard values are (English units):

- addendum $a - 0.8/P_d$,
- minimum dedendum $b = 1/P_d$.

For SI units the standard values for full-depth involute teeth with a pressure angle of 20° are

- addendum $a = m$,
- minimum dedendum $b = 1.25m$.

6.3 Interference and Contact Ratio

The contact of segments of tooth profiles, which are not conjugate, is called *interference*. The involute tooth form is only defined outside the base circle. In some cases, the dedendum will extend below the base circle. In such cases the portion of tooth below the base circle will not be an involute and will interfere with the tip of the tooth on the meshing gear, which is an involute. Interference will occur, preventing rotation of the meshing gears, if either of the addendum

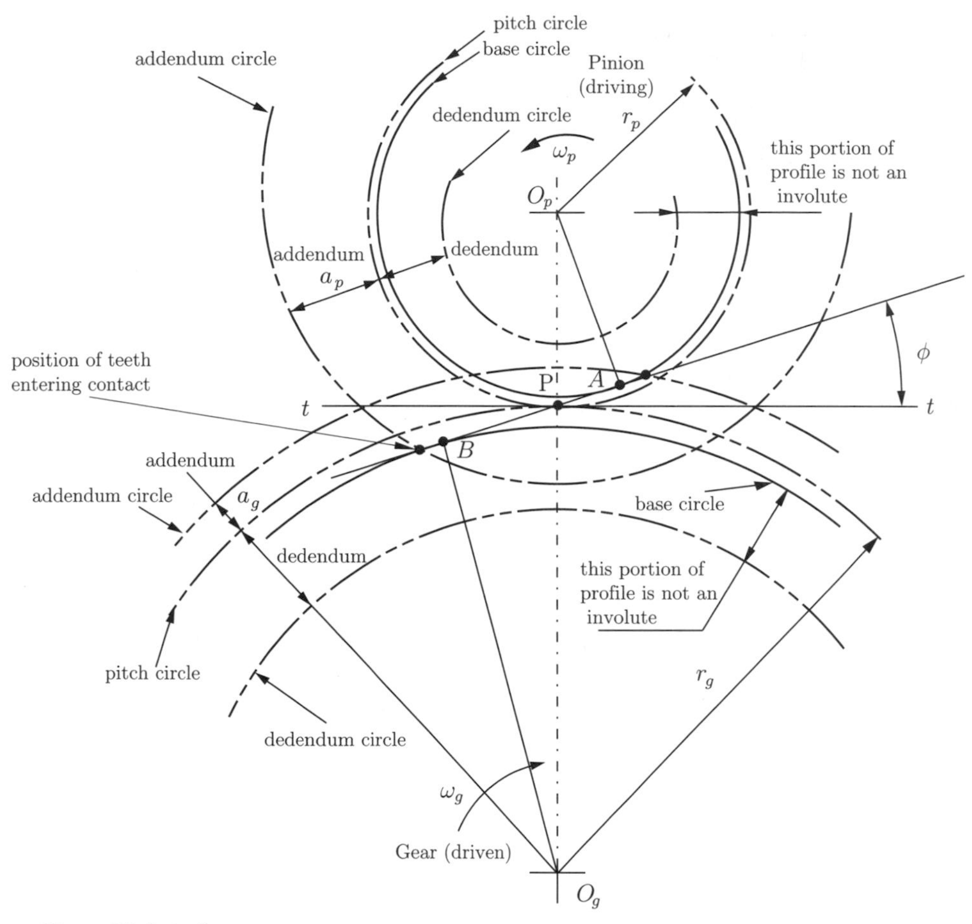

Figure 6.5. Interference.

circles extends beyond tangent points A and B, which are called interference points, as shown in Fig. 6.5. In Fig. 6.5 both addendum circles extend beyond the interference points.

The maximum possible addendum circle radius, of pinion or gear, without interference is

$$r_{a(\max)} = \sqrt{r_b^2 + c^2 \sin^2 \phi}, \tag{6.7}$$

where $r_b = r \cos \phi$ is the base circle radius of the pinion or gear. The base circle diameter is

$$d_b = d \cos \phi. \tag{6.8}$$

The average number of teeth in contact as the gears rotate together is the contact ratio CR, which is calculated from the following equation:

$$\mathrm{CR} = \frac{\sqrt{r_{ap}^2 - r_{bp}^2} + \sqrt{r_{ag}^2 - r_{bg}^2} - c \sin \phi}{p_b}, \tag{6.9}$$

where r_{ap}, r_{ag} are the addendum radii of the meshing pinion and gear, and r_{bp}, r_{bg} are the base circle radii of the meshing pinion and gear The base pitch p_b is computed with

$$p_b = \pi d_b / N = p \cos \phi. \tag{6.10}$$

The base pitch is similar to the circular pitch except that it represents an arc of the base circle rather than an arc of the pitch circle.

6.4 Linkage Transformation

The half-joint at the contact point of two gears in motion can be substituted for two full joints A and B and an extra link 3, between gears 1 and 2, as shown in Fig. 6.6. The mechanism still has one DOF, and the two-gear system (0, 1, and 2) is in fact a four-bar mechanism (0, 1, 2, and 3) in another disguise.

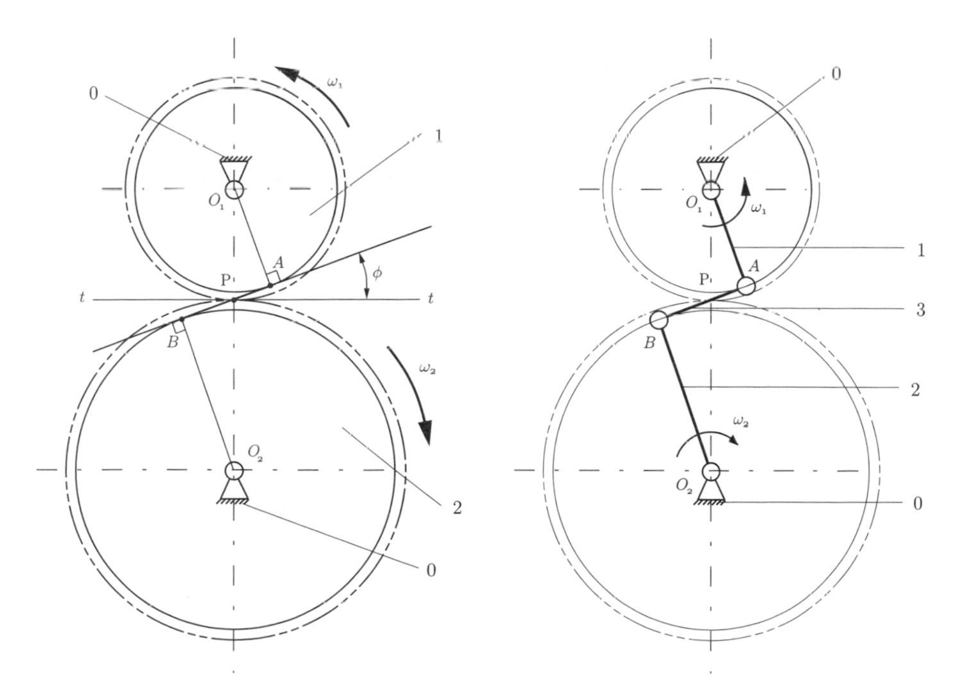

Figure 6.6. Linkage transformation.

The following relations can be written:

$$O_1O_2 = \frac{m}{2}(N_1 + N_2),$$
$$O_1A = r_1 \cos\phi,$$
$$O_2B = r_2 \cos\phi,$$
$$AB = AP + PB = \frac{mN_1}{2}\sin\phi + \frac{mN_2}{2}\sin\phi = \frac{m}{2}(N_1 + N_2)\sin\phi.$$

Because m, N_1, N_2, and ϕ are constants, the links of the four-bar mechanism are constant too.

6.5 Ordinary Gear Trains

A gear train is any collection of two or more meshing gears. Figure 6.7(a) shows a simple gear train with three gears in series. The train ratio can be computed with the relation

$$i_{13} = \frac{\omega_1}{\omega_3} = \frac{\omega_1}{\omega_2}\frac{\omega_2}{\omega_3} = \left(-\frac{N_2}{N_1}\right)\left(-\frac{N_3}{N_2}\right) = \frac{N_3}{N_1}. \tag{6.11}$$

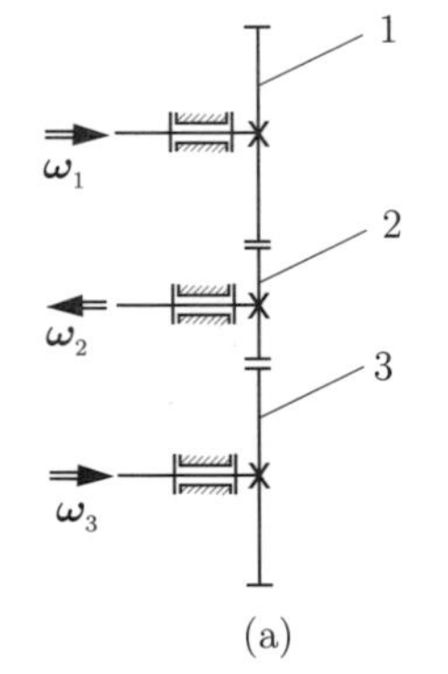

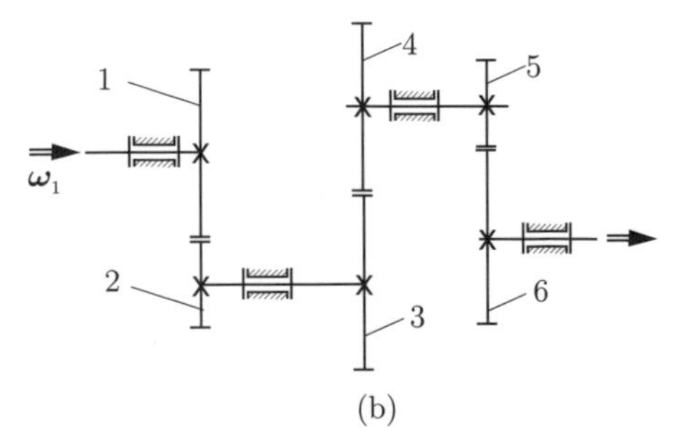

Figure 6.7. Gear train: (a) simple, with three gears in series and (b) compound.

Only the sign of the overall ratio is affected by intermediate gear 2. Gear 2 is called an *idler*.

Figure 6.7(b) shows a compound gear train, without idler gears, with the train ratio

$$i_{16} = \frac{\omega_1}{\omega_2}\frac{\omega_3}{\omega_4}\frac{\omega_5}{\omega_6} = \left(-\frac{N_2}{N_1}\right)\left(-\frac{N_4}{N_3}\right)\left(-\frac{N_6}{N_5}\right) = -\frac{N_2 N_4 N_6}{N_1 N_3 N_5}. \tag{6.12}$$

6.6 Epicyclic Gear Trains

When at least one of the gear axes rotates relative to the frame, in addition to the gear's own rotation about its own axes, the train is called a *planetary gear train* or an *epicyclic gear train*. The term *epicyclic* comes from the fact that points on gears with moving axes of rotation describe epicyclic paths. When a generating circle (a planet gear) rolls on the outside of another circle, called the directing circle (a sun gear), each point on the generating circle describes an epicycloid, as shown in Fig. 6.8.

Generally, the greater the number of planet gears, the greater the torque capacity of the system. For better load balancing, new designs have two sun gears and up to 12 planetary assemblies in one casing.

In the case of simple and compound gears it is not difficult to visualize the motion of the gears, and the determination of the speed ratio is relatively easy. In the case of epicyclic gear trains it is often difficult to visualize the motion of the gears. A systematic procedure using the contour method is presented below. The contour method is applied to determine the distribution of velocities for several epicyclic gear trains.

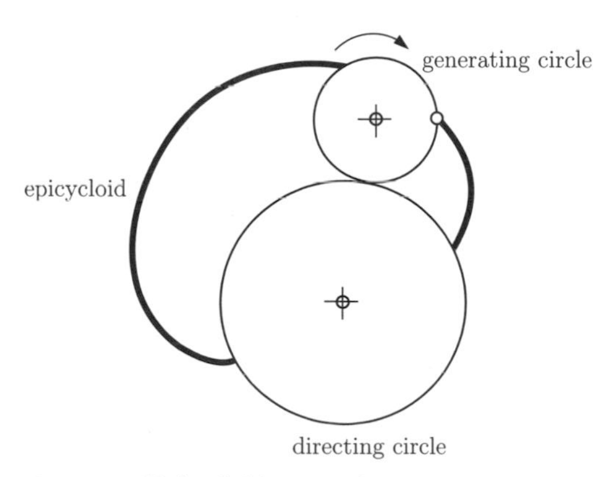

Figure 6.8. Epicycloid generation.

The velocity equations for a simple closed kinematic chain are

$$\sum_{(i)} \omega_{i,i-1} = \mathbf{0},$$

$$\sum_{(i)} \mathbf{AA}_i \times \omega_{i,i-1} + \sum_{(i)} \mathbf{v}_{A_{i,i-1}} = \mathbf{0}, \tag{6.13}$$

where $\omega_{i,i-1}$ is the relative angular velocity of rigid body (i) with respect to rigid body $(i-1)$, $\mathbf{AA}_i$ is the position vector of the kinematic pair, A_i, between rigid body (i) and rigid body $(i-1)$ with respect to a "fixed" reference frame, and $\mathbf{v}_{A_{i,i-1}}$ is the relative velocity of link (i) with respect to link $(i-1)$, permitted by the joint at A_i.

EXAMPLE 1

The first epicyclic (planetary) gear train considered is shown in Fig. 6.9. The system consists of a central gear 2 (a sun gear) and another gear 3 in mesh with 2 (a planet gear) at B. Gear 3 is carried by the arm 1 hinged at A, as shown. The ring gear 4 meshes with planet gear 3 and pivots at A, so it can be easily tapped as the output member. The sun gear and the ring gear are concentric. The sun gear, the ring gear, and the arm can be accessed so that their angular velocities and torques can be tapped either as inputs or outputs.

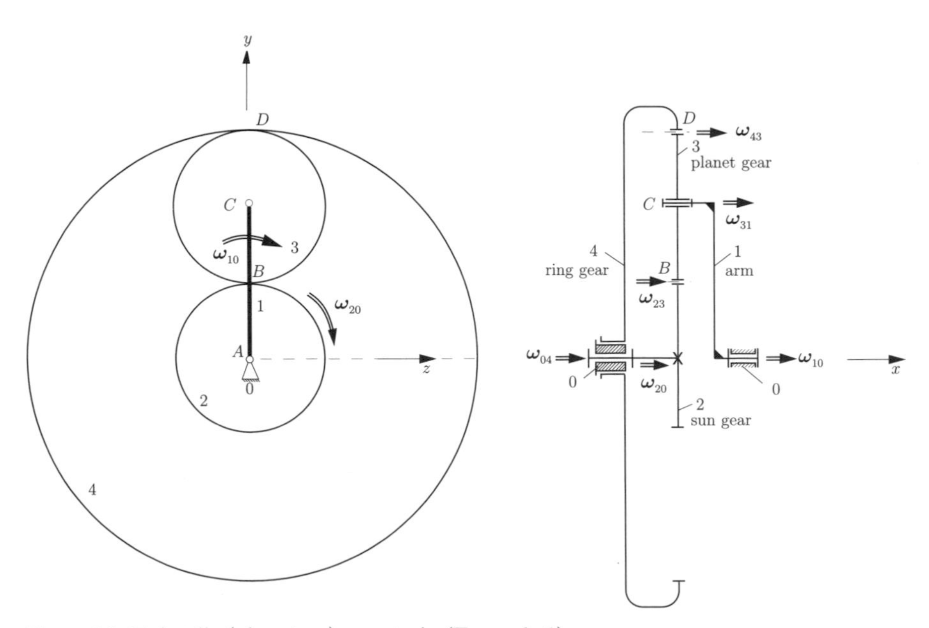

Figure 6.9. Epicyclic (planetary) gear train (Example 1).

There are four moving bodies 1, 2, 3, and 4, ($n = 4$) connected by

- four full joints ($c_5 = 4$): one hinge between arm 1 and planet gear 3 at C, one hinge between frame 0 and the shaft of sun gear 2 at A, one hinge between frame 0 and ring gear 4 at A, and one hinge between frame 0 and arm 1 at A;
- two half-joints ($c_4 = 2$): one between sun gear 2 and planet gear 3, and one between planet gear 3 and ring gear 4. The system possesses two DOF:

$$M = 3n - 2c_5 - c_4 = 3 \cdot 4 - 2 \cdot 4 - 2 = 2. \tag{6.14}$$

The sun gear has $N_2 = 40$ external gear teeth, the planet gear has $N_3 = 20$ external gear teeth, and the ring gear has $N_4 = 80$ internal gear teeth. If the arm and the sun gear rotate with input angular speeds, $n_1 = 200$ rpm and $n_2 = 100$ rpm, find the absolute output angular velocity of the ring gear.

SOLUTION

The velocity analysis is carried out by using the contour method. The system shown in Fig. 6.9 has a total of five elements ($p = 5$): frame 0 and four moving links 1, 2, 3, and 4. There are six joints ($l = 6$), four full joints and two half-joints. The number of independent loops is given by

$$n_c = l - p + 1 - 6 - 5 + 1 = 2.$$

This gear system has two independent contours. The diagram representing the kinematic chain is shown in Fig. 6.10.

The angular speeds of the arm and the sun gear, expressed in radians per second, are

$$\omega_1 = \omega_{10} = \frac{\pi n_1}{30} = \frac{20\pi}{3} \text{ rad/s},$$

$$\omega_2 = \omega_{20} = \frac{\pi n_2}{30} = \frac{10\pi}{3} \text{ rad/s}.$$

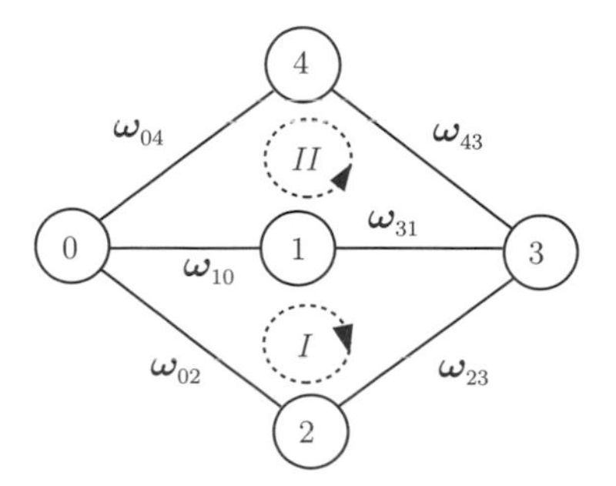

Figure 6.10. Diagram representing the planetary gear train (Example 1).

First Contour

The first contour is formed by elements 0, 1, 3, 2, and 0 (following a clockwise path). For the velocity analysis, the following vectorial equations can be written:

$$\boldsymbol{\omega}_{10} + \boldsymbol{\omega}_{31} + \boldsymbol{\omega}_{23} + \boldsymbol{\omega}_{02} = \mathbf{0},$$
$$\mathbf{AC} \times \boldsymbol{\omega}_{31} + \mathbf{AB} \times \boldsymbol{\omega}_{23} = \mathbf{0}, \tag{6.15}$$

where the input angular velocities are

$$\boldsymbol{\omega}_{10} = [\omega_{10}, 0, 0] = [\omega_1, 0, 0] = \omega_1\mathbf{ı} + 0\mathbf{ȷ} + 0\mathbf{k},$$
$$\boldsymbol{\omega}_{02} = [\omega_{02}, 0, 0] = [-\omega_2, 0, 0] = -\omega_2\mathbf{ı} + 0\mathbf{ȷ} + 0\mathbf{k},$$

and the unknown angular velocities are

$$\boldsymbol{\omega}_{31} = [\omega_{31}, 0, 0],$$
$$\boldsymbol{\omega}_{23} = [\omega_{23}, 0, 0].$$

The sign of the relative angular velocities is selected to be positive, and then the numerical computation will give the true orientation of the vectors.

The vectors **AB**, **AC**, and **AD** are defined as follows:

$$\mathbf{AB} = [x_B, y_B, 0], \qquad \mathbf{AC} = [x_C, y_C, 0], \qquad \mathbf{AD} = [x_D, y_D, 0], \tag{6.16}$$

where

$$y_B = r_2 = mN_2/2,$$
$$y_C = r_2 + r_3 = m(N_2 + N_3)/2,$$
$$y_D = r_2 + 2r_3 = mN_2/2 + mN_3.$$

The module of the gears is m. Equation (6.15) becomes

$$\omega_1\mathbf{ı} + \omega_{31}\mathbf{ı} + \omega_{23}\mathbf{ı} - \omega_2\mathbf{ı} = \mathbf{0},$$
$$\begin{vmatrix} \mathbf{ı} & \mathbf{ȷ} & \mathbf{k} \\ x_C & y_C & 0 \\ \omega_{31} & 0 & 0 \end{vmatrix} + \begin{vmatrix} \mathbf{ı} & \mathbf{ȷ} & \mathbf{k} \\ x_B & y_B & 0 \\ \omega_{23} & 0 & 0 \end{vmatrix} = \mathbf{0}. \tag{6.17}$$

Equation (6.17) can be projected on the "fixed" reference frame $xOyz$:

$$\omega_1 + \omega_{31} + \omega_{23} - \omega_2 = 0,$$
$$y_C\omega_{31} + y_B\omega_{23} = 0.$$

Equation (6.18) represents a system of two equations with two unknowns, ω_{31} and ω_{23}. When the algebraic equations are solved, the values of the relative angular velocities are obtained as follows:

$$\omega_{31} = N_2(\omega_1 - \omega_2)/N_3 = 20\pi/3 \text{ rad/s},$$
$$\omega_{23} = -\omega_1 + \omega_2 - N_2(\omega_1 - \omega_2)/N_3 = -10\pi \text{ rad/s}.$$

Second Contour

The second closed contour contains elements 0, 1, 3, 4, and 0, as shown in Fig. 6.10. The contour velocity equations can be written as (using the counter-clockwise path)

$$\boldsymbol{\omega}_{10} + \boldsymbol{\omega}_{31} + \boldsymbol{\omega}_{43} + \boldsymbol{\omega}_{04} = \mathbf{0},$$
$$\mathbf{AC} \times \boldsymbol{\omega}_{31} + \mathbf{AD} \times \boldsymbol{\omega}_{43} = \mathbf{0}, \tag{6.18}$$

where the known angular velocities are $\boldsymbol{\omega}_{10}$, $\boldsymbol{\omega}_{31}$, and the unknown angular velocities are

$$\boldsymbol{\omega}_{43} = [\omega_{43}, 0, 0],$$
$$\boldsymbol{\omega}_{04} = [\omega_{04}, 0, 0].$$

Equation (6.18) can be written as

$$\omega_1 \mathbf{1} + \omega_{31} \mathbf{1} + \omega_{43} \mathbf{1} + \omega_{04} \mathbf{1} = \mathbf{0},$$
$$\begin{vmatrix} \mathbf{1} & \mathbf{J} & \mathbf{k} \\ x_C & y_C & 0 \\ \omega_{31} & 0 & 0 \end{vmatrix} + \begin{vmatrix} \mathbf{1} & \mathbf{J} & \mathbf{k} \\ x_D & y_D & 0 \\ \omega_{43} & 0 & 0 \end{vmatrix} = \mathbf{0}. \tag{6.19}$$

From Eq. (6.19) the absolute angular velocity of the ring gear is

$$\omega_{40} = -\omega_{04} = \frac{2N_2\omega_1 + 2N_3\omega_1 - N_2\omega_2}{N_2 + 2N_3} = 25\pi/3 \text{ rad/s},$$

or $n_4 = 250$ rpm.

■ EXAMPLE 2

The second planetary gear train considered is shown in Fig. 6.11. The system consists of an input sun gear 1 and a planet gear 2 in mesh with 1 at B. Gear 2 is carried by arm S fixed on the shaft of gear 3, as shown. Gear 3 meshes with output gear 4 at F. The fixed ring gear 4 meshes with planet gear 2 at D.

There are four moving gears (1, 2, 3, and 4) connected by

- four full joints ($c_5 = 4$): one at A, between frame 0 and sun gear 1; one at C, between arm S and planet gear 2; one at E, between frame 0 and gear 3, and another at G, between frame 0 and gear 3;

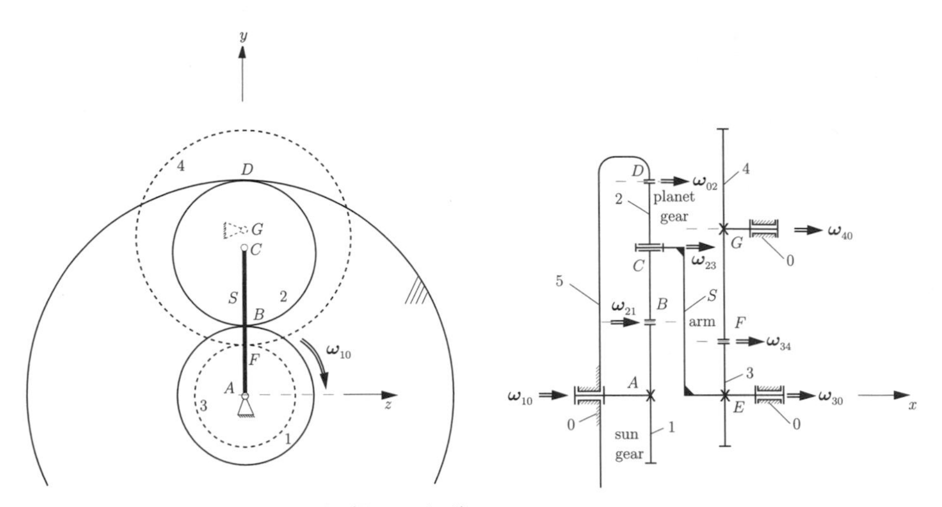

Figure 6.11. Planetary gear train (Example 2).

- three half-joints ($c_4 = 3$): one at B, between sun gear 1 and planet gear 2; one at D, between planet gear 2 and the ring gear; and another at F, between gear 3 and output gear 4. The module of the gears is $m = 5$ mm.

The system possesses one degree of freedom:

$$M = 3n - 2c_5 - c_4 = 3 \cdot 4 - 2 \cdot 4 - 3 = 1. \tag{6.20}$$

The sun gear has $N_1 = 19$ external gear teeth, the planet gear has $N_2 = 28$ external gear teeth, and the ring gear has $N_5 = 75$ internal gear teeth. The gear 3 has $N_3 = 18$ external gear teeth, and the output gear has $N_4 = 36$ external gear teeth. The sun gear rotates with an input angular speed $n_1 = 2970$ rpm ($\omega_1 = \omega_{10} = \pi n_1/30 = 311.018$ rad/s). Find the absolute output angular velocity of gear 4, the velocities of pitch points B and F, and the velocity of joint C.

SOLUTION

The velocity analysis is carried out by using the contour equation method. The system shown in Fig. 6.11 has five elements (0, 1, 2, 3, 4) and seven joints. The number of independent loops is given by

$$n_c = l - p + 1 = 7 - 5 + 1 = 3.$$

This gear system has three independent contours. The diagram representing the kinematic chain and the independent contours is shown in Fig. 6.12.

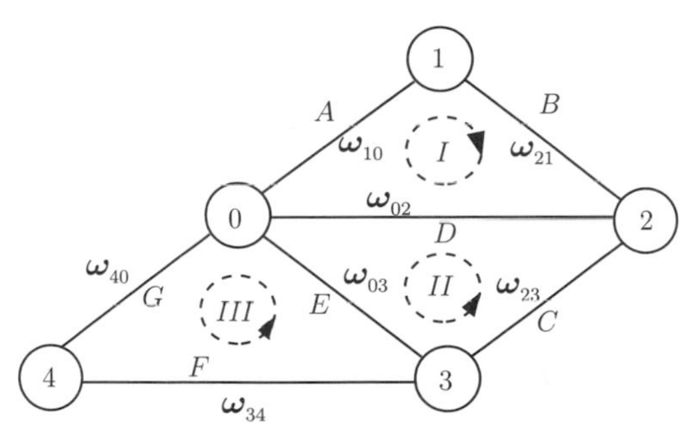

Figure 6.12. Diagram representing the planetary gear train (Example 2).

The position vectors **AB**, **AC**, **AD**, **AF**, and **AG** are defined as follows:

$$\begin{aligned}
\mathbf{AB} &= [x_B, y_B, 0] = [x_B, r_1, 0] = [x_B, mN_1/2, 0],\\
\mathbf{AC} &= [x_C, y_C, 0] = [x_C, r_1 + r_2, 0] = [x_C, m(N_1 + N_2)/2, 0],\\
\mathbf{AD} &= [x_D, y_D, 0] = [x_D, r_1 + 2r_2, 0] = [x_D, m(N_1 + 2N_2)/2, 0],\\
\mathbf{AF} &= [x_F, y_F, 0] = [x_F, r_3, 0] = [x_F, mN_3/2, 0],\\
\mathbf{AG} &= [x_G, y_G, 0] = [x_C, r_3 + r_4, 0] = [x_G, m(N_3 + N_4)/2, 0].
\end{aligned}$$

First Contour

The first closed contour contains elements 0, 1, 2, and 0 (following the clockwise path). For the velocity analysis, the following vectorial equations can be written:

$$\begin{aligned}
&\boldsymbol{\omega}_{10} + \boldsymbol{\omega}_{21} + \boldsymbol{\omega}_{02} = \mathbf{0},\\
&\mathbf{AB} \times \boldsymbol{\omega}_{21} + \mathbf{AD} \times \boldsymbol{\omega}_{02} = \mathbf{0},
\end{aligned} \tag{6.21}$$

where the input angular velocity is

$$\boldsymbol{\omega}_{10} = [\omega_{10}, 0, 0] = [\omega_1, 0, 0],$$

and the unknown angular velocities are

$$\begin{aligned}
\boldsymbol{\omega}_{21} &= [\omega_{21}, 0, 0],\\
\boldsymbol{\omega}_{02} &= [\omega_{02}, 0, 0].
\end{aligned}$$

The sign of the relative angular velocities is selected to be positive, and then the numerical results give the real orientation of the vectors.

Equation (6.21) becomes

$$\omega_1 \mathbf{\imath} + \omega_{21} \mathbf{\imath} + \omega_{02} \mathbf{\imath} = \mathbf{0},$$

$$\begin{vmatrix} \mathbf{\imath} & \mathbf{\jmath} & \mathbf{k} \\ x_B & y_B & 0 \\ \omega_{21} & 0 & 0 \end{vmatrix} + \begin{vmatrix} \mathbf{\imath} & \mathbf{\jmath} & \mathbf{k} \\ x_D & y_D & 0 \\ \omega_{02} & 0 & 0 \end{vmatrix} = \mathbf{0}. \tag{6.22}$$

Equation (6.22) projected on a "fixed" reference frame $xOyz$ is

$$\begin{aligned} \omega_1 + \omega_{21} + \omega_{02} &= 0, \\ y_B \omega_{21} + y_D \omega_{02} &= 0. \end{aligned} \tag{6.23}$$

Equation (6.23) represents a system of two equations with two unknowns, ω_{21} and ω_{02}. When the algebraic equations are solved, the following value is obtained for the absolute angular velocity of planet gear 2:

$$\omega_{20} = -\omega_{02} = -\frac{N_1 \omega_1}{2N_2} = -105.524 \text{ rad/s}. \tag{6.24}$$

Second Contour

The second closed contour contains elements 0, 3, 2, and 0 (following the counterclockwise path). For the velocity analysis, the following vectorial equations can be written:

$$\begin{aligned} &\boldsymbol{\omega}_{30} + \boldsymbol{\omega}_{23} + \boldsymbol{\omega}_{02} = \mathbf{0}, \\ &\mathbf{AE} \times \boldsymbol{\omega}_{30} + \mathbf{AC} \times \boldsymbol{\omega}_{23} + \mathbf{AD} \times \boldsymbol{\omega}_{02} = \mathbf{0}. \end{aligned} \tag{6.25}$$

The unknown angular velocities are

$$\begin{aligned} \boldsymbol{\omega}_{30} &= [\omega_{21}, 0, 0], \\ \boldsymbol{\omega}_{23} &= [\omega_{23}, 0, 0]. \end{aligned}$$

When Eq. (6.25) is solved, the following value is obtained for the absolute angular velocity of gear 3 and arm S:

$$\omega_{30} = \frac{N_1 \omega_1}{2(N_1 + N_2)} = 62.865 \text{ rad/s}. \tag{6.26}$$

Third Contour

The third closed contour contains links 0, 4, 3, and 0 (following the counter-clockwise path). The velocity vectorial equations are

$$\boldsymbol{\omega}_{40} + \boldsymbol{\omega}_{34} + \boldsymbol{\omega}_{03} = \mathbf{0},$$
$$\mathbf{AG} \times \boldsymbol{\omega}_{40} + \mathbf{AF} \times \boldsymbol{\omega}_{34} + \mathbf{AE} \times \boldsymbol{\omega}_{03} = \mathbf{0}, \tag{6.27}$$

or

$$\omega_{40}\mathbf{1} + \omega_{34}\mathbf{1} - \omega_{30}\mathbf{1} = \mathbf{0},$$
$$\begin{vmatrix} \mathbf{1} & \mathbf{J} & \mathbf{k} \\ x_G & y_G & 0 \\ \omega_{40} & 0 & 0 \end{vmatrix} + \begin{vmatrix} \mathbf{1} & \mathbf{J} & \mathbf{k} \\ x_F & y_F & 0 \\ \omega_{34} & 0 & 0 \end{vmatrix} = \mathbf{0}. \tag{6.28}$$

The unknown angular velocities are

$$\boldsymbol{\omega}_{40} = [\omega_{40}, 0, 0],$$
$$\boldsymbol{\omega}_{34} = [\omega_{34}, 0, 0].$$

The absolute angular velocity of output gear 4 is

$$\omega_{40} = -\frac{N_1 N_3 \omega_1}{2(N_1 + N_2)N_4} = -31.432 \text{ rad/s}. \tag{6.29}$$

Linear Velocities of Pitch Points

The velocity of pitch point B is

$$v_B = \omega_{10} r_1 = 14.773 \text{ m/s},$$

and the velocity of pitch point F is

$$v_F = \omega_{40} r_4 = 2.828 \text{ m/s}.$$

The velocity of joint C is

$$v_C = \omega_{30}(r_1 + r_2) = 7.386 \text{ m/s}.$$

Gear Geometrical Dimensions

For standard external gear teeth, the addendum is $a = m$.

Gear 1

- pitch circle diameter $d_1 = mN_1 = 95.0$ mm;
- addendum circle diameter $d_{a1} = m(N_1 + 2) = 105.0$ mm; and
- dedendum circle diameter $d_{d1} = m(N_1 - 2.5) = 82.5$ mm.

Gear 2

- pitch circle diameter $d_2 = mN_2 = 140.0$ mm;
- addendum circle diameter $d_{a2} = m(N_2 + 2) = 150.0$ mm; and
- dedendum circle diameter $d_{d2} = m(N_2 - 2.5) = 127.5$ mm.

Gear 3

- pitch circle diameter $d_3 = mN_3 = 90.0$ mm;
- addendum circle diameter $d_{a3} = m(N_3 + 2) = 100.0$ mm; and
- dedendum circle diameter $d_{d3} = m(N_3 - 2.5) = 77.5$ mm.

Gear 4

- pitch circle diameter $d_4 = mN_4 = 180.0$ mm;
- addendum circle diameter $d_{a4} = m(N_4 + 2) = 190.0$ mm; and
- dedendum circle diameter $d_{d4} = m(N_4 - 2.5) = 167.5$ mm.

Gear 5 (Internal Gear)

- pitch circle diameter $d_5 = mN_5 = 375.0$ mm;
- addendum circle diameter $d_{a5} = m(N_5 - 2) = 365.0$ mm; and
- dedendum circle diameter $d_{d5} = m(N_5 + 2.5) = 387.5$ mm.

Number of Planet Gears

The number of necessary planet gears k is given by the assembly condition,

$$(N_1 + N_5)/k = \text{integer},$$

and for the planetary gear train, $k = 2$ planet gears. The vicinity condition between the sun gear and the planet gear,

$$m(N_1 + N_2)\sin(\pi/k) > d_{a2},$$

is thus verified.

EXAMPLE 3

The third planetary gear train considered is shown in Fig. 6.13. The system consists of an input sun gear 1 and a planet gear 2 in mesh with 1 at B. Gear $2'$

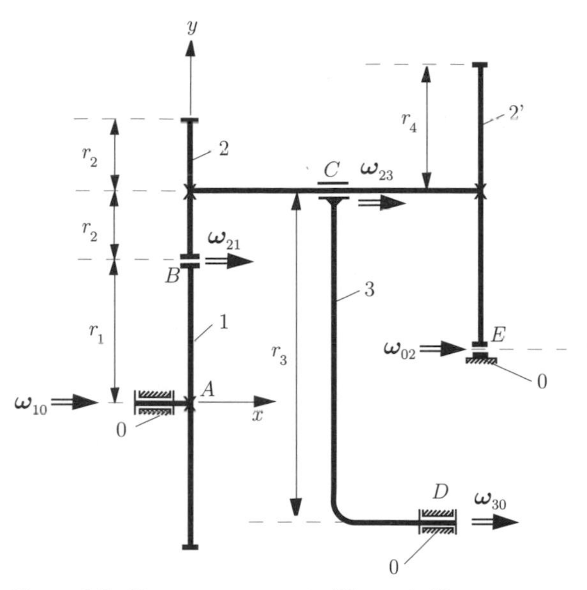

Figure 6.13. Planetary gear train (Example 3).

is fixed on the shaft of gear 2. The system of gears 2 and 2′ is carried by arm 3. Gear 2′ meshes with fixed frame 0 at E.

There are three moving gears (1, 2, and 3) connected by

- three full joints ($c_5 = 3$): one at A, between frame 0 and sun gear 1; one at C, between arm 3 and planet gear system 2; and another at D, between frame 0 and arm 3;
- two half-joints ($c_4 = 2$): one at B, between sun gear 1 and planet gear 2, and another at E, between planet gear 2′ and frame 0.

The system possesses one DOF:

$$M = 3n - 2c_5 - c_4 = 3 \cdot 3 - 2 \cdot 3 - 2 = 1. \tag{6.30}$$

The sun gear has a radius of the pitch circle equal to r_1, planet gear 2 has a radius of the pitch circle equal to r_2, arm 3 has a length equal to r_3, and planet gear 2′ has a radius of the pitch circle equal to r_4, as shown in Fig. 6.13.

The sun gear rotates with the input angular velocity ω_1. Find the speed ratio i_{13} between sun gear 1 and arm 3.

SOLUTION

The system shown in Fig. 6.13 has four elements (0, 1, 2, 3) and five kinematic pairs. The number of independent loops is given by

$$n_c = l - p + 1 = 5 - 4 + 1 = 2.$$

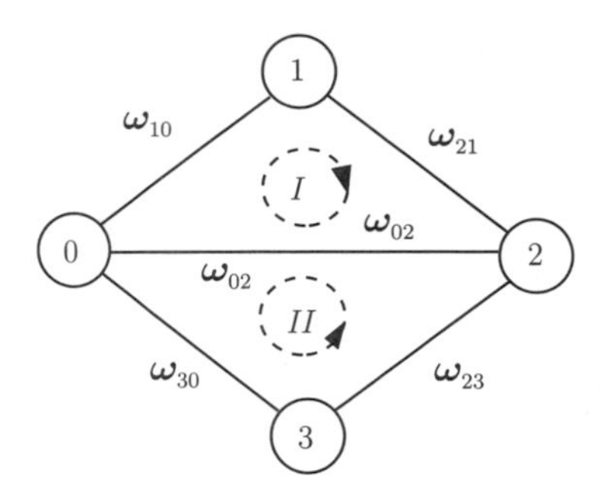

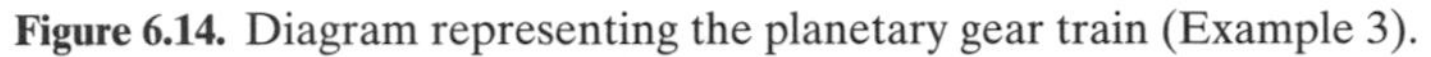

Figure 6.14. Diagram representing the planetary gear train (Example 3).

This gear system has two independent contours. The diagram representing the kinematic chain and the independent contours is shown in Fig. 6.14.

The position vectors **AB**, **AC**, **AD**, and **AE** are defined as follows:

$$\begin{aligned}
\mathbf{AB} &= [x_B, y_B, 0] = [x_B, r_1, 0],\\
\mathbf{AC} &= [x_C, y_C, 0] = [x_C, r_1 + r_2, 0],\\
\mathbf{AD} &= [x_D, y_D, 0] = [x_D, r_1 + r_2 - r_3, 0],\\
\mathbf{AE} &= [x_E, y_E, 0] = [x_E, r_1 + r_2 - r_4, 0].
\end{aligned} \tag{6.31}$$

First Contour

The first closed contour contains elements 0, 1, 2, and 0 (following the clockwise path). For the velocity analysis, the following vectorial equations can be written:

$$\begin{aligned}
&\boldsymbol{\omega}_{10} + \boldsymbol{\omega}_{21} + \boldsymbol{\omega}_{02} = \mathbf{0},\\
&\mathbf{AB} \times \boldsymbol{\omega}_{21} + \mathbf{AE} \times \boldsymbol{\omega}_{02} = \mathbf{0},
\end{aligned} \tag{6.32}$$

where the input angular velocity is

$$\boldsymbol{\omega}_{10} = [\omega_{10}, 0, 0] = [\omega_1, 0, 0],$$

and the unknown angular velocities are

$$\begin{aligned}
\boldsymbol{\omega}_{21} &= [\omega_{21}, 0, 0],\\
\boldsymbol{\omega}_{02} &= [\omega_{02}, 0, 0].
\end{aligned}$$

Equation (6.32) becomes

$$\begin{aligned}
&\omega_1 \mathbf{i} + \omega_{21} \mathbf{i} + \omega_{02} \mathbf{i} = \mathbf{0},\\
&\begin{vmatrix} \mathbf{i} & \mathbf{j} & \mathbf{k} \\ x_B & y_B & 0 \\ \omega_{21} & 0 & 0 \end{vmatrix} + \begin{vmatrix} \mathbf{i} & \mathbf{j} & \mathbf{k} \\ x_E & y_E & 0 \\ \omega_{02} & 0 & 0 \end{vmatrix} = \mathbf{0}.
\end{aligned} \tag{6.33}$$

Equation (6.33) projected onto a "fixed" reference frame $xOyz$ is

$$\omega_1 + \omega_{21} + \omega_{02} = 0,$$
$$y_B\omega_{21} + y_E\omega_{02} = 0. \tag{6.34}$$

Equation (6.34) represents a system of two equations with two unknowns, ω_{21} and ω_{02}. When the algebraic equations are solved, the following value is obtained for the absolute angular velocity of planet gear 2:

$$\omega_{20} = -\omega_{02} = -\frac{r_1\omega_1}{r_2 - r_4}. \tag{6.35}$$

Second Contour

The second closed contour contains elements 0, 3, 2, and 0 (counterclockwise path). For the velocity analysis, the following vectorial equations can be written:

$$\boldsymbol{\omega}_{30} + \boldsymbol{\omega}_{23} + \boldsymbol{\omega}_{02} = \mathbf{0},$$
$$\mathbf{AD} \times \boldsymbol{\omega}_{30} + \mathbf{AC} \times \boldsymbol{\omega}_{23} + \mathbf{AE} \times \boldsymbol{\omega}_{02} = \mathbf{0}. \tag{6.36}$$

The unknown angular velocities are

$$\boldsymbol{\omega}_{30} = [\omega_{21}, 0, 0],$$
$$\boldsymbol{\omega}_{23} = [\omega_{23}, 0, 0].$$

When Eq. (6.36) is solved, the following value is obtained for the absolute angular velocity of arm 3:

$$\omega_{30} = \frac{r_1 r_4 \omega_1}{r_3(-r_2 + r_4)}. \tag{6.37}$$

The speed ratio is

$$i_{13} = \frac{\omega_{10}}{\omega_{30}} = \frac{\omega_1}{\omega_{30}} = \frac{r_3(-r_2 + r_4)}{r_1 r_4}. \tag{6.38}$$

6.7 Differential

Figure 6.15 is a schematic drawing of an ordinary bevel-gear automotive differential. The drive shaft pinion 1 and the ring gear 2 are normally hypoid gears. Ring gear 2 acts as the planet carrier for planet gear 3, and its speed can be calculated in the case of a simple gear train when the speed of the drive shaft is given. Sun gears 4 and 5 are connected, respectively, to each rear wheel.

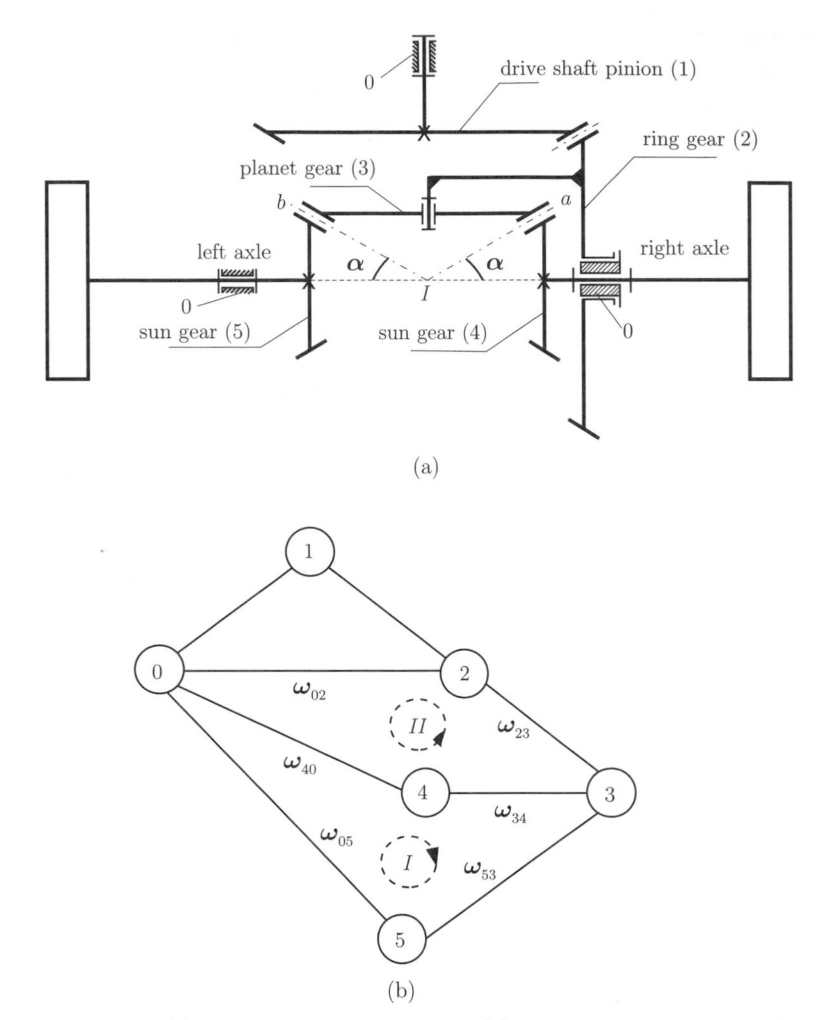

Figure 6.15. (a) Automotive differential and (b) diagram representing the differential.

When the car is traveling in a straight line, the two sun gears rotate in the same direction with exactly the same speed. Thus for straight-line motion of the car, there is no relative motion between planet gear 3 and ring gear 2. Planet gear 3, in effect, serves only as key to transmit motion from the planet carrier to both wheels.

When the vehicle is making a turn, the wheel on the inside of the turn makes fewer revolutions than the wheel with a larger turning radius. Unless this difference in speed is accommodated in some manner, one or both of the tires would have to slip in order to make the turn. The differential permits the two wheels to rotate at different angular velocities while at the same time delivering power to both. During a turn, planet gear 3 rotates about its own axis, thus permitting gears 4 and 5 to revolve at different velocities. The purpose of a differential is to differentiate between the speeds of the two wheels. In the usual passenger-car differential, the torque is divided equally between the two wheels whether the car is traveling in a straight line or on a curve. Sometimes the road conditions are such that the tractive effort developed by the two wheels is unequal. In this case the total tractive effort available will be only twice that on the wheel having the least traction, because the differential divides the torque equally. If one wheel should happen to be resting on snow or ice, the total tractive effort available is very small, and only a small torque will be required to cause the wheel to spin. Thus the car will sit there with one wheel spinning and the other at rest with no tractive effort. And, if the car is in motion and encounters slippery surfaces, then all traction, as well as control, is lost.

It is possible to overcome the disadvantages of the simple bevel-gear differential by adding a coupling unit that is sensitive to wheel speeds. The object of such a unit is to cause most of the torque to be directed to the slowly moving wheel. Such a combination is then called a limited-slip differential.

Diagram for Angular Velocities

The velocity analysis is carried out by using the contour equation method and the graphical diagram for angular velocities.

There are five moving elements (1, 2, 3, 4, and 5) connected by

- five full joints ($c_5 = 5$): one between frame 0 and drive shaft pinion gear 1; one between frame 0 and ring gear 2; one between planet carrier arm 2 and planet gear 3; one between frame 0 and sun gear 4; and another between frame 0 and sun gear 5;
- three half-joints ($c_4 = 3$): one between drive shaft pinion gear 1 and ring gear 2; one between planet gear 3 and sun gear 4; and another between planet gear 3 and sun gear 5.

The system possesses two DOF:

$$M = 3n - 2c_5 - c_4 = 3 \cdot 5 - 2 \cdot 5 - 3 = 2.$$

The input data are the absolute angular velocities ω_{40} and ω_{50} of the two wheels.

The system shown in Fig. 6.15(a) has six elements (0, 1, 2, 3, 4, and 5) and eight joints ($c_4 + c_5$). The number of independent loops is given by

$$n_c = 8 - p + 1 = 8 - 6 + 1 = 3.$$

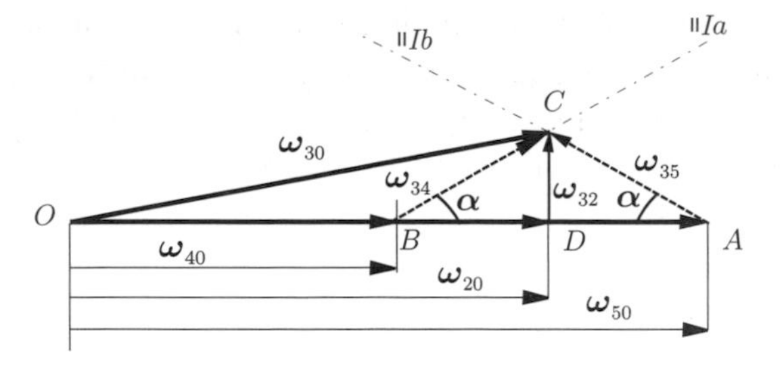

Figure 6.16. Angular velocities diagram.

This gear system has three independent contours. The diagram representing the kinematic chain and the independent contours is shown in Fig. 6.15(b).

The first closed contour contains elements 0, 4, 3, 5, and 0 (clockwise path). For the velocity analysis, the following vectorial equations can be written:

$$\boldsymbol{\omega}_{40} + \boldsymbol{\omega}_{34} + \boldsymbol{\omega}_{53} + \boldsymbol{\omega}_{05} = \mathbf{0},$$

or

$$\boldsymbol{\omega}_{40} + \boldsymbol{\omega}_{34} = \boldsymbol{\omega}_{50} + \boldsymbol{\omega}_{35}. \tag{6.39}$$

The unknown angular velocities are $\boldsymbol{\omega}_{34}$ and $\boldsymbol{\omega}_{35}$. The relative angular velocity of the planet gear 3 with respect to sun gear 4 is parallel to line Ia, and the relative angular velocity of planet gear 3 with respect to sun gear 5 is parallel to line Ib. Equation (6.39) can be solved graphically, as shown in Fig. 6.16. Vectors OA and OB represent velocities $\boldsymbol{\omega}_{50}$ and $\boldsymbol{\omega}_{40}$. At A and B we draw parallels to Ib and Ia. The intersection between the two lines is point C. Vector BC represents the relative angular velocity of planet gear 3 with respect to sun gear 4, and vector AC represents the relative angular velocity of planet gear 3 with respect to sun gear 5.

The absolute angular velocity of planet gear 3 is

$$\boldsymbol{\omega}_{30} = \boldsymbol{\omega}_{40} + \boldsymbol{\omega}_{34}.$$

Vector OC represents the absolute angular velocity of the planet gear.

The second closed contour contains elements 0, 4, 3, 2, and 0 (counterclockwise path). For the velocity analysis, the following vectorial equations can be written:

$$\boldsymbol{\omega}_{40} + \boldsymbol{\omega}_{34} + \boldsymbol{\omega}_{23} + \boldsymbol{\omega}_{02} = \mathbf{0}. \tag{6.40}$$

With the use of the velocity diagram, Fig. 6.16, vector DC represents the relative angular velocity of planet gear 3 with respect to ring gear 2, ω_{23}, and OD represents the absolute angular velocity of ring gear 2, ω_{20}.

From Fig. 6.16 one can write

$$\omega_{20} = |OD| = 1/2(\omega_{40} + \omega_{50}),$$
$$\omega_{32} = |DC| = 1/2(\omega_{50} - \omega_{40}) \tan\alpha. \tag{6.41}$$

When the car is traveling in a straight line, the two sun gears rotate in the same direction with exactly the same speed, $\omega_{50} = \omega_{40}$, and there is no relative motion between the planet gear and the ring gear, $\omega_{32} = 0$. When the wheels are jacked up, $\omega_{50} = -\omega_{40}$, and the absolute angular velocity of ring gear 2 is zero.

6.8 Gear Force Analysis

Basic Relations

The force between meshing teeth (neglecting the sliding friction) can be resolved at the pitch point (P in Fig. 6.17) into two force components:

- tangential component F_t, which accounts for the power transmitted; and
- radial component F_r, which does no work but tends to push the gears apart.

The relationship between these components is

$$F_r = F_t \tan\phi, \tag{6.42}$$

where ϕ is the pressure angle.

The pitch line velocity in feet per minute is equal to

$$V = \pi d n/12 \text{ [ft/min]}, \tag{6.43}$$

where d is the pitch diameter in inches of the gear rotating at n rpm.

In SI units,

$$V = \pi d n/60{,}000 \text{ [m/s]}, \tag{6.44}$$

where d is the pitch diameter in millimeters of the gear rotating at n rpm.

The transmitted power in horsepower is

$$H = F_t V/33{,}000 \text{ [hp]}, \tag{6.45}$$

where F_t is in pounds and V is in feet per minute.

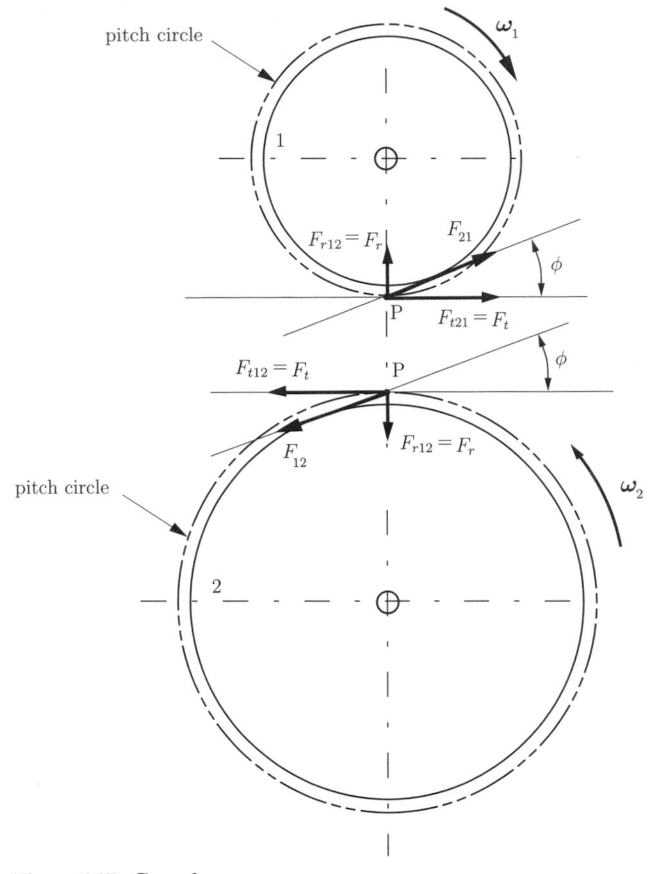

Figure 6.17. Gear forces.

In SI units the transmitted power in watts is

$$H = F_t V \text{ [W]}, \tag{6.46}$$

where F_t is in newtons and V is in meters per second.

Joint Reactions for Planetary Gear Trains

The planetary gear train considered is shown in Fig. 6.18. The sun gear has $N_1 = 19$ external gear teeth, the planet gear has $N_2 = N_{2'} = 28$ external gear teeth, and the ring gear has $N_5 = 75$ internal gear teeth. Gear 3 has $N_3 = 18$ external gear teeth, and the output gear has $N_4 = 36$ external gear teeth. The module of the gears is $m = 5$ mm, and the pressure angle is $\phi = 20°$. The driven torque is $\mathbf{M}_4 = M_4\mathbf{\imath}$, where $M_4 = 500$ N m, and it is opposed to the angular velocity of the output gear, $\boldsymbol{\omega}_{40} = \omega_4\mathbf{\imath}$, $\omega_4 < 0$, as shown in Fig. 6.19. The joints are assumed to be frictionless.

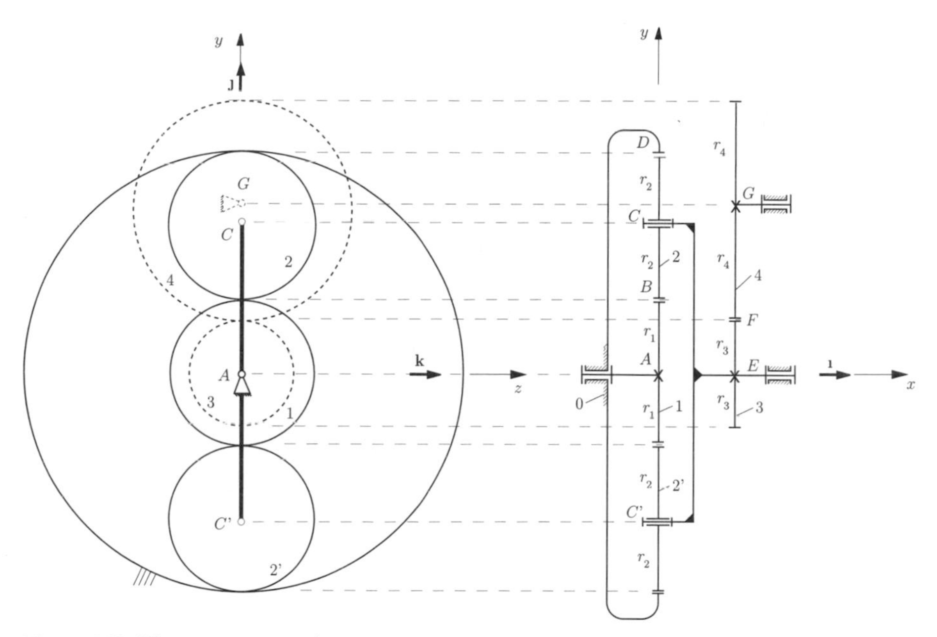

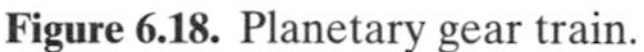
Figure 6.18. Planetary gear train.

The position vectors of the joints are defined as follows (Fig. 6.18):

$$
\begin{aligned}
\mathbf{r}_A &= [0, 0, 0],\\
\mathbf{r}_B &= [0, r_1, 0] = [0, mN_1/2, 0],\\
\mathbf{r}_C &= [0, r_1 + r_2, 0] = [0, m(N_1 + N_2)/2, 0],\\
\mathbf{r}_{C'} &= [0, -r_1 - r_2, 0] = [0, -m(N_1 + N_2)/2, 0],\\
\mathbf{r}_D &= [0, r_1 + 2r_2, 0] = [0, m(N_1 + 2N_2)/2, 0],\\
\mathbf{r}_E &= [\#, 0, 0],\\
\mathbf{r}_F &= [\#, r_3, 0] = [\#, mN_3/2, 0],\\
\mathbf{r}_G &= [\#, r_3 + r_4, 0] = [\#, m(N_3 + N_4)/2, 0].
\end{aligned}
\tag{6.47}
$$

The x parameter # is not important for the calculation.

Gear 4

The force of gear 3 that acts on gear 4 at pitch point F is denoted by F_{34}. The force between meshing teeth can be resolved at the pitch point into two components, a tangential component $F_{t34} = F_{34}\cos\phi$, and a radial component $F_{r34} = F_{34}\sin\phi$ or

$$\mathbf{F}_{34} = [0, F_{r34}, F_{t34}] = F_{34}\sin\phi\,\mathbf{J} + F_{34}\cos\phi\,\mathbf{k}. \tag{6.48}$$

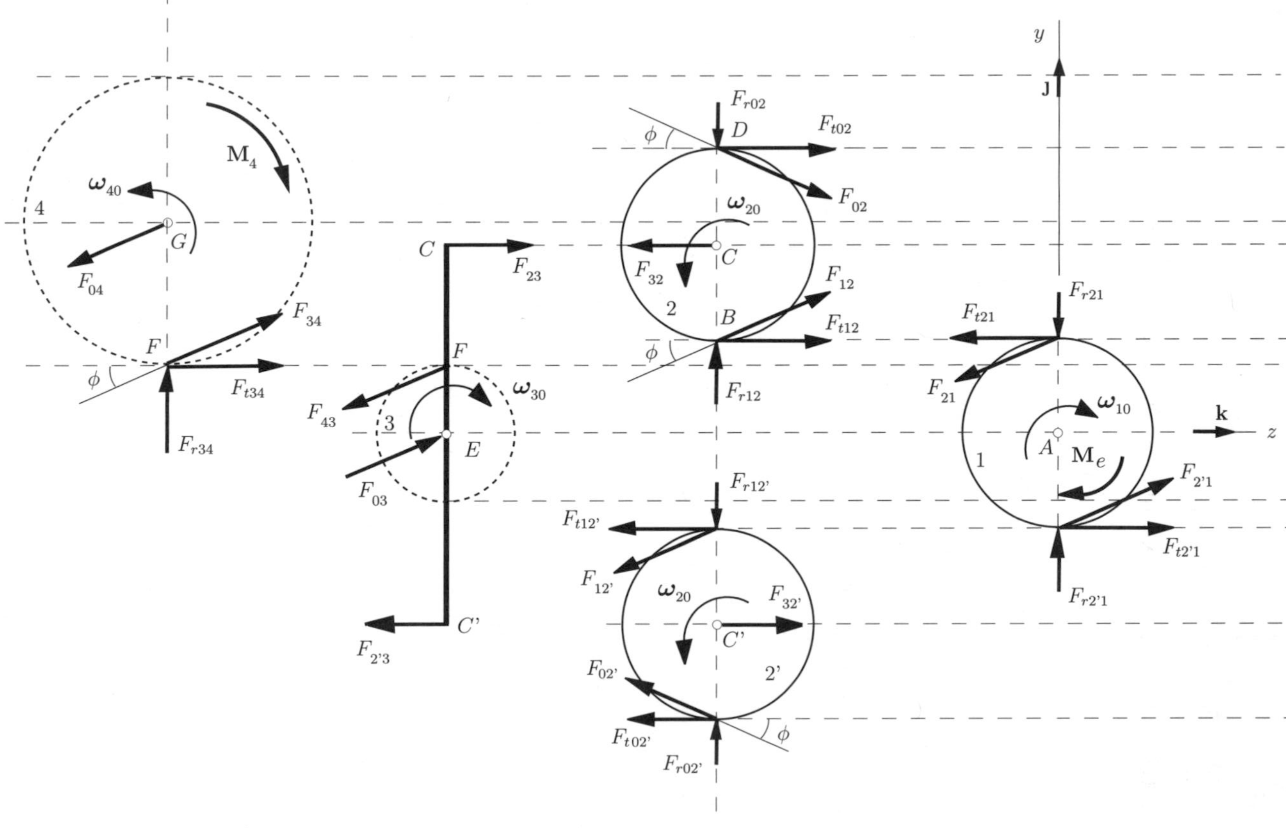

Figure 6.19. Free-body diagrams for the planetary gear train.

The equilibrium of moments for gear 4 with respect to its center G can be written as

$$\sum \mathbf{M}_G^{(\text{gear }4)} = \mathbf{M}_4 + \mathbf{GF} \times \mathbf{F}_{34} = \mathbf{0}, \tag{6.49}$$

where $\mathbf{GF} = \mathbf{r}_F - \mathbf{r}_G = -r_4\mathbf{j}$. Equation (6.49) can be written as

$$M_4\mathbf{i} + \begin{vmatrix} \mathbf{i} & \mathbf{j} & \mathbf{k} \\ 0 & -r_4 & 0 \\ 0 & F_{34}\sin\phi & F_{34}\cos\phi \end{vmatrix} = \mathbf{0}. \tag{6.50}$$

When Eq. (6.50) is solved, the reaction F_{34} is obtained:

$$F_{34} = \frac{2M_4}{mN_4\cos\phi} = \frac{2 \cdot 500}{0.005 \cdot 36 \cdot \cos 20^\circ} = 5912.1 \text{ N}. \tag{6.51}$$

The reaction of ground 0 on gear 4 at G is

$$\mathbf{F}_{04} = -\mathbf{F}_{34}.$$

Link 3

Link 3 is composed of gear 3 and the planetary arm. The reaction of gear 4 on gear 3 at F is known:

$$\mathbf{F}_{43} = -\mathbf{F}_{34} = -F_{34}\sin\phi\mathbf{j} - F_{34}\cos\phi\mathbf{k}.$$

The unknowns are the reactions of planet gears 2 and 2′ on the planet arm at C and C':

$$\mathbf{F}_{23} = F_{23r}\mathbf{j} + F_{23t}\mathbf{k},$$
$$\mathbf{F}_{2'3} = -F_{23r}\mathbf{j} - F_{23t}\mathbf{k}. \tag{6.52}$$

The reaction of ground 0 on gear 3 at E is

$$\mathbf{F}_{03} = -\mathbf{F}_{43}.$$

From the free-body diagram of link 3 (Fig. 6.19), the tangential component of force F_{23t} can be computed by writing a moment equation with respect to the center of gear 3, E:

$$\begin{aligned}\sum \mathbf{M}_E^{(\text{link }3)} &= (\mathbf{r}_F - \mathbf{r}_E) \times \mathbf{F}_{43} + (\mathbf{r}_C - \mathbf{r}_E) \times \mathbf{F}_{23} + (\mathbf{r}_{C'} - \mathbf{r}_E) \times \mathbf{F}_{2'3} \\ &= \begin{vmatrix} \mathbf{i} & \mathbf{j} & \mathbf{k} \\ 0 & r_3 & 0 \\ 0 & -F_{34}\sin\phi & -F_{34}\cos\phi \end{vmatrix} + \begin{vmatrix} \mathbf{i} & \mathbf{j} & \mathbf{k} \\ \# & r_1 + r_2 & 0 \\ 0 & F_{23r} & F_{23t} \end{vmatrix} \\ &\quad + \begin{vmatrix} \mathbf{i} & \mathbf{j} & \mathbf{k} \\ \# & -r_1 - r_2 & 0 \\ 0 & -F_{23r} & -F_{23t} \end{vmatrix} \\ &= -F_{34}r_3\cos\phi\mathbf{i} + 2F_{23t}(r_1 + r_2)\mathbf{i} = \mathbf{0}.\end{aligned} \tag{6.53}$$

The force F_{23t} is

$$F_{23t} = \frac{M_4 r_3}{2(r_1 + r_2) r_4} = 1063.83 \text{ N}. \tag{6.54}$$

Gear 2

The forces that act on gear 2 are

- $\mathbf{F}_{32} = -F_{23r}\mathbf{J} - F_{23t}\mathbf{k}$, the reaction of the arm on planet 2 at C; the tangential component $F_{32t} = -F_{23t}$ is known;
- $\mathbf{F}_{12} = F_{12} \sin\phi\mathbf{J} + F_{12} \cos\phi\mathbf{k}$, the reaction of sun gear 1 on planet 2 at B, is unknown; and
- $\mathbf{F}_{02} = -F_{02} \sin\phi\mathbf{J} + F_{02} \cos\phi\mathbf{k}$, the reaction of ring gear 0 on planet 2 at D, is unknown.

Two vectorial equilibrium equations can be written.

1. The sum of moments that act on gear 2 with respect to center C is zero.

$$\begin{aligned}\sum \mathbf{M}_C^{(\text{gear 2})} &= (\mathbf{r}_D - \mathbf{r}_C) \times \mathbf{F}_{02} + (\mathbf{r}_B - \mathbf{r}_C) \times \mathbf{F}_{12} \\ &= \begin{vmatrix} \mathbf{I} & \mathbf{J} & \mathbf{k} \\ 0 & r_2 & 0 \\ 0 & -F_{02}\sin\phi & F_{02}\cos\phi \end{vmatrix} + \begin{vmatrix} \mathbf{I} & \mathbf{J} & \mathbf{k} \\ 0 & -r_2 & 0 \\ 0 & F_{12}\sin\phi & F_{12}\cos\phi \end{vmatrix} \\ &= \mathbf{0}. \end{aligned} \tag{6.55}$$

2. The sum of all the forces that act on gear 2 is zero.

$$\begin{aligned}\sum \mathbf{F}^{(\text{gear 2})} = \mathbf{F}_{02} + \mathbf{F}_{12} + \mathbf{F}_{32} &= (-F_{02}\sin\phi\mathbf{J} + F_{02}\cos\phi\mathbf{k}) \\ &+ (F_{12}\sin\phi\mathbf{J} + F_{12}\cos\phi\mathbf{k}) + (F_{23r}\mathbf{J} + F_{23t}\mathbf{k}) = \mathbf{0}. \end{aligned} \tag{6.56}$$

Solving the system of equations, Eqs. (6.55) and (6.56), gives the result

$$F_{32r} = 0, \qquad F_{12} = F_{02} = \frac{M_4 r_3 \sec\phi}{4(r_1 + r_2) r_4} = 566.052 \text{ N}. \tag{6.57}$$

Gear 1

The motor torque $\mathbf{M} = M\mathbf{I}$ that acts on the input sun gear 1 is computed from the moment equation with respect to center A:

$$\sum \mathbf{M}_A^{(\text{gear 1})} = \mathbf{M} + 2\mathbf{r}_B \times \mathbf{F}_{21}, \tag{6.58}$$

and

$$M = \frac{M_4 r_1 r_3}{2(r_1 r_4 + r_2 r_4)} = 50.531 \text{ Nm}. \tag{6.59}$$

Motor torque $\mathbf{M}$ has the same direction and orientation as angular velocity ω_{10}.

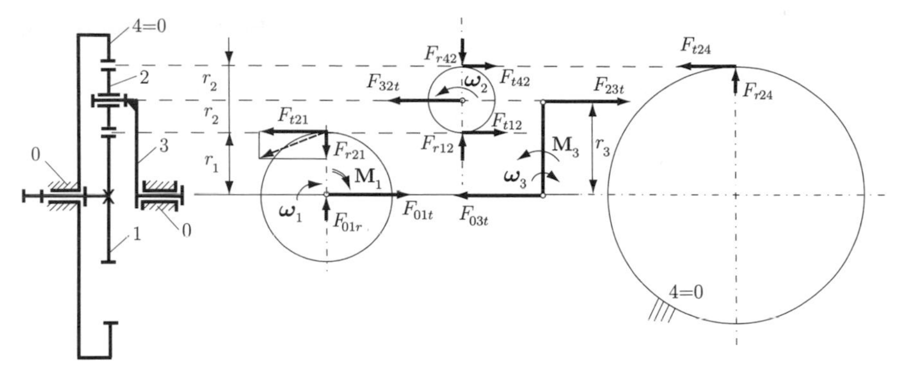

Figure 6.20. Planetary gear train with single planet and free-body diagrams.

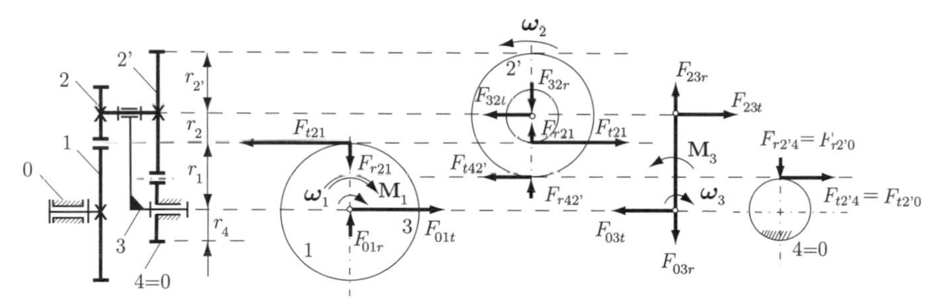

Figure 6.21. Planetary gear train with double planet and free-body diagrams.

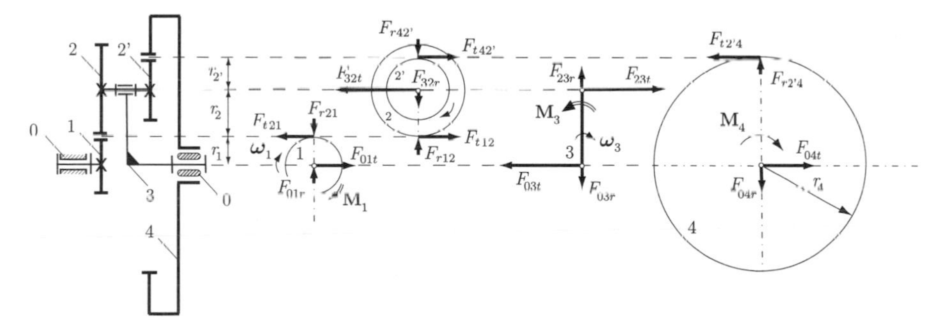

Figure 6.22. Planetary gear train with double planet and free-body diagrams.

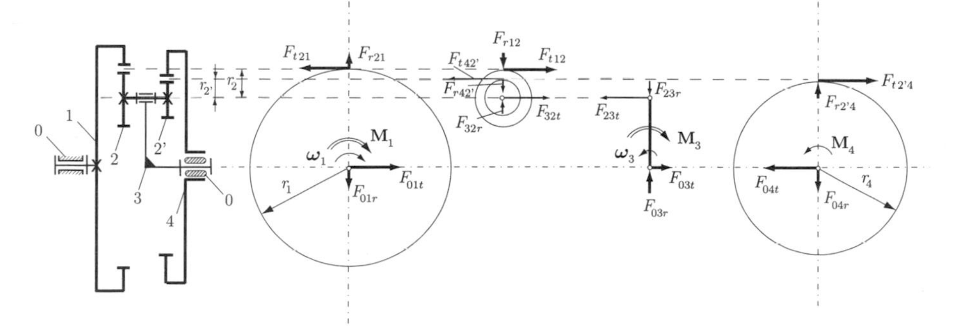

Figure 6.23. Planetary gear train with double planet and free-body diagrams.

Forces Diagrams

Figures 6.20, 6.21, 6.22, and 6.23 give free-body diagrams for different types of planetary gear trains. The torque on the sun gear is M_1 and the torque on the planet arm is M_3. The tangential force that acts on sun gear 1 at the pitch point is

$$F_{t21} = -F_{t12} = M_1/r_1, \tag{6.60}$$

and the radial force is

$$F_{r21} = -F_{r12} = F_{t21} \tan\phi, \tag{6.61}$$

where ϕ is the pressure angle. The reactions of the ground on the sun gear are

$$F_{01r} = -F_{r21}, \qquad F_{01t} = -F_{t21}. \tag{6.62}$$

Figure 6.20 shows a planetary gear train with a single planet. For the planetary gear trains with a double planet system (Figs. 6.21, 6.22, and 6.23), the tangential force of the planet system that acts on the arm is

$$F_{23t} = -F_{32t} = [M_3/(r_1 + r_2)]. \tag{6.63}$$

The output torque on the ring gear is

$$M_4 = F_{t2'4} r_4. \tag{6.64}$$

7 Open Kinematic Chains

In this chapter Kane's approach is used to formulate the equations of motion for open kinematic chains. A detailed dynamic analysis of a three DOF open kinematic chain (Kane and Levinson) is presented.

7.1 Kinematics of Open Kinematic Chains

Figure 7.1 is a schematic representation of an open kinematic chain (robot arm) consisting of three links 1, 2, and 3. The last link 3 grips firmly the rigid body RB. Link 1 can be rotated at A in a "fixed" reference frame (0) of unit vectors $[\mathbf{i}_0, \mathbf{j}_0, \mathbf{k}_0]$ about a vertical axis $\mathbf{i}_0$. The unit vector $\mathbf{i}_0$ is fixed in 1. Link 1 is connected to link 2 at pin joint B. Element 2 rotates relative to 1 about a horizontal axis fixed in both 1 and 2, passing through B, and perpendicular to the axis of 1. The last link 3 is connected to 2 by means of a slider joint. The mass centers of links 1, 2, and 3 are C_1, C_2, and C_3, respectively. The mass center of rigid body RB is C_R. The distances $L_1 = AC_1$, $L_2 = BC_2$, and $L_B = AB$ are indicated in Fig. 7.1. The reference frame (1) of the unit vectors $[\mathbf{i}_1, \mathbf{j}_1, \mathbf{k}_1]$ is attached to link 1, and the reference frame (2) of the unit vectors $[\mathbf{i}_2, \mathbf{j}_2, \mathbf{k}_2]$ is attached to link 2, as shown in Fig. 7.1.

The position vector from point C_3 to point C_R is $\mathbf{r}_{C_3C_R} = p_x\mathbf{i}_2 + p_y\mathbf{j}_2 + p_z\mathbf{k}_2$, and its components on reference frame (2) are

$$p_x = \mathbf{r}_{C_3C_R} \cdot \mathbf{i}_2, \qquad p_y = \mathbf{r}_{C_3C_R} \cdot \mathbf{j}_2, \qquad p_z = \mathbf{r}_{C_3C_R} \cdot \mathbf{k}_2.$$

To characterize the instantaneous configuration of the arm, *generalized coordinates* $q_1(t)$, $q_2(t)$, and $q_3(t)$ are employed. The generalized coordinates are quantities associated with the position of the system. The first generalized coordinate q_1 denotes the radian measure of the angle between the axes of 1 and 2 ($s_1 = \sin q_1$, $c_1 = \cos q_1$), and q_2 is the distance from C_2 to C_3. The last

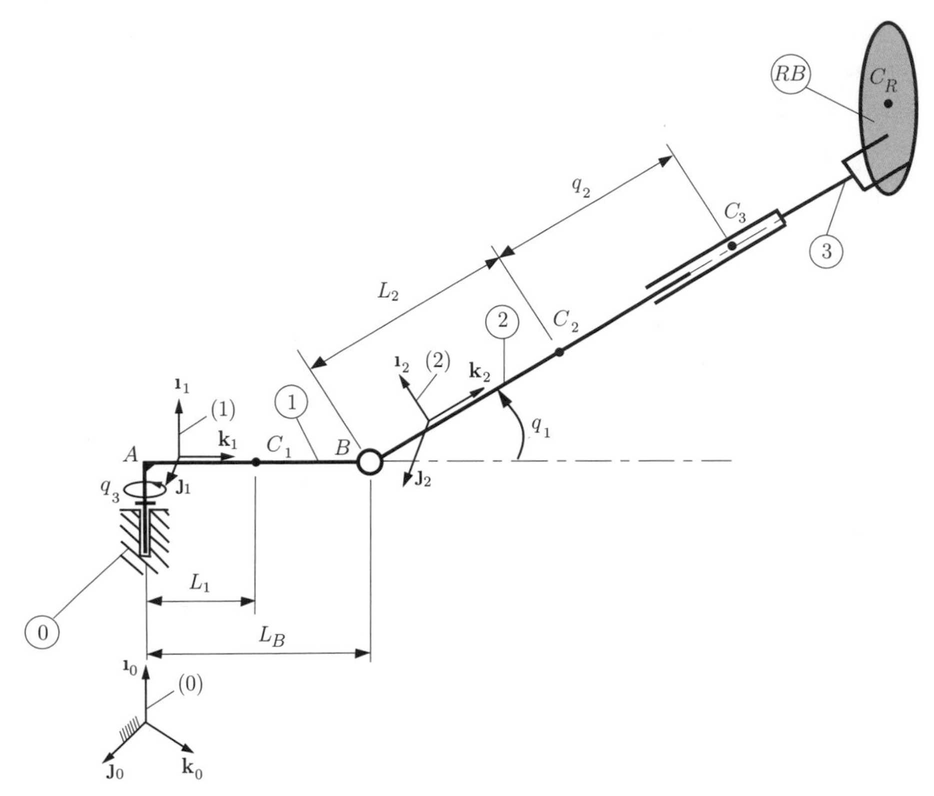

Figure 7.1. Three DOF open kinematic chain.

generalized coordinate, q_3 ($s_3 = \sin q_3, c_3 = \cos q_3$), also designates a radian measure of the rotation angle between 1 and 0.

Just as important as generalized coordinates are *generalized speeds*; these are quantities associated with the motion of a system. The generalized speeds $u_1(t), \ldots, u_n(t)$, where n is the number of generalized coordinates, can be introduced as

$$u_r = \sum_{s=1}^{n} A_{rs}\,\dot{q}_s + B_r, \qquad r = 1, \ldots, n, \tag{7.1}$$

where A_{rs} and B_r are functions of the generalized coordinates $q_1, \ldots, q_n$, and the time t. Functions A_{rs} and B_r $(r, s = 1, \ldots, n)$ are chosen in such a way that Eq. (7.1) can be solved uniquely for $q_1, \ldots, q_n$. The generalized speeds $u_1, \ldots, u_n$ serve as variables on an equal footing with the generalized coordinates $q_1, \ldots, q_n$. Their introduction can enable one to take advantage of special features of a given physical system to bring the equations of motion into a particularly simple form. Generally, this is accomplished by taking u_r to be an angular velocity measure number, a velocity measure number, or simply $\dot{q}_r$. Consider, for example, the robotic arm. One can introduce the generalized speeds u_1, u_2, and u_3 as

$$u_1 = \boldsymbol{\omega}_{10} \cdot \mathbf{\imath}_1, \qquad u_2 = \boldsymbol{\omega}_{21} \cdot \mathbf{\jmath}_2, \qquad u_3 = \mathbf{v}_{C_3} \cdot \mathbf{k}_2, \tag{7.2}$$

where $\boldsymbol{\omega}_{10}$ is the angular velocity of link 1 in fixed reference frame (0), $\boldsymbol{\omega}_{21}$ is the angular velocity of link 2 with respect to reference frame (1), and $\mathbf{v}_{C_3}$ is the velocity of C_3.

In the case of the three DOF robot arm, $\dot{q}_3 = u_1$, $\dot{q}_1 = u_2$, and $\dot{q}_2 = u_3$ or

$$u_1 = \dot{q}_3, \qquad u_2 = \dot{q}_1, \qquad u_3 = \dot{q}_2. \tag{7.3}$$

Equation (7.3) can be solved uniquely for $\dot{q}_1$, $\dot{q}_2$, and $\dot{q}_3$.

Angular Velocities

Next, the angular velocity of links 1, 2, and 3 will be expressed in fixed reference frame (0). One can express the angular velocity of 1 in (0) as

$$\boldsymbol{\omega}_{10} = \dot{q}_3 \mathbf{\imath}_1 = u_1 \mathbf{\imath}_1. \tag{7.4}$$

The angular velocity of link 2 with respect to (1) is

$$\boldsymbol{\omega}_{21} = \dot{q}_1 \mathbf{\jmath}_2, \tag{7.5}$$

and the angular velocity of link 2 with respect to fixed reference frame (0) is

$$\boldsymbol{\omega}_{20} = \boldsymbol{\omega}_{10} + \boldsymbol{\omega}_{21} = \dot{q}_3 \mathbf{\imath}_1 + \dot{q}_1 \mathbf{\jmath}_2. \tag{7.6}$$

Unit vector $\mathbf{\imath}_0$ can be expressed as (Fig. 7.2)

$$\mathbf{\imath}_0 = \mathbf{\imath}_1 = c_1 \mathbf{\imath}_2 + s_1 \mathbf{k}_2. \tag{7.7}$$

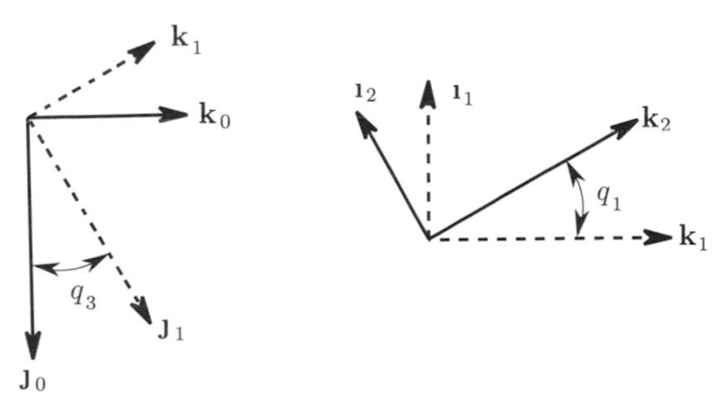

Figure 7.2. Reference frame (0) of unit vectors $[\mathbf{\imath}_0, \mathbf{\jmath}_0, \mathbf{k}_0]$, reference frame (1) of unit vectors $[\mathbf{\imath}_1, \mathbf{\jmath}_1, \mathbf{k}_1]$, and reference frame (2) of unit vectors $[\mathbf{\imath}_2, \mathbf{\jmath}_2, \mathbf{k}_2]$.

The angular velocity of link 2 in reference frame (0) written in terms of reference frame (2) is

$$\boldsymbol{\omega}_{20} = u_1(c_1\mathbf{\imath}_2 + s_1\mathbf{k}_2) + u_2\mathbf{\jmath}_2 = u_1c_1\mathbf{\imath}_2 + u_2\mathbf{\jmath}_2 + u_1s_1\mathbf{k}_2, \tag{7.8}$$

or

$$\boldsymbol{\omega}_{20} = Z_1\mathbf{\imath}_2 + u_2\mathbf{\jmath}_2 + Z_2\mathbf{k}_2, \tag{7.9}$$

where $Z_1 = u_1c_1$ and $Z_2 = u_1s_1$. The quantities Z_i are introduced to minimize the length of the equations. Link 3 and rigid body RB have the same rotational motion as link 2; that is, $\boldsymbol{\omega}_{30} = \boldsymbol{\omega}_{R0} = \boldsymbol{\omega}_{20}$.

Angular Accelerations

The angular acceleration of link 1 in reference frame (0) can be expressed as

$$\boldsymbol{\alpha}_{10} = \ddot{q}_3\mathbf{\imath}_1 = \dot{u}_1\mathbf{\imath}_1. \tag{7.10}$$

The angular velocity of link 2 with respect to reference frame (0) is

$$\begin{aligned}\boldsymbol{\alpha}_{20} &= \frac{d}{dt}\boldsymbol{\omega}_{20} = \frac{^{(2)}\partial}{\partial t}\boldsymbol{\omega}_{20} = (\dot{u}_1c_1 - u_1\dot{q}_1s_1)\mathbf{\imath}_2 + \dot{u}_2\mathbf{\jmath}_2 + (\dot{u}_1s_1 + u_1\dot{q}_1c_1)\mathbf{k}_2\\ &= (\dot{u}_1c_1 - u_1u_2s_1)\mathbf{\imath}_2 + \dot{u}_2\mathbf{\jmath}_2 + (\dot{u}_1s_1 + u_1u_2c_2)\mathbf{k}_2\\ &= (\dot{u}_1c_1 + Z_3)\mathbf{\imath}_2 + \dot{u}_2\mathbf{\jmath}_2 + (\dot{u}_1s_1 + Z_4)\mathbf{k}_2,\end{aligned} \tag{7.11}$$

where $^{(2)}\partial/\partial t$ represents the partial derivative with respect to time in reference frame (2), $[\mathbf{\imath}_2, \mathbf{\jmath}_2, \mathbf{k}_2]$, $Z_3 = -Z_1u_2$, and $Z_4 = Z_1u_2$. Link 3 and rigid body RB have the same angular acceleration as link 2; that is; $\boldsymbol{\alpha}_{30} = \boldsymbol{\alpha}_{20}$.

Linear Velocities

The position vector of C_1, the mass center of link 1, is

$$\mathbf{r}_{C_1} = L_1\mathbf{k}_1, \tag{7.12}$$

and the velocity of C_1 in (0) is

$$\begin{aligned}\mathbf{v}_{C_1} &= \frac{d}{dt}\mathbf{r}_{C_1} = \frac{^{(1)}\partial}{\partial t}\mathbf{r}_1 + \boldsymbol{\omega}_{10} \times \mathbf{r}_{C_1}\\ &= \mathbf{0} + \begin{vmatrix}\mathbf{\imath}_1 & \mathbf{\jmath}_1 & \mathbf{k}_1\\ u_1 & 0 & 0\\ 0 & 0 & L_1\end{vmatrix} = -u_1L_1\mathbf{\jmath}_1 = Z_5\mathbf{\jmath}_1,\end{aligned} \tag{7.13}$$

where $Z_5 = -u_1L_1$.

The position vector of C_2, the mass center of link 2, is

$$\begin{aligned}\mathbf{r}_{C_2} &= L_B\mathbf{k}_1 + L_2\mathbf{k}_2 = L_B(-s_1\mathbf{\imath}_2 + c_1\mathbf{k}_2) + L_2\mathbf{k}_2\\ &= -L_Bs_1\mathbf{\imath}_2 + (L_Bc_1 + L_2)\mathbf{k}_2.\end{aligned} \tag{7.14}$$

The velocity of C_2 in (0) is

$$
\begin{aligned}
\mathbf{v}_{C_2} &= \frac{d}{dt}\mathbf{r}_{C_2} = \frac{^{(2)}\partial}{\partial t}\mathbf{r}_{C_2} + \boldsymbol{\omega}_{20} \times \mathbf{r}_{C_2} \\
&= -L_B c_1 u_2 \mathbf{ı}_2 - L_B c_1 u_2 \mathbf{k}_2 + \begin{vmatrix} \mathbf{ı}_2 & \mathbf{ȷ}_2 & \mathbf{k}_2 \\ u_1 c_1 & u_2 & u_1 s_1 \\ -L_B s_1 & 0 & L_B c_1 + L_2 \end{vmatrix} \\
&= L_2 u_2 \mathbf{ı}_2 - (L_B + L_2 c_1) u_1 \mathbf{ȷ}_2 = L_2 u_2 \mathbf{ı}_2 + Z_6 u_1 \mathbf{ȷ}_2 \\
&= Z_7 \mathbf{ı}_2 + Z_8 \mathbf{ȷ}_2,
\end{aligned} \tag{7.15}
$$

where $Z_6 = -(L_B + L_2 c_1)$, $Z_7 = L_2 u_2$, and $Z_8 = Z_6 u_1$.

The position vector of C_3 with respect to reference frame (0) is

$$
\begin{aligned}
\mathbf{r}_{C_3} &= \mathbf{r}_{C_2} + q_2 \mathbf{k}_2 \\
&= -L_B s_1 \mathbf{ı}_2 + (L_B c_1 + L_2 + q_2) \mathbf{k}_2,
\end{aligned} \tag{7.16}
$$

and the velocity of this mass center in (0) is

$$
\begin{aligned}
\mathbf{v}_{C_3} &= \frac{d}{dt}\mathbf{r}_{C_3} = \frac{^{(2)}\partial}{\partial t}\mathbf{r}_{C_3} + \boldsymbol{\omega}_{20} \times \mathbf{r}_{C_3} \\
&= -L_B c_1 u_2 \mathbf{ı}_2 - (L_B c_1 u_2 + u_3) \mathbf{k}_2 + \begin{vmatrix} \mathbf{ı}_2 & \mathbf{ȷ}_2 & \mathbf{k}_2 \\ u_1 c_1 & u_2 & u_1 s_1 \\ -L_B s_1 & 0 & L_B c_1 + L_2 + q_2 \end{vmatrix} \\
&= (L_2 + q_2) u_2 \mathbf{ı}_2 - (L_B + L_2 c_1 + c_1 q_2) u_1 \mathbf{ȷ}_2 + u_3 \mathbf{k}_2 \\
&= u_2 Z_9 \mathbf{ı}_2 + u_1 Z_{10} \mathbf{ȷ}_2 + u_3 \mathbf{k}_2 \\
&= Z_{11} \mathbf{ı}_2 + Z_{12} \mathbf{ȷ}_2 + u_3 \mathbf{k}_2,
\end{aligned} \tag{7.17}
$$

where $Z_9 = L_2 + q_2$, $Z_{10} = Z_6 + q_2 c_1$, $Z_{11} = u_2 Z_9$, and $Z_{12} = Z_{10} u_1$. The position vector of mass center C_R of rigid body RB is

$$
\begin{aligned}
\mathbf{r}_{C_R} &= \mathbf{r}_{C_3} + \mathbf{r}_{C_3 C_R} \\
&= (p_x - L_B s_1) \mathbf{ı}_2 + p_y \mathbf{ȷ}_2 + (L_B c_1 + L_2 + q_2 + p_z) \mathbf{k}_2.
\end{aligned} \tag{7.18}
$$

The velocity of C_R in (0) is

$$
\begin{aligned}
\mathbf{v}_{C_R} &= \frac{d}{dt}\mathbf{r}_{C_R} = \frac{^{(2)}\partial}{\partial t}\mathbf{r}_{C_R} + \boldsymbol{\omega}_{20} \times \mathbf{r}_{C_R} \\
&= -L_B c_1 u_2 \mathbf{ı}_2 - (L_B c_1 u_2 + u_3) \mathbf{k}_2 \\
&\quad + \begin{vmatrix} \mathbf{ı}_2 & \mathbf{ȷ}_2 & \mathbf{k}_2 \\ u_1 c_1 & u_2 & u_1 s_1 \\ p_x - L_B s_1 & p_y & L_B c_1 + L_2 + q_2 + p_z \end{vmatrix} \\
&= (u_1 Z_{13} + u_2 Z_{14}) \mathbf{ı}_2 + u_1 Z_{15} \mathbf{ȷ}_2 + (Z_16 u_1 - u_2 p_x + u_3) \mathbf{k}_2 \\
&= Z_{17} \mathbf{ı}_2 + Z_{18} \mathbf{ȷ}_2 + Z_{19} \mathbf{k}_2,
\end{aligned} \tag{7.19}
$$

where $Z_{13} = -s_1 p_y$, $Z_{14} = Z_9 + p_z$, $Z_{15} = Z_{10} + s_1 p_x - c_1 p_z$, $Z_{16} = c_1 p_y$, $Z_{17} = u_1 Z_{13} + u_2 Z_{14}$, $Z_{18} = u_1 Z_{15}$, and $Z_{19} = Z_{16} u_1 - u_2 p_x + u_3$.

Linear Accelerations

The acceleration of C_1 is

$$\begin{aligned}
\mathbf{a}_{C_1} &= \frac{d}{dt}\mathbf{v}_{C_1} = \frac{^{(1)}\partial}{\partial t}\mathbf{v}_{C_1} + \boldsymbol{\omega}_{10} \times \mathbf{v}_{C_1} \\
&= -L_1 \dot{u}_1 \mathbf{J} + \begin{vmatrix} \mathbf{\imath}_1 & \mathbf{J}_1 & \mathbf{k}_1 \\ u_1 & 0 & 0 \\ 0 & -L_1 u_1 & 0 \end{vmatrix} \\
&= -L_1 \dot{u}_1 \mathbf{J}_1 - L_1 u_1^2 \mathbf{k}_1 \\
&= -L_1 \dot{u}_1 \mathbf{J}_1 + Z_{20} \mathbf{k}_1,
\end{aligned} \tag{7.20}$$

where $Z_{20} = -L_1 u_1^2 = u_1 Z_5$.

The linear acceleration of the mass center C_2 is

$$\begin{aligned}
\mathbf{a}_{C_2} &= \frac{d}{dt}\mathbf{v}_{C_2} = \frac{^{(2)}\partial}{\partial t}\mathbf{v}_{C_2} + \boldsymbol{\omega}_{20} \times \mathbf{v}_{C_2} \\
&= (\dot{u}_2 L_2 - Z_2 Z_8)\mathbf{\imath}_2 + (Z_6 \dot{u}_1 + L_2 s_1 u_2 u_1 + Z_2 Z_7)\mathbf{J}_2 + (Z_1 Z_8 - u_2 Z_7)\mathbf{k}_2 \\
&= (\dot{u}_2 L_2 + Z_{22})\mathbf{\imath}_2 + (Z_6 \dot{u}_1 + Z_{23})\mathbf{J}_2 + Z_{24}\mathbf{k}_2,
\end{aligned} \tag{7.21}$$

where $Z_{21} = L_2 s_1 u_2$, $Z_{22} = -Z_2 Z_8$, $Z_{23} = Z_{21} u_1 + Z_2 Z_7$, and $Z_{24} = Z_1 Z_8 - u_2 Z_7$.

The acceleration of C_3 is

$$\begin{aligned}
\mathbf{a}_{C_3} &= \frac{d}{dt}\mathbf{v}_{C_3} = \frac{^{(2)}\partial}{\partial t}\mathbf{v}_{C_3} + \boldsymbol{\omega}_{20} \times \mathbf{v}_{C_3} \\
&= (\dot{u}_2 Z_9 + Z_{26})\mathbf{\imath}_2 + (\dot{u}_1 Z_{10} + Z_{27})\mathbf{J}_2 + (\dot{u}_3 + Z_{28})\mathbf{k}_2,
\end{aligned} \tag{7.22}$$

where $Z_{25} = Z_{21} - u_3 c_1 + q_2 s_1 u_2$, $Z_{26} = 2u_2 u_3 - Z_2 Z_{12}$, $Z_{27} = Z_{25} u_1 + Z_2 Z_{11} - Z_1 u_3$, and $Z_{28} = Z_1 Z_{12} - u_2 Z_{11}$.

The acceleration of C_R is

$$\begin{aligned}
\mathbf{a}_{C_R} &= \frac{d}{dt}\mathbf{v}_{C_R} = \frac{^{(2)}\partial}{\partial t}\mathbf{v}_{C_R} + \boldsymbol{\omega}_{20} \times \mathbf{v}_{C_R} = (\dot{u}_1 Z_{13} + \dot{u}_2 Z_{14} + Z_{32})\mathbf{\imath}_2 \\
&\quad + (\dot{u}_1 Z_{15} + Z_{33})\mathbf{J}_2 + (\dot{u}_1 Z_{16} - p_x \dot{u}_2 + \dot{u}_3 + Z_{34})\mathbf{k}_2,
\end{aligned} \tag{7.23}$$

where $Z_{29} = -Z_{16} u_2$, $Z_{30} = Z_{25} + u_2(c_1 p_x + s_1 p_z)$, $Z_{31} = Z_{13} u_2$, $Z_{32} = Z_{29} u_1 + u_2(u_3 + Z_{19}) - Z_2 Z_{18}$, $Z_{33} = Z_{30} u_1 + Z_2 Z_{17} - Z_1 Z_{19}$, and $Z_{34} = Z_{31} u_1 + Z_1 Z_{18} - u_2 Z_{17}$.

7.2 Generalized Inertia Forces

As a way to explain what the *generalized inertia forces* are, a system $\{S\}$ formed by ν particles $P_1, \ldots, P_\nu$ and having masses $m_1, \ldots, m_\nu$ is considered. Suppose that n generalized speeds u_r $(r = 1, \ldots, n)$ have been introduced. Let $\mathbf{v}_{P_j}$ and $\mathbf{a}_{P_j}$ denote, respectively, the velocity of P_j and the acceleration of P_j in a reference frame (0).

Define $\mathbf{F}_{\text{in}j}$, the inertia force for the particle P_j, as

$$\mathbf{F}_{\text{in}j} = -m_j \mathbf{a}_{P_j}. \tag{7.24}$$

The quantities $F_1^*, \ldots, F_n^*$, defined as

$$F_r^* = \sum_{j=1}^{\nu} \frac{\partial \mathbf{v}_{P_j}}{\partial u_r} \cdot \mathbf{a}_{P_j}, \qquad r = 1, \ldots, n, \tag{7.25}$$

are called *generalized inertia forces* for $\{S\}$.

The contribution to F_r^*, made by the particles of a rigid body RB belonging to $\{S\}$, are

$$(F_r^*)_R = \frac{\partial \mathbf{v}_C}{\partial u_r} \cdot \mathbf{F}_{\text{in}} + \frac{\partial \omega}{\partial u_r} \cdot \mathbf{T}_{\text{in}}, \qquad r = 1, \ldots, n, \tag{7.26}$$

where $\mathbf{v}_C$ is the velocity of the center of gravity of RB in (0), and $\omega = \omega_x \mathbf{\imath} + \omega_y \mathbf{\jmath} + \omega_z \mathbf{k}$ is the angular velocity of RB in (0).

The inertia force for rigid body RB is

$$\mathbf{F}_{\text{in}} = -m\mathbf{a}_C, \tag{7.27}$$

where m is the mass of RB, and $\mathbf{a}_C$ is the acceleration of the mass center of RB in the fixed reference frame. The inertia torque $\mathbf{T}_{\text{in}}$ for RB is

$$\mathbf{T}_{\text{in}} = -\alpha \cdot \bar{I} - \omega \times (\bar{I} \cdot \omega), \tag{7.28}$$

where $\alpha = \dot{\omega} = \alpha_x \mathbf{\imath} + \alpha_y \mathbf{\jmath} + \alpha_z \mathbf{k}$ is the angular acceleration of RB in (0), and $\bar{I} = (I_x \mathbf{\imath})\mathbf{\imath} + (I_y \mathbf{\jmath})\mathbf{\jmath} + (I_z \mathbf{k})\mathbf{k}$ is the central inertia dyadic of RB. The central principal axes of RB are parallel to $\mathbf{\imath}, \mathbf{\jmath}, \mathbf{k}$ and the associated moments of inertia have the values $I_x,\ I_y,\ I_z$, respectively. The inertia matrix associated with $\bar{I}$ is

$$\bar{I} \to \begin{bmatrix} I_x & 0 & 0 \\ 0 & I_y & 0 \\ 0 & 0 & I_z \end{bmatrix}. \tag{7.29}$$

The dot product of the vector α with the dyadic $\bar{I}$ is

$$\alpha \cdot \bar{I} = \bar{I} \cdot \alpha = \alpha_x I_x \,\mathbf{\imath} + \alpha_y I_y \,\mathbf{\jmath} + \alpha_z I_z \,\mathbf{k}, \tag{7.30}$$

and the cross product between a vector and a dyadic is

$$\boldsymbol{\omega} \times (\bar{I} \cdot \boldsymbol{\omega}) = \begin{vmatrix} \mathbf{\imath} & \mathbf{\jmath} & \mathbf{k} \\ \omega_x & \omega_y & \omega_z \\ \omega_x I_x & \omega_y I_y & \omega_z I_z \end{vmatrix}$$
$$= -\omega_y \omega_z (I_y - I_z)\mathbf{\imath} - \omega_z \omega_x (I_z - I_x)\mathbf{\jmath} - \omega_x \omega_y (I_x - I_y)\mathbf{k}. \tag{7.31}$$

Referring to the three DOF robot arm, let m_1, m_2, m_3, m_R be the masses of 1, 2, 3, RB, respectively. Links 1, 2, 3, and rigid body RB have the following mass distribution properties. The central principal axes of 1 are parallel to $\mathbf{\imath}_1$, $\mathbf{\jmath}_1$, $\mathbf{k}_1$, as shown in Fig. 7.1, and the associated moments of inertia have the values A_x, A_y, A_z, respectively. The central inertia dyadic of 1 is

$$\bar{I}_1 = (A_x \mathbf{\imath}_1)\mathbf{\imath}_1 + (A_y \mathbf{\jmath}_1)\mathbf{\jmath}_1 + (A_z \mathbf{k}_1)\mathbf{k}_1. \tag{7.32}$$

The central principal axes of 2 and 3 are parallel to $\mathbf{\imath}_2$, $\mathbf{\jmath}_2$, $\mathbf{k}_2$ and the associated moments of inertia have values B_x, B_y, B_z and C_x, C_y, C_z, respectively.

The central inertia dyadic of 2 is

$$\bar{I}_2 = (B_x \mathbf{\imath}_2)\mathbf{\imath}_2 + (B_y \mathbf{\jmath}_2)\mathbf{\jmath}_2 + (B_z \mathbf{k}_2)\mathbf{k}_2, \tag{7.33}$$

and the central inertia dyadic of 3 is

$$\bar{I}_3 = (C_x \mathbf{\imath}_2)\mathbf{\imath}_2 + (C_y \mathbf{\jmath}_2)\mathbf{\jmath}_2 + (C_z \mathbf{k}_2)\mathbf{k}_2. \tag{7.34}$$

The central inertia dyadic of rigid body RB is

$$\bar{I}_R = (D_{11}\mathbf{\imath}_2 + D_{12}\mathbf{\jmath}_2 + D_{13}\mathbf{k}_2)\mathbf{\imath}_2 + (D_{21}\mathbf{\imath}_2 + D_{22}\mathbf{\jmath}_2 + D_{23}\mathbf{k}_2)\mathbf{\jmath}_2$$
$$+ (D_{31}\mathbf{\imath}_2 + D_{32}\mathbf{\jmath}_2 + D_{33}\mathbf{k}_2)\mathbf{k}_2. \tag{7.35}$$

To make the results more compact, the following notation is introduced:

$$\begin{array}{llll} k_1 = B_y - B_z, & k_2 = B_z - B_x, & k_3 = B_x - B_y, & k_4 = C_y - C_z, \\ k_5 = C_z - C_x, & k_6 = C_x - C_y, & k_7 = D_{33} - D_{22}, & k_8 = D_{11} - D_{33}, \\ k_9 = D_{22} - D_{11}, & k_{10} = B_x + k_1, & k_{11} = B_z - k_3, & k_{12} = C_x + k_4, \\ k_{13} = C_z - k_6, & k_{14} = D_{11} - k_7, & k_{15} = D_{31} + k_{13}, & k_{16} = D_{33} + k_9. \end{array}$$

The inertia torque of 1 in (0) can be written as

$$\mathbf{T}_{\text{in}1} = -\boldsymbol{\alpha}_{10} \cdot \bar{I}_1 - \boldsymbol{\omega}_{10} \times (\bar{I}_1 \cdot \boldsymbol{\omega}_{10}) = -A_x \ddot{q}_3 \mathbf{\imath}_1 = -A_x \dot{u}_1 \mathbf{\imath}_1. \tag{7.36}$$

The inertia torque of 2 in (0) is

$$\mathbf{T}_{\text{in}2} = -\boldsymbol{\alpha}_{20} \cdot \bar{I}_2 - \boldsymbol{\omega}_{20} \times (\bar{I}_2 \cdot \boldsymbol{\omega}_{20}),$$

or

$$\mathbf{T}_{\text{in}2} = -(\dot{u}_1 Z_{35} + Z_{36})\mathbf{\imath}_2 - (\dot{u}_2 B_y + Z_{38})\mathbf{\jmath}_2 - (\dot{u}_1 Z_{39} + Z_{40})\mathbf{k}_2, \qquad (7.37)$$

where $Z_{35} = c_1 B_x$, $Z_{36} = Z_3 k_{10}$, $Z_{37} = Z_2 Z_1$, $Z_{38} = -Z_{37} k_2$, $Z_{39} = s_1 B_z$, and $Z_{40} = Z_4 k_{11}$.

Similarly, the inertia torque of 3 in (0) is

$$\mathbf{T}_{\text{in}3} = -(\dot{u}_1 Z_{41} + Z_{42})\mathbf{\imath}_2 - (\dot{u}_2 C_y + Z_{43})\mathbf{\jmath}_2 - (\dot{u}_1 Z_{44} + Z_{45})\mathbf{k}_2, \qquad (7.38)$$

where $Z_{40} = Z_4 k_{11}$, $Z_{41} = c_1 C_x$, $Z_{42} = Z_3 k_{12}$, $Z_{43} = -Z_{37} k_5$, $Z_{44} = s_1 C_z$, and $Z_{45} = Z_4 k_{13}$.

The inertia torque of RB in (0) is

$$\mathbf{T}_{\text{in}R} = -(\dot{u}_1 Z_{49} + \dot{u}_2 D_{12} + Z_{50})\mathbf{\imath}_2 - (\dot{u}_1 Z_{51} + \dot{u}_2 D_{22} + Z_{52})\mathbf{\jmath}_2$$
$$- (\dot{u}_1 Z_{53} + \dot{u}_2 D_{32} + Z_{54})\mathbf{k}_2, \qquad (7.39)$$

where $Z_{46} = Z_1^2$, $Z_{47} = u_2^2$, $Z_{48} = Z_2^2$, $Z_{49} = D_{11} c_1 + D_{13} s_1$, $Z_{50} = k_{14} Z_3 + k_{15} Z_4 - D_{12} Z_{37} + D_{23}(Z_{47} - Z_{48})$, $Z_{51} = D_{23} s_1 + D_{21} c_1$, $Z_{52} = D_{31}(Z_{48} - Z_{46}) + k_8 Z_{37}$, $Z_{53} = D_{33} s_1 + D_{31} c_1$, and $Z_{54} = k_{15} Z_3 + k_{16} Z_4 + D_{23} Z_{37} + D_{12}(Z_{46} - Z_{47})$.

The inertia force for link $j = 1, 2, 3$ is

$$\mathbf{F}_{\text{in}j} = -m_j \mathbf{a}_{C_j}, \qquad (7.40)$$

and the inertia force for rigid body RB is

$$\mathbf{F}_{\text{in}R} = -m_R \mathbf{a}_{C_R}. \qquad (7.41)$$

The contribution of link $j = 1, 2, 3$ to generalized inertia force F_r^* is

$$(F_r^*)_j = \frac{\partial \mathbf{v}_{C_j}}{\partial u_r} \cdot \mathbf{F}_{\text{in}j} + \frac{\partial \boldsymbol{\omega}_{j0}}{\partial u_r} \cdot \mathbf{T}_{\text{in}j}, \qquad r = 1, 2, 3, \qquad (7.42)$$

and the contribution of rigid body RB to the generalized inertia force F_r^* is

$$(F_r^*)_R = \frac{\partial \mathbf{v}_{C_R}}{\partial u_r} \cdot \mathbf{F}_{\text{in}R} + \frac{\partial \boldsymbol{\omega}_{20}}{\partial u_r} \cdot \mathbf{T}_{\text{in}R}. \qquad (7.43)$$

The three generalized inertia forces can be computed as

$$F_r^* = \sum_{j=1}^{3} (F_r^*)_j + (F_r^*)_R = \sum_{j=1}^{3} \left(\frac{\partial \mathbf{v}_{C_j}}{\partial u_r} \cdot \mathbf{F}_{\text{in}j} + \frac{\partial \boldsymbol{\omega}_{j0}}{\partial u_r} \cdot \mathbf{T}_{\text{in}j} \right)$$
$$+ \frac{\partial \mathbf{v}_{C_R}}{\partial u_r} \cdot \mathbf{F}_{\text{in}R} + \frac{\partial \boldsymbol{\omega}_{20}}{\partial u_r} \cdot \mathbf{T}_{\text{in}R}, \qquad r = 1, 2, 3. \qquad (7.44)$$

7.3 Generalized Active Forces

As a way to explain what the generalized active forces are, again the system $\{S\}$ of ν particles is considered. The force $\mathbf{R}_i$ is the resultant of all contact and body forces acting on a generic particle P_i of $\{S\}$. The rth *generalized active force*, F_r, for the system $\{S\}$ is defined as

$$F_r = \sum_{i=1}^{\nu} \left(\frac{\partial \mathbf{v}_{P_i}}{\partial u_i} \cdot \mathbf{R}_i \right), \qquad r = 1, \ldots, n. \tag{7.45}$$

The task of constructing expressions for F_r frequently is facilitated by the following facts. Many forces that contribute to the resultant force $\mathbf{R}_i$ make no contributions to the generalized active force F_r. For example, if RB is a rigid body belonging to $\{S\}$, the total contribution to F_r of all gravitational forces exerted by particles of RB on each other is equal to zero. Furthermore, if a set of contact and/or body forces acting on RB is equivalent to a couple of torque $\mathbf{T}$ together with force $\mathbf{R}$ applied at a point Q of RB, then $(F_r)_R$, the contribution of this set of forces to F_r, is given by

$$(F_r)_R = \frac{\partial \omega}{\partial u_r} \cdot \mathbf{T} + \frac{\partial \mathbf{v}_Q}{\partial u_r} \cdot \mathbf{R}, \qquad r = 1, \ldots, n, \tag{7.46}$$

where ω is the angular velocity of RB in (0), and $\mathbf{v}_Q$ is the velocity of Q in (0).

In the case of the robot arm, there are two kinds of forces that contribute to the generalized active forces F_1, F_2, and F_3, namely contact forces applied in order to drive links 1, 2, 3 and the rigid body RB, and gravitational forces exerted on 1, 2, 3, and RB by the Earth. Consider first the contact forces, as shown in Fig. 7.3. The set of such forces transmitted from 0 to 1 (through the bearings and by means of the motor) can be replaced with a couple of torque $\mathbf{T}_{01}$ together with a force $\mathbf{F}_{01}$ applied to 1 at A. Similarly, the set of contact forces transmitted from 1 to 2 can be replaced with a couple of torque $\mathbf{T}_{12}$ together with a force $\mathbf{F}_{12}$ applied to 2 at B. The law of action and reaction then guarantees that the set of contact forces transmitted from 1 to 2 is equivalent to a couple of torque $-\mathbf{T}_{12}$ together with the force $-\mathbf{F}_{12}$ applied to 1 and B.

Next, the set of contact forces exerted on link 2 by link 3 can be replaced with a couple of torque $\mathbf{T}_{23}$ together with a force $\mathbf{F}_{23}$ applied to 3 at C_3. The law of action and reaction guarantees that the set of contact forces transmitted from 3 to 2 is equivalent to a couple of torque $-\mathbf{T}_{23}$ together with the force $-\mathbf{F}_{23}$ applied to 2 at C_{32}. The point C_{32} ($C_{32} \in$ link 2) instantaneously coincides with C_3, ($C_3 \in$ link 3).

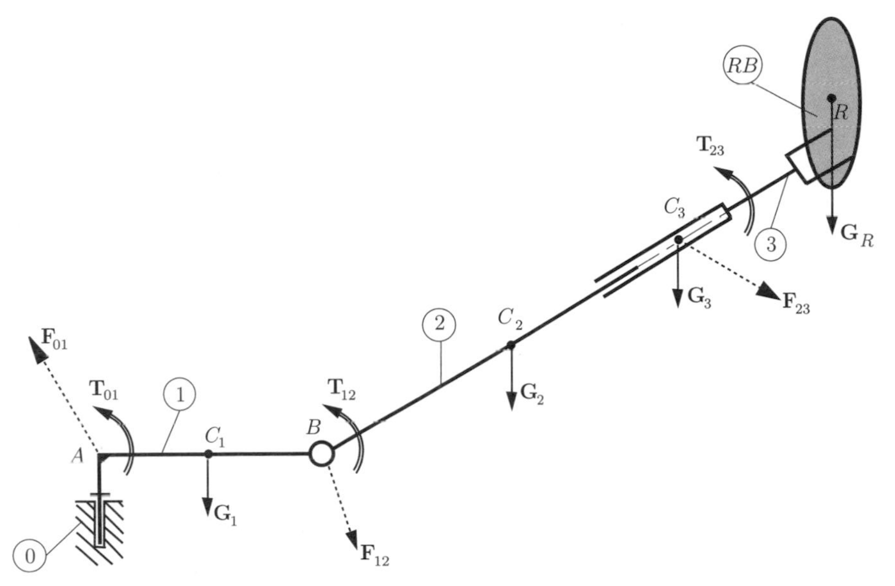

Figure 7.3. Contact forces.

The expressions for $\mathbf{T}_{01}$, $\mathbf{F}_{01}$, $\mathbf{T}_{12}$, $\mathbf{F}_{12}$, $\mathbf{T}_{23}$, and $\mathbf{F}_{23}$ are as follows:

$$\begin{aligned}
\mathbf{T}_{01} &= T_{01x}\mathbf{i}_1 + T_{01y}\mathbf{j}_1 + T_{01z}\mathbf{k}_1, \quad \mathbf{F}_{01} = F_{01x}\mathbf{i}_1 + F_{01y}\mathbf{j}_1 + F_{01z}\mathbf{k}_1,\\
\mathbf{T}_{12} &= T_{12x}\mathbf{i}_2 + T_{12y}\mathbf{j}_2 + T_{12z}\mathbf{k}_2, \quad \mathbf{F}_{12} = F_{12x}\mathbf{i}_2 + F_{12y}\mathbf{j}_2 + F_{12z}\mathbf{k}_2,\\
\mathbf{T}_{23} &= T_{23x}\mathbf{i}_2 + T_{23y}\mathbf{j}_2 + T_{23z}\mathbf{k}_2, \quad \mathbf{F}_{23} = F_{23x}\mathbf{i}_2 + F_{23y}\mathbf{j}_2 + F_{23z}\mathbf{k}_2. \qquad (7.47)
\end{aligned}$$

The external gravitational forces exerted on links 1, 2, 3, and rigid body RB by the Earth, can be denoted by $\mathbf{G}_1$, $\mathbf{G}_2$, $\mathbf{G}_3$, and $\mathbf{G}_R$, respectively, and they can be expressed as

$$\begin{aligned}
\mathbf{G}_1 &= -m_1 g \mathbf{i}_1,\\
\mathbf{G}_2 &= -m_2 g \mathbf{i}_1 = -m_2 g(c_1\mathbf{i}_2 + s_1\mathbf{k}_2),\\
\mathbf{G}_3 &= -m_3 g \mathbf{i}_1 = -m_3 g(c_1\mathbf{i}_2 + s_1\mathbf{k}_2),\\
\mathbf{G}_R &= -m_R g \mathbf{i}_1 = -m_R g(c_1\mathbf{i}_2 + s_1\mathbf{k}_2). \qquad (7.48)
\end{aligned}$$

The reason for replacing $\mathbf{i}_1$ with $c_1\mathbf{i}_2 + s_1\mathbf{k}_2$ in connection with the forces $\mathbf{G}_2$, $\mathbf{G}_3$, and $\mathbf{G}_R$ is that they are soon to be dot multiplied with $\partial \mathbf{v}_{C_2}/\partial u_r$, $\partial \mathbf{v}_{C_3}/\partial u_r$, and $\partial \mathbf{v}_{C_R}/\partial u_r$, which have been expressed in terms of $\mathbf{i}_2, \mathbf{j}_2, \mathbf{k}_2$.

One can express $(F_r)_1$, the contribution to the generalized active force F_r of all the forces and torques acting on the particles of link 1, as

$$(F_r)_1 = \frac{\partial \boldsymbol{\omega}_{10}}{\partial u_r}\cdot(\mathbf{T}_{01}-\mathbf{T}_{12}) + \frac{\partial \mathbf{v}_{C_1}}{\partial u_r}\cdot\mathbf{G}_1 + \frac{\partial \mathbf{v}_B}{\partial u_r}\cdot(-\mathbf{F}_{12}), \qquad r = 1,\ 2,\ 3. \qquad (7.49)$$

The contribution $(F_r)_2$ to the generalized active force of all the forces and torques acting on link 2 is

$$(F_r)_2 = \frac{\partial \omega_{20}}{\partial u_r} \cdot (\mathbf{T}_{12} - \mathbf{T}_{23}) + \frac{\partial \mathbf{v}_B}{\partial u_r} \cdot \mathbf{F}_{12} + \frac{\partial \mathbf{v}_{C_2}}{\partial u_r} \cdot \mathbf{G}_2 + \frac{\partial \mathbf{v}_{C_{32}}}{\partial u_r} \cdot (-\mathbf{F}_{23}), \qquad r = 1,\ 2,\ 3. \tag{7.50}$$

The contribution $(F_r)_3$ to the generalized active force of all the forces and torques acting on link 3 is

$$(F_r)_3 = \frac{\partial \omega_{20}}{\partial u_r} \cdot \mathbf{T}_{23} + \frac{\partial \mathbf{v}_{C_3}}{\partial u_r} \cdot \mathbf{G}_3 + \frac{\partial \mathbf{v}_{C_3}}{\partial u_r} \cdot \mathbf{F}_{23}, \qquad r = 1,\ 2,\ 3. \tag{7.51}$$

The contribution $(F_r)_R$ to the generalized active force of all the forces and torques acting on rigid body RB is

$$(F_r)_R = \frac{\partial \mathbf{v}_{C_R}}{\partial u_r} \cdot \mathbf{G}_R, \qquad r = 1,\ 2,\ 3. \tag{7.52}$$

The generalized active forces F_r of all the forces and torques acting on links 1, 2, 3, and rigid body RB are

$$F_r = (F_r)_1 + (F_r)_2 + (F_r)_3 + (F_r)_R, \qquad r = 1, 2, 3, \tag{7.53}$$

or

$$\begin{aligned} F_1 &= T_{12x}, \\ F_2 &= T_{12y} - g\,[(m_2 L_2 + m_3 Z_9 + m_r Z_{14})\, c_1 - m_R p_x s_1], \\ F_3 &= F_{23z} - g\,(m_3 + m_R)\, s_1. \end{aligned} \tag{7.54}$$

To arrive at the dynamical equations governing the robot arm, all that remains to be done is to substitute into Kane's dynamical equations (Appendix 9), namely,

$$F_r^* + F_r = 0, \qquad r = 1,\ 2,\ 3, \tag{7.55}$$

or

$$\begin{aligned} X_{11}\dot{u}_1 + X_{12}\dot{u}_2 + X_{13}\dot{u}_3 &= Y_1, \\ X_{21}\dot{u}_1 + X_{22}\dot{u}_2 + X_{23}\dot{u}_3 &= Y_2, \\ X_{31}\dot{u}_1 + X_{32}\dot{u}_2 + X_{33}\dot{u}_3 &= Y_3, \end{aligned} \tag{7.56}$$

where

$$
\begin{aligned}
X_{11} &= -\big[A_x + c_1(Z_{35} + Z_{41} + Z_{49}) + s_1(Z_{39} + Z_{44} + Z_{53}) \\
&\quad + m_1 L_1^2 + m_2 Z_6^2 + m_3 Z_{10}^2 + m_R\big(Z_{13}^2 + Z_{15}^2 + Z_{16}^2\big)\big], \\
X_{12} &= X_{21} = -[Z_{51} + m_R(Z_{13}Z_{14} - Z_{16}p_x)], \\
X_{13} &= X_{31} = -m_R Z_{16}, \\
Y_1 &= c_1(Z_{36} + Z_{42} + Z_{50}) + s_1(Z_{40} + Z_{45} + Z_{54}) + m_2 Z_6 Z_{23} \\
&\quad + m_3 Z_{10} Z_{27} + m_R(Z_{13}Z_{32} + Z_{15}Z_{33} + Z_{16}Z_{34}) - T_{01x}, \\
X_{22} &= -\big[B_y + C_y + D_{22} + m_2 L_2^2 + m_3 Z_9^2 + m_R\big(Z_{14}^2 + p_x^2\big)\big], \\
X_{23} &= X_{32} = m_R p_x, \\
Y_2 &= Z_{38} + Z_{42} + Z_{52} + m_2 L_2 Z_{22} + m_3 Z_9 Z_{26} + m_R(Z_{14}Z_{32} - p_x Z_{34}) \\
&\quad - T_{12y} + g[m_2 L_2 + m_3 Z_9 + m_R Z_{14})c_1 - m_R p_x s_1], \\
X_{33} &= -(m_3 + m_R), \\
Y_3 &= m_3 Z_{28} + m_R Z_{34} - F_{23z} + g(m_3 + m_R)s_2.
\end{aligned}
$$

7.4 Numerical Simulation

The robot arm is characterized by the following geometry (Kane and Levinson): $L_1 = 0.3$ m, $L_2 = 0.5$ m, $L_B = 1.1$ m, $p_x = 0.2$ m, $p_y = 0.4$ m, $p_z = 0.6$ m, $A_x = 11$ kg m^2, $B_x = 7$ kg m^2, $B_y = 6$ kg m^2, $B_z = 2$ kg m^2, $C_x = 5$ kg m^2, $C_y = 4$ kg m^2, $C_z = 1$ kg m^2, $D_{11} = 2$ kg m^2, $D_{22} = 2.5$ kg m^2, $D_{33} = 1.3$ kg m^2, $D_{12} = D_{21} = 0.6$ kg m^2, $D_{13} = D_{31} = -1.1$ kg m^2, and $D_{32} = D_{23} = 0.75$ kg m^2. The masses of the rigid bodies are $m_1 = 87$ kg, $m_2 = 63$ kg, $m_3 = 42$ kg, and $m_R = 50$ kg, and the gravitational acceleration is assumed to be $g = 9.81$ m/s^2.

The initial conditions, at $t = 0$ s, are $q_1(0) = \pi/6$ rad, $q_2(0) = 0.1$ m, $q_3(0) = \pi/18$ rad, and $\dot{q}_1(0) = \dot{q}_2(0) = \dot{q}_3(0) = 0$.

The robot arm can be brought from an initial state of rest in reference frame (0) to a final state of rest in (0) in such a way that q_1, q_2, and q_3 have specified values q_{1f}, q_{2f}, and q_{3f}, respectively, by using the following feedback control laws:

$$
\begin{aligned}
T_{01x} &= -\beta_{01}\dot{q}_3 - \gamma_{01}(q_3 - q_{3f}), \\
T_{12y} &= -\beta_{12}\dot{q}_1 - \gamma_{12}(q_1 - q_{1f}) + g[(m_2 L_2 + m_3 Z_9 + m_R Z_{14})c_1 \\
&\quad - m_R p_x s_1], \\
F_{23z} &= -\beta_{23}\dot{q}_2 - \gamma_{23}(q_2 - q_{2f}) + g(m_3 + m_R)s_3.
\end{aligned}
$$

The constant gains are $\beta_{01} = 464$ N m s/rad, $\gamma_{01} = 306$ N m/rad, $\beta_{12} = 216$ N m s/rad, $\gamma_{12} = 285$ N m/rad, $\beta_{23} = 169$ N s/m, and $\gamma_{23} = 56$ N/m. The values specified for the generalized coordinates are $q_{1f} = \pi/3$ rad, $q_{2f} = 0.4$ m,

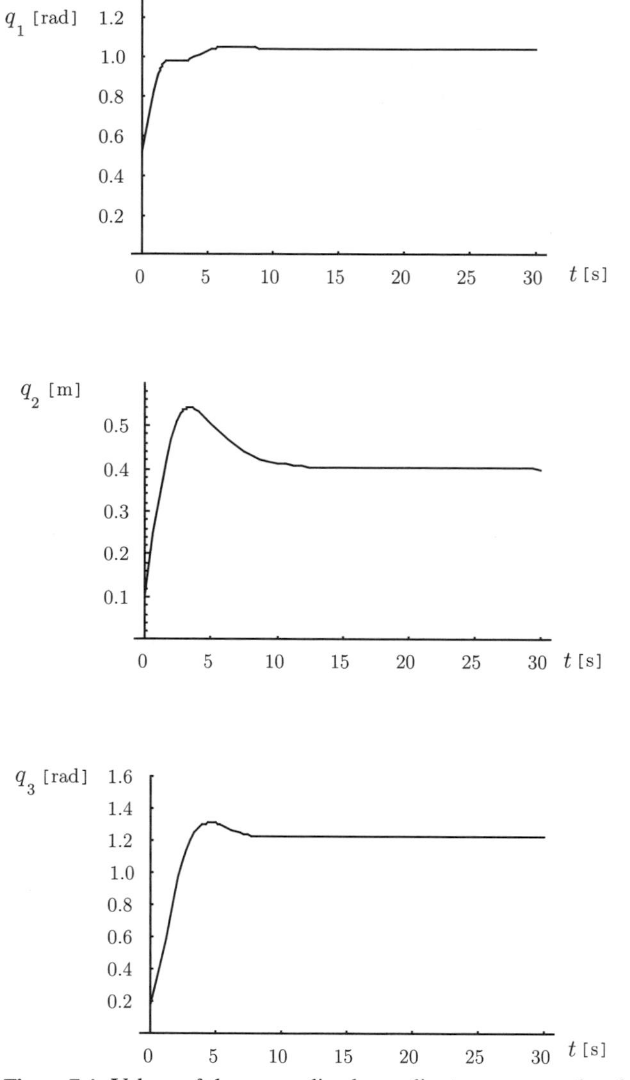

Figure 7.4. Values of the generalized coordinates q_1, q_2, and q_3 from time $t = 0$ to $t = 30$ s.

and $q_{3f} = 7\pi/18$ rad. Figure 7.4 represents the values of q_1, q_2, and q_3 from $t = 0$ to $t = 30$ s, and the *Mathematica* program that was used to compute these is given in Appendix 10.

7.5 Kinetic Energy

The total kinetic energy of the robot arm in reference frame (0) is

$$T = \sum_{i=1}^{3} T_i + T_R. \tag{7.57}$$

The kinetic energy of link i, $i = 1, 2, 3$, is

$$T_i = 1/2 m_i \mathbf{v}_{C_i} \cdot \mathbf{v}_{C_i} + 1/2 \boldsymbol{\omega}_{i0} \cdot (\bar{I}_i \cdot \boldsymbol{\omega}_{i0}). \tag{7.58}$$

The kinetic energy of rigid body RB is

$$T_R = 1/2 m_R \mathbf{v}_R \cdot \mathbf{v}_R + 1/2 \boldsymbol{\omega}_{20} \cdot (\bar{I}_R \cdot \boldsymbol{\omega}_{20}). \tag{7.59}$$

The generalized inertia forces can be computed also with the formula

$$F_r^* = \frac{d}{dt}\left(\frac{\partial T}{\partial \dot{q}_r}\right) - \frac{\partial T}{\partial q_r}, \qquad r = 1,\ 2,\ 3. \tag{7.60}$$

8 Kinematic Chains with Continuous Flexible Links

8.1 Transverse Vibrations of a Flexible Link

Often a kinematic chain consists in part of continuous elastic components supported by rigid bodies. Small motions of the components relative to the rigid bodies generally are governed by partial differential equations. In such cases, these equations cannot be solved by the method of separation of variables.

Figure 8.1 shows a cantilever beam B of length L, with constant flexural rigidity EI and constant mass per unit length ρ. When B is supported by a rigid body fixed in a frame, the small flexural vibrations of B are governed by the equation

$$EI\frac{\partial^4 y(x,t)}{\partial x^4} + \rho\frac{\partial^2 y(x,t)}{\partial t^2} = 0, \tag{8.1}$$

and by the boundary conditions

$$y(0,t) = y'(0,t) = y''(L,t) = y'''(L,t) = 0. \tag{8.2}$$

The general solution of Eq. (8.1) that satisfies Eq. (8.2) can be expressed as

$$y(x,t) = \sum_{i=1}^{\infty} \Phi_i(x)\, q_i(t), \tag{8.3}$$

where $\Phi_i(x)$ and $q_i(t)$ are functions of x and t, respectively, defined as

$$\Phi_i = \cosh\frac{\lambda_i x}{L} - \cos\frac{\lambda_i x}{L} - \frac{\cosh\lambda_i + \cos\lambda_i}{\sinh\lambda_i + \sin\lambda_i}\left(\sinh\frac{\lambda_i x}{L} - \sin\frac{\lambda_i x}{L}\right), \tag{8.4}$$

and

$$q_i = \alpha_i \cos p_i t + \beta_i \sin p_i t, \tag{8.5}$$

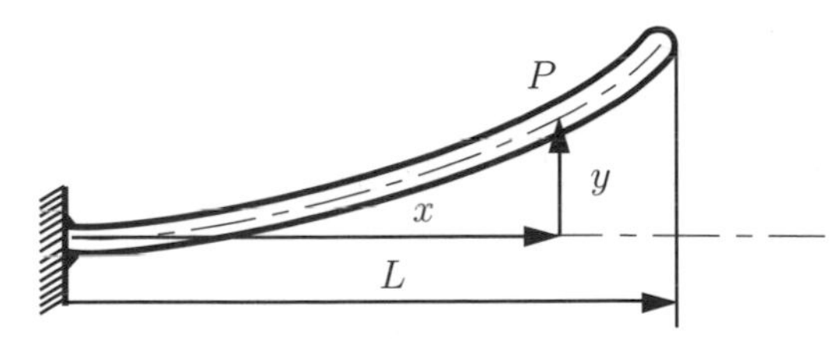

Figure 8.1. Cantilever beam.

where λ_i, $i = 1, \ldots, \infty$ are the consecutive roots of the transcendental equation

$$\cos\lambda \cosh\lambda + 1 = 0, \tag{8.6}$$

and where

$$p_i = \left(\frac{\lambda_i}{L}\right)^2 \left(\frac{EI}{\rho}\right)^{1/2}, \tag{8.7}$$

and α_i and β_i are constants that depend upon initial conditions.

Functions $\Phi_i(x)$ satisfy the orthogonality relations

$$\int_0^L \Phi_i \Phi_j \, \rho \, dx = m\delta_{ij}, \qquad (i, j = 1, \ldots, \infty), \tag{8.8}$$

$$EI \int_0^L \Phi_i'' \, \Phi_j'' \, dx = p_i^2 \, m\delta_{ij}, \qquad (i, j = 1, \ldots, \infty), \tag{8.9}$$

where m is the mass of the beam and δ_{ij} is the Kronecker delta.

8.2 Equations of Motion for a Flexible Link

In Fig. 8.2, a schematic representation of a kinematic chain is given. The system is formed by a rigid body RB that supports a uniform cantilever beam B of length L, flexural rigidity EI, and mass per unit length ρ. Only planar motions of the kinematic chain in a fixed reference frame (0) with unit vectors $[\mathbf{\imath}_0, \mathbf{\jmath}_0, \mathbf{k}_0]$ will be considered.

To characterize the instantaneous configuration of rigid body RB, generalized coordinates q_1, q_2, q_3 are employed. The first generalized coordinate q_1 denotes the distance from C_R, the mass center of RB, to the horizontal axis of reference frame (0). The generalized coordinate q_2 denotes the distance from C_R to the vertical axis of (0). The last generalized coordinate, q_3 ($s_3 = \sin q_3$, $c_3 = \cos q_3$), designates the radian measure of the rotation angle between RB and the horizontal axis.

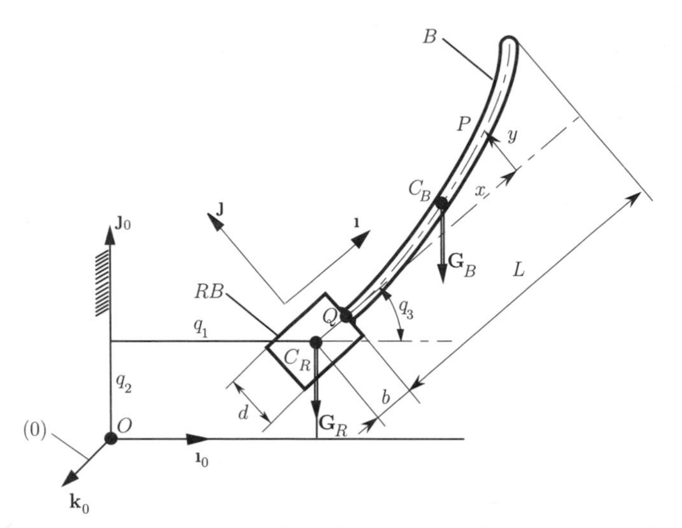

Figure 8.2. System formed by a rigid body RB that supports a uniform cantilever beam B.

The generalized speeds u_1, u_2, and u_3, used to characterize the motion of RB in (0), are defined as

$$u_1 = \mathbf{v}_{C_R} \cdot \mathbf{\imath}, \qquad u_2 = \mathbf{v}_{C_R} \cdot \mathbf{\jmath}, \qquad u_3 = \boldsymbol{\omega}_{R0} \cdot \mathbf{k}, \tag{8.10}$$

where $\mathbf{v}_{C_R}$ is the velocity in (0) of the mass center C_R of RB, $\boldsymbol{\omega}_{R0}$ is the angular velocity of RB in (0), and $[\mathbf{\imath}, \mathbf{\jmath}, \mathbf{k}]$ form a dextral set of mutually perpendicular unit vectors fixed in RB and directed as shown in Fig. 8.2. It follows immediately that

$$\mathbf{v}_{C_R} = u_1 \mathbf{\imath} + u_2 \mathbf{\jmath}, \qquad \boldsymbol{\omega}_{R0} = u_3 \mathbf{k}. \tag{8.11}$$

The unit vectors $\mathbf{\imath}_0, \mathbf{\jmath}_0, \mathbf{k}_0$ can be expressed as

$$\begin{aligned} \mathbf{\imath}_0 &= c_3 \mathbf{\imath} - s_3 \mathbf{\jmath}, \\ \mathbf{\jmath}_0 &= s_3 \mathbf{\imath} + c_3 \mathbf{\jmath}, \\ \mathbf{k}_0 &= \mathbf{k}. \end{aligned} \tag{8.12}$$

The velocity of C_R in (0) is

$$\begin{aligned} \mathbf{v}_{C_R} &= \dot{q}_1 \mathbf{\imath}_0 + \dot{q}_2 \mathbf{\jmath}_0 \\ &= (\dot{q}_1 c_3 + \dot{q}_2 s_3) \mathbf{\imath} + (-\dot{q}_1 s_3 + \dot{q}_2 c_3) \mathbf{\jmath}. \end{aligned} \tag{8.13}$$

From Eqs. (8.11) and (8.13) it follows that

$$\begin{aligned} u_1 &= \dot{q}_1 c_3 + \dot{q}_2 s_3, \\ u_2 &= -\dot{q}_1 s_3 + \dot{q}_2 c_3, \\ u_3 &= \dot{q}_3. \end{aligned} \tag{8.14}$$

Equation (8.14) can be solved uniquely for $\dot{q}_1, \dot{q}_2, \dot{q}_3$, and thus u_1, u_2, u_3 form a set of generalized speeds for RB.

Kinematics

The deformations of the cantilever beam B can be discussed in terms of the displacement $y(x, t)$ of a generic point P on beam B. Point P is situated at a distance x from point Q, the point at which B is attached to RB. The displacement y can be expressed as

$$y(x, t) = \sum_{i=1}^{n} \Phi_i(x)\, q_{3+i}(t), \tag{8.15}$$

where $\Phi_i(x)$ is a totally unrestricted function of x, $q_{3+i}(t)$ is an equally unrestricted function of t, and n is any positive integer. Generalized speeds u_{3+i}, $i = 1, \ldots, n$ are introduced as

$$u_{3+i} = \dot{q}_{3+i}, \qquad i = 1, \ldots, n. \tag{8.16}$$

The velocity of Q in (0) is

$$\begin{aligned}\mathbf{v}_Q &= \mathbf{v}_{C_R} + \boldsymbol{\omega}_{R0} \times b\,\mathbf{i}\\ &= u_1\mathbf{i} + u_2\mathbf{j} + \begin{vmatrix} \mathbf{i} & \mathbf{j} & \mathbf{k}\\ 0 & 0 & u_3\\ b & 0 & 0 \end{vmatrix} = u_1\mathbf{i} + (u_2 + bu_3)\mathbf{j},\end{aligned} \tag{8.17}$$

where b is the distance from C_R to Q.

The velocity of any point P of the elastic link B in (0) is

$$\begin{aligned}\mathbf{v}_P &= \mathbf{v}_{C_R} + \frac{\partial}{\partial t}[(b + x)\mathbf{i} + y\,\mathbf{j}] + \boldsymbol{\omega}_{R0} \times [(b + x)\mathbf{i} + y\,\mathbf{j}]\\ &= u_1\mathbf{i} + u_2\mathbf{j} + \dot{y}\,\mathbf{j} + \begin{vmatrix} \mathbf{i} & \mathbf{j} & \mathbf{k}\\ 0 & 0 & u_3\\ b + x & y & 0 \end{vmatrix}\\ &= (u_1 - u_3\, y)\mathbf{i} + [u_2 + (b + x)\, u_3 + \dot{y}]\,\mathbf{j}\\ &= \left(u_1 - u_3 \sum_{i=1}^{n} \Phi_i q_{3+i}\right)\mathbf{i}\\ &\quad + \left[u_2 + (b + x)\, u_3 + \sum_{i=1}^{n} \Phi_i u_{3+i}\right]\mathbf{j}.\end{aligned} \tag{8.18}$$

The velocity of the midpoint C_B of the uniform elastic link B in (0) is

$$\mathbf{v}_{C_B} = \mathbf{v}_P\left(x = \frac{L}{2}, t\right) = \left(u_1 - u_3 \sum_{i=1}^{n} \Phi_i q_{3+i}\right)\mathbf{\imath} + \left[u_2 + (b + 0.5\,L)\,u_3 + \sum_{i=1}^{n} \Phi_i u_{3+i}\right]\mathbf{\jmath}. \tag{8.19}$$

The angular acceleration of RB in reference frame (0) is

$$\boldsymbol{\alpha}_{R0} = \dot{\boldsymbol{\omega}}_{R0} = \dot{u}_3\mathbf{k}. \tag{8.20}$$

The linear acceleration of C_R in reference frame (0) is

$$\begin{aligned}
\mathbf{a}_{C_R} &= \frac{\partial}{\partial t}\mathbf{v}_{C_R} + \boldsymbol{\omega}_{R0} \times \mathbf{v}_{C_R} \\
&= \dot{u}_1\mathbf{\imath} + \dot{u}_2\mathbf{\jmath} + \begin{vmatrix} \mathbf{\imath} & \mathbf{\jmath} & \mathbf{k} \\ 0 & 0 & u_3 \\ u_1 & u_2 & 0 \end{vmatrix} \\
&= (\dot{u}_1 - u_2 u_3)\mathbf{\imath} + (\dot{u}_2 + u_3 u_1)\mathbf{\jmath}.
\end{aligned}$$

The acceleration of point P in the reference frame (0) is

$$\begin{aligned}
\mathbf{a}_P &= \frac{\partial}{\partial t}\mathbf{v}_P + \boldsymbol{\omega}_{R0} \times \mathbf{v}_P \\
&= \left[\dot{u}_1 - u_2 u_3 - (b + x)u_3^2 - \sum_{i=1}^{n} \Phi_i(\dot{u}_3 q_{3+i} + 2u_3 u_{3+i})\right]\mathbf{\imath} \\
&\quad + \left[\dot{u}_2 + u_3 u_1 + (b + x)\dot{u}_3 + \sum_{i=1}^{n} \Phi_i\left(\dot{u}_{3+i} - u_3^2 q_{3+i}\right)\right]\mathbf{\jmath}.
\end{aligned} \tag{8.21}$$

Generalized Inertia Forces

If m_R and I_z are the mass of RB and the moment of inertia of RB about a line passing through C_R and parallel to $\mathbf{k}$, then the generalized inertia force F_r^* is given by

$$\begin{aligned}
F_r^* &= \frac{\partial \boldsymbol{\omega}_{R0}}{\partial u_r} \cdot (-I_z\, \boldsymbol{\alpha}_{R0}) + \frac{\partial \mathbf{v}_{C_R}}{\partial u_r} \cdot (-m_R \mathbf{a}_{C_R}) \\
&\quad + \int_o^L \frac{\partial \mathbf{v}_P}{\partial u_r} \cdot (-\mathbf{a}_P)\rho\, dx, \qquad r = 1, \ldots, 3 + n.
\end{aligned} \tag{8.22}$$

The constants m_B, e_B, I_B, E_i, F_i, and G_{ij} are defined as

$$m_B = \int_0^L \rho\, dx, \qquad e_B = \int_0^L x\rho\, dx, \qquad I_B = \int_0^L x^2 \rho\, dx,$$

$$E_i = \int_0^L \Phi_i\, \rho\, dx, \qquad F_i = \int_0^L x\Phi_i\, \rho\, dx, \qquad G_{ij} = \int_0^L \Phi_i\, \Phi_j \rho\, dx,$$

$$i, j = 1, \ldots, n.$$

Equation (8.22) then leads to

$$\begin{aligned}
F_1^* &= -(m_R + m_B)(\dot{u}_1 - u_2 u_3) + \dot{u}_3 \sum_{i=1}^{n} E_i q_{3+i} \\
&\quad + 2u_3 \sum_{i=1}^{n} E_i u_{3+i} + u_3^2(b m_B + e_B), \\
F_2^* &= -(m_R + m_B)(\dot{u}_2 + u_3 u_1) - \sum_{i=1}^{n} E_i \dot{u}_{3+i} \\
&\quad - \dot{u}_3(b m_B + e_B) + u_3^2 \sum_{i=1}^{n} E_i q_{3+i}, \\
F_3^* &= (\dot{u}_1 - u_2 u_3) \sum_{i=1}^{n} E_i q_{3+i} - (\dot{u}_2 + u_3 u_1)(b m_B + e_B) \\
&\quad - \dot{u}_3\left(b^2 m_B + 2b e_B + I_B + I_z\right) - \sum_{i=1}^{n} \dot{u}_{3+i}(b E_i + F_i) \\
&\quad - 2u_3 \sum_{i=1}^{n} \sum_{k=1}^{n} G_{ik} q_{3+i} u_{3+k} - \dot{u}_3 \sum_{i=1}^{n} \sum_{k=1}^{n} G_{ik} q_{3+i} q_{3+k}, \\
F_{3+j}^* &= -\dot{u}_2 E_j - \sum_{i=1}^{n} G_{ij} \dot{u}_{3+i} - \dot{u}_3(b E_j + F_j) - u_3 u_1 E_j \\
&\quad + u_3^2 \sum_{i=1}^{n} G_{ij} q_{3+i}, \qquad j = 1, \ldots, n.
\end{aligned} \tag{8.23}$$

Generalized Active Forces

The contributions to the generalized active forces are made by the internal forces, and by the gravitational forces exerted. The internal forces are considered first. The force $d\mathbf{F}$ is the force exerted on a generic differential element of B,

$$d\mathbf{F} = -\frac{\partial V(x, t)}{\partial x} dx\, \mathbf{J}, \tag{8.24}$$

where $V(x,t)$ is the shear at point P. If rotary inertia is neglected, then $V(x,t)$ may be expressed in terms of the bending moment $M(x,t)$ as

$$V(x,t) = [\partial M(x,t)/\partial x]. \tag{8.25}$$

Since

$$M = EI(\partial^2 y/\partial x^2), \tag{8.26}$$

Eqs. (8.24) and (8.26) yield

$$d\mathbf{F} = -\frac{\partial^2}{\partial x^2}\left(EI\frac{\partial^2 y}{\partial x^2}\right) dx\ \mathbf{J}. \tag{8.27}$$

The system of forces exerted on rigid body RB by elastic beam B is equivalent to a couple of torque $M(0,t)\mathbf{k}$ together with a force $-V(0,t)\mathbf{J}$ applied at point Q. Hence, $(F_r)_I$, the contribution of the internal forces to the generalized active force F_r, is given by

$$\begin{aligned}(F_r)_I &= \frac{\partial \boldsymbol{\omega}_{R0}}{\partial u_r}\cdot M(0,t)\mathbf{k} - \frac{\partial \mathbf{v}_Q}{\partial u_r}\cdot V(0,t)\mathbf{J}\\ &\quad - \int_0^L \frac{\partial \mathbf{v}_P}{\partial u_r}\cdot \mathbf{J}\frac{\partial^2}{\partial x^2}\left(EI\frac{\partial^2 y}{\partial x^2}\right) dx\\ &= EI\left[\frac{\partial \boldsymbol{\omega}_{R0}}{\partial u_r}\cdot \mathbf{k}\frac{\partial^2 y(0,t)}{\partial x^2} - \frac{\partial \mathbf{v}_Q}{\partial u_r}\cdot \mathbf{J}\frac{\partial^3 y(0,t)}{\partial x^3}\right]\\ &\quad - \int_0^L \frac{\partial \mathbf{v}_P}{\partial u_r}\cdot \mathbf{J}\frac{\partial^2}{\partial x^2}\left[EI\frac{\partial^2 y(x,t)}{\partial x^2}\right]dx, \qquad r = 1,\ldots,3+n,\end{aligned} \tag{8.28}$$

which leads to

$$\begin{aligned}(F_1)_I &= 0,\\ (F_2)_I &= -\sum_{i=1}^{n} q_{3+i}\left[(EI\Phi_i''')_{x=0} + \int_0^L (EI\Phi_i'')''\,dx\right],\\ (F_3)_I &= b\,(F_2)_I + \sum_{i=1}^{n} q_{3+i}\left[(EI\Phi_i'')_{x=0} - \int_0^L x(EI\Phi_i'')''\,dx\right],\\ (F_{3+j})_I &= -\sum_{i=1}^{n} q_{3+i}\int_0^L \Phi_j(EI\Phi_i'')''\,dx, \qquad j = 1,\ldots,n.\end{aligned} \tag{8.29}$$

The restrictions on Φ_i to ensure that y and y' vanish at $x = 0$ while M and V vanish at $x = L$ are

$$\Phi_i(0) = \Phi_i'(0) = \Phi_i''(L) = \Phi_i'''(L) = 0, \qquad i = 1,\ldots,n. \tag{8.30}$$

When the integrations are carried out, the following expressions result for the contribution of the internal forces to the generalized active forces:

$$(F_1)_I = (F_2)_I = (F_3)_I = 0,$$
$$(F_{3+j})_I = -\sum_{i=1}^{n} H_{ij} q_{3+i}, \qquad j = 1, \ldots n, \tag{8.31}$$

where H_{ij} is defined as

$$H_{ij} = \int_0^L EI\Phi_i''\Phi_j''\,dx, \qquad i, j = 1, \ldots, n. \tag{8.32}$$

The gravity forces exerted on RB and B by the Earth are denoted by $\mathbf{G}_R$, and $\mathbf{G}_B$, and they can be expressed as

$$\mathbf{G}_R = -m_R g\,\mathbf{J}_0 = -m_R g\,(s_3\mathbf{\imath} + c_3\,\mathbf{J}),$$
$$\mathbf{G}_B = -m_B g\,\mathbf{J}_0 = -m_B g(s_3\mathbf{\imath} + c_3\,\mathbf{J}). \tag{8.33}$$

The contribution to F_r of the gravitational forces is

$$(F_r)_G = \frac{\partial \mathbf{v}_{C_R}}{\partial u_r} \cdot \mathbf{G}_R + \frac{\partial \mathbf{v}_{C_B}}{\partial u_r} \cdot \mathbf{G}_B, \qquad r = 1, \ldots, 3+n. \tag{8.34}$$

The generalized active forces are

$$F_r = (F_r)_I + (F_r)_G, \qquad r = 1, \ldots, 3+n. \tag{8.35}$$

To arrive at the dynamical equations governing the system, all that remains to be done is to substitute into Kane's dynamical equations, namely,

$$F_r^* + F_r = 0, \quad r = 1, \ldots, 3+n. \tag{8.36}$$

EXAMPLE

For the link illustrated in Fig. 8.2, the following data are given: the length of the elastic link $L = 1$ m. The distance between the center of mass of the rigid body C_R and the point where the beam is cantilevered Q is $b = 0.1$ m. The rigid body has a mass $m_R = 1$ kg, a width $d = 0.3$ m, and a length $c = 2b = 0.2$ m. The constant mass per unit length of the elastic link is $\rho = 0.21991$ kg/m, and the constant flexural rigidity is $EI = 35.3$ N m^2. The universal gravitational constant is $g = 9.81$ m/s^2.

The elastic deflection of the beam is computed as

$$y(x,t) = \sum_{i=1}^{n} \Phi_i(x) q_{3+i}(t) = \Phi_1 q_4,$$

where $n = 1$. Only one elastic coordinate $q_4(t)$ is used for this example.

The shape function is

$$\Phi_1 = \cosh\frac{\lambda_1 x}{L} - \cos\frac{\lambda_1 x}{L} - \frac{\cosh\lambda_1 + \cos\lambda_1}{\sinh\lambda_1 + \sin\lambda_1}\left(\sinh\frac{\lambda_1 x}{L} - \sin\frac{\lambda_1 x}{L}\right),$$

where $\lambda_1 = 1.875$. The velocity of C_R in (0) is

$$\mathbf{v}_{C_R} = u_1\mathbf{\imath} + u_2\mathbf{\jmath},$$

and the angular velocity of rigid body RB in (0) is

$$\boldsymbol{\omega}_{R0} = u_3\mathbf{k},$$

where $u_3 = \dot{q}_3$.

The velocity of Q in (0) has the following expression:

$$\mathbf{v}_Q = \mathbf{v}_{C_R} + \boldsymbol{\omega}_{R0} \times b\mathbf{\imath} = u_1\mathbf{\imath} + (u_2 + bu_3)\mathbf{\jmath}.$$

The velocity of any point P of the elastic link in (0) is

$$\begin{aligned}\mathbf{v}_P &= \mathbf{v}_{C_R} + \frac{\partial}{\partial t}[(b+x)\mathbf{\imath} + y\mathbf{\jmath}] + \boldsymbol{\omega}_{R0} \times [(b+x)\mathbf{\imath} + y\mathbf{\jmath}]\\ &= (u_1 - u_3\Phi_1 q_4)\mathbf{\imath} + [u_2 + (b+x)u_3 + \Phi_1 u_4]\mathbf{\jmath},\end{aligned}$$

where $u_4 = \dot{q}_4$.

The velocity of the midpoint C_B of the elastic link in (0) is

$$\begin{aligned}\mathbf{v}_{C_B} = \mathbf{v}_P(x = 0.5L, t) &= (u_1 - u_3\Phi_{CB}q_4)\mathbf{\imath}\\ &+ [u_2 + (b + 0.5L)u_3 + \Phi_{CB}u_4]\mathbf{\jmath},\end{aligned}$$

where $\Phi_{CB} = \Phi_1(x = 0.5L)$ is the shape function corresponding to the mass center C_B $(x = L/2)$.

The angular acceleration of RB in reference frame (0) is

$$\boldsymbol{\alpha}_{R0} = \dot{\boldsymbol{\omega}}_{R0} = \dot{u}_3\mathbf{k}.$$

The linear acceleration of C_R in reference frame (0) is

$$\mathbf{a}_{C_R} = \frac{\partial}{\partial t}\mathbf{v}_{C_R} + \boldsymbol{\omega}_{R0} \times \mathbf{v}_{C_R} = (\dot{u}_1 - u_2u_3)\mathbf{\imath} + (\dot{u}_2 + u_3u_1)\mathbf{\jmath},$$

and the acceleration of point P in reference frame (0) is

$$\begin{aligned}\mathbf{a}_P = \frac{\partial}{\partial t}\mathbf{v}_P + \boldsymbol{\omega}_{R0} \times \mathbf{v}_P &= \left[\dot{u}_1 - u_2u_3 - (b+x)u_3^2 - \Phi_1(\dot{u}_3q_4 + 2u_3u_4)\right]\mathbf{\imath}\\ &+ \left[\dot{u}_2 + u_3u_1 + (b+x)\dot{u}_3 + \Phi_1\left(\dot{u}_4 - u_3^2q_4\right)\right]\mathbf{\jmath}.\end{aligned}$$

The generalized inertia forces F_r^* are

$$F_r^* = \frac{\partial \boldsymbol{\omega}_{R0}}{\partial u_r} \cdot (-I_z\boldsymbol{\alpha}_{R0}) + \frac{\partial \mathbf{v}_{C_R}}{\partial u_r} \cdot (-m_R\mathbf{a}_{C_R}) + \int_0^L \frac{\partial \mathbf{v}_P}{\partial u_r} \cdot (-\mathbf{a}_P)\rho\, dx, \quad r = 1, \ldots, 4.$$

The mass moment of inertia I_z is the moment of inertia of RB about a line passing through C_R and parallel to $\mathbf{k}_0$, and it is calculated with the formula

$$I_z = {}^1/_{12}\, m_R(c^2 + d^2) = {}^1/_{12}\,(1)[(0.2)^2 + (0.3)^2] = 0.01 \text{ kg m}^2.$$

In this case, the generalized inertia forces are

$$\begin{aligned}
F_1^* &= -(m_R + m_B)(\dot{u}_1 - u_2u_3) + \dot{u}_3E_1q_4 + 2u_3E_1u_4 + u_3^2(bm_B + e_B)\\
&= 0.13194u_3^2 + 0.34434u_3u_4 - 1.21991(u_2u_3 - \dot{u}_1) + 0.17217q_4\dot{u}_3,\\
F_2^* &= -(m_R + m_B)(\dot{u}_2 + u_3u_1) - E_1\dot{u}_4 - \dot{u}_3(bm_B + e_B) + u_3^2E_1q_4\\
&= 0.17217q_4u_3^2 - 1.21991(u_1u_3 + \dot{u}_2) - 0.13194\dot{u}_3 - 0.17217\dot{u}_4,\\
F_3^* &= (\dot{u}_1 - u_2u_3)E_1q_4 - (\dot{u}_2 + u_3u_1)(bm_B + e_B)\\
&\quad - \dot{u}_3\big(b^2m_B + 2be_B + I_B + J_3\big) - \dot{u}_4(bE_1 + F_1)\\
&\quad - G_{11}q_4(2u_3u_4 + q_4\dot{u}_3)\\
&= -0.13194u_1u_3 + q_4(-0.17217u_2u_3 - 0.43974u_3u_4 + 0.17217\dot{u}_1)\\
&\quad - 0.13194\dot{u}_2 - 0.10832\dot{u}_3 - 0.14229\dot{u}_4 - 0.219873q_4^2\dot{u}_3,\\
F_4^* &= -\dot{u}_2E_1 - G_{11}\dot{u}_4 - \dot{u}_3(bE_1 + F_1) - u_3u_1E_1 + u_3^2G_{11}q_4\\
&= -0.17217u_1u_3 + 0.21987q_4u_3^2 - 0.17217\dot{u}_2\\
&\quad - 0.14229\dot{u}_3 - 0.21991\dot{u}_4,
\end{aligned}$$

where the constants m_B, e_B, I_B, E_1, F_1, and G_{11} are

$$\begin{aligned}
m_B &= \int_0^L \rho\,dx = 0.21991 \text{ kg}, & e_B &= \int_0^L x\rho\,dx = 0.109955 \text{ kg m},\\
I_B &= \int_0^L x^2\rho\,dx = 0.0733033 \text{ kg m}^2, & E_1 &= \int_0^L \Phi_1\rho\,dx = 0.172173 \text{ kg},\\
F_1 &= \int_0^L x\Phi_1\rho\,dx = 0.12508 \text{ kg m}, & G_{11} &= \int_0^L \Phi_1^2\rho\,dx = 0.219873 \text{ kg}.
\end{aligned}$$

The contribution of internal forces to the generalized active force is given by

$$\begin{aligned}
(F_r)_I &= \frac{\partial \boldsymbol{\omega}_{R0}}{\partial u_r}\cdot M(0,t)\mathbf{k} - \frac{\partial \mathbf{v}_Q}{\partial u_r}\cdot V(0,t)\mathbf{J} - \int_0^L \frac{\partial \mathbf{v}_P}{\partial u_r}\cdot \mathbf{J}\frac{\partial^2}{\partial x^2}\\
&\quad \left[EI\frac{\partial^2 y(x,t)}{\partial x^2}\right]dx\\
&= EI\left[\frac{\partial \boldsymbol{\omega}_{R0}}{\partial u_r}\cdot \mathbf{k}\frac{\partial^2 y(0,t)}{\partial x^2} - \frac{\partial \mathbf{v}_Q}{\partial u_r}\cdot \mathbf{J}\frac{\partial^3 y(0,t)}{\partial x^3}\right]\\
&\quad - \int_0^L \frac{\partial \mathbf{v}_P}{\partial u_r}\cdot \mathbf{J}\frac{\partial^2}{\partial x^2}\left[EI\frac{\partial^2 y(x,t)}{\partial x^2}\right]dx, \qquad r = 1,\ldots,4,
\end{aligned}$$

where $M(x,t) = EI(\partial^2 y/\partial x^2)$ is the bending moment, and $V(x,t) = [\partial M(x,t)/\partial x]$ is the shear. This leads to

$$(F_1)_I = 0,$$

$$(F_2)_I = -q_4\left[(EI\Phi_1''')_{x=0} + \int_0^L (EI\Phi_1'')''dx\right],$$

$$(F_3)_I = b(F_2)_I + q_4\left[(EI\Phi_1'')_{x=0} - \int_0^L x(EI\Phi_1'')''dx\right],$$

$$(F_4)_I = -q_4\int_0^L EI\Phi_1\Phi_1''''dx = -q_4 H_{11},$$

where $H_{11} = \int_0^L EI(\Phi_1'')^2\,dx = 436.295$ N m^3.

After integrations, the results are

$$(F_1)_I = (F_2)_I = (F_3)_I = 0,$$
$$(F_4)_I = -436.295 q_4.$$

The contribution to F_r of the gravity forces are

$$(F_r)_G = \frac{\partial \mathbf{v}_{C_R}}{\partial u_r}\cdot \mathbf{G}_R + \frac{\partial \mathbf{v}_{C_B}}{\partial u_r}\cdot \mathbf{G}_B, \qquad r = 1,\ldots,4,$$

or

$$\begin{aligned}
(F_1)_G &= \frac{\partial \mathbf{v}_{C_R}}{\partial u_1}\cdot \mathbf{G}_R + \frac{\partial \mathbf{v}_{C_B}}{\partial u_1}\cdot \mathbf{G}_B \\
&= \frac{\partial \mathbf{v}_{C_R}}{\partial \dot{q}_1}\cdot\left(\frac{\partial \dot{q}_1}{\partial u_1}\mathbf{G}_R\right) + \frac{\partial \mathbf{v}_{C_B}}{\partial \dot{q}_1}\cdot\left(\frac{\partial \dot{q}_1}{\partial u_1}\mathbf{G}_B\right) = -11.9673\sin q_3, \\
(F_2)_G &= \frac{\partial \mathbf{v}_{C_R}}{\partial u_2}\cdot \mathbf{G}_R + \frac{\partial \mathbf{v}_{C_B}}{\partial u_2}\cdot \mathbf{G}_B \\
&= \frac{\partial \mathbf{v}_{C_R}}{\partial \dot{q}_2}\cdot\left(\frac{\partial \dot{q}_2}{\partial u_2}\mathbf{G}_R\right) + \frac{\partial \mathbf{v}_{C_B}}{\partial \dot{q}_2}\cdot\left(\frac{\partial \dot{q}_2}{\partial u_2}\mathbf{G}_B\right) \\
&= -gm_B\cos q_3 - gm_R\cos q_3 = -11.9673\cos q_3, \\
(F_3)_G &= \frac{\partial \mathbf{v}_{C_R}}{\partial u_3}\cdot \mathbf{G}_R + \frac{\partial \mathbf{v}_{C_B}}{\partial u_3}\cdot \mathbf{G}_B \\
&= \frac{\partial \mathbf{v}_{C_R}}{\partial \dot{q}_3}\cdot\left(\frac{\partial \dot{q}_3}{\partial u_3}\mathbf{G}_R\right) + \frac{\partial \mathbf{v}_{C_B}}{\partial \dot{q}_3}\cdot\left(\frac{\partial \dot{q}_3}{\partial u_3}\mathbf{G}_B\right) \\
&= -g(b + L/2)m_B\cos q_3 + gm_B\Phi_{CB}q_4\sin q_3 \\
&= -1.29439\cos q_3 + 1.46478 q_4\sin q_3, \\
(F_4)_G &= \frac{\partial \mathbf{v}_{C_R}}{\partial u_4}\cdot \mathbf{G}_R + \frac{\partial \mathbf{v}_{C_B}}{\partial u_4}\cdot \mathbf{G}_B \\
&= \frac{\partial \mathbf{v}_{C_R}}{\partial \dot{q}_4}\cdot\left(\frac{\partial \dot{q}_4}{\partial u_4}\mathbf{G}_R\right) + \frac{\partial \mathbf{v}_{C_B}}{\partial \dot{q}_4}\cdot\left(\frac{\partial \dot{q}_4}{\partial u_4}\mathbf{G}_B\right) \\
&= -gm_B\cos\ q_3\Phi_{CB} = -1.46478\cos q_3,
\end{aligned}$$

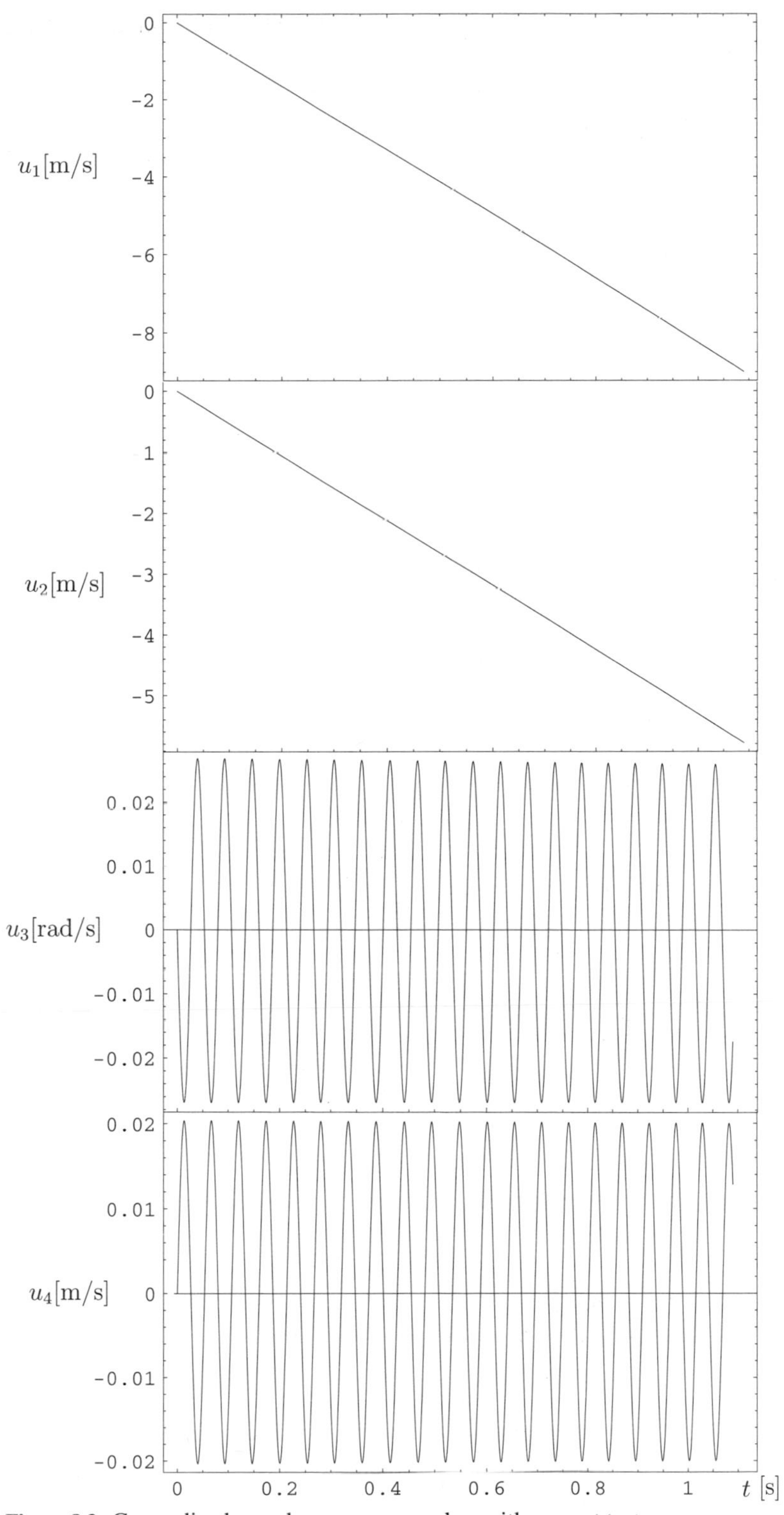

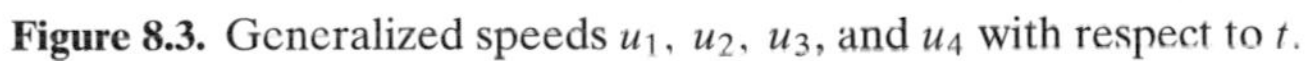

Figure 8.3. Generalized speeds u_1, u_2, u_3, and u_4 with respect to t.

where the gravitational forces exerted on RB and B are

$$\begin{aligned}\mathbf{G}_R &= -m_R g \mathbf{J}_0 = -m_R g(\sin q_3 \mathbf{I} + \cos q_3 \mathbf{J})\\ &= -9.81 \sin q_3 \mathbf{I} - 9.81 \cos q_3 \mathbf{J},\\ \mathbf{G}_B &= -m_B g \mathbf{J}_0 = -m_B g(\sin q_3 \mathbf{I} + \cos q_3 \mathbf{J})\\ &= -2.15732 \sin q_3 \mathbf{I} - 2.15732 \cos q_3 \mathbf{J}.\end{aligned}$$

The generalized active forces are

$$F_r = (F_r)_I + (F_r)_G, \qquad r = 1, \ldots, 4.$$

Finally, Kane's dynamical equations are

$$F_r^* + (F_r)_I + (F_r)_G = 0, \qquad r = 1, \ldots, 4,$$

which gives

$$\begin{aligned}&-11.9673 \sin q_3 + 0.13194 u_3^2 + 0.34434 u_3 u_4 - 1.21991(-u_2 u_3 + \dot{u}_1)\\ &\quad + 0.17217 q_4 \dot{u}_3 = 0,\\ &-11.9673 \cos q_3 + 0.17217 q_4 u_3^2 - 1.21991(u_1 u_3 + \dot{u}_2)\\ &\quad - 0.13194 \dot{u}_3 - 0.17217 \dot{u}_4 = 0,\\ &-1.29439 \cos q_3 + 1.46478 q_4 \sin q_3 + 0.17217 q_4(-u_2 u_3 + \dot{u}_1)\\ &\quad -0.13194(u_1 u_3 + \dot{u}_2) - 0.10832 \dot{u}_3 - 0.14229 \dot{u}_4\\ &\quad -0.43974 q_4 u_3 u_4 - 0.21987 q_4^2 \dot{u}_3 = 0,\\ &-1.46478 \cos q_3 - 436.295 q_4 - 0.17217 u_1 u_3 + 0.21991 q_4 {u_3}^2\\ &\quad -0.17217 \dot{u}_2 - 0.14229 \dot{u}_3 - 0.21991 \dot{u}_4 = 0.\end{aligned}$$

The kinematical equations are

$$u_3 = \dot{q}_3, \qquad u_4 = \dot{q}_4.$$

The initial conditions for the system of ordinary differential equations (dynamical equations and kinematical equations) at $t = 0$ are

$$\begin{aligned}&u_1(0) = 0, \quad u_2(0) = 0, \quad u_3(0) = 0, \quad u_4(0) = 0, \quad q_3(0) = 1,\\ &q_4(0) = 0.0001.\end{aligned}$$

The solution of the system $\{u_1, u_2, u_3, u_4 q_3, q_4\}$ is computed for the interval of time $0 < t < 1$ s, and Fig. 8.3 represents the generalized speeds u_1, u_2, u_3, and u_4 with respect to t.

9 Problems

9.1 Introduction

1. Determine the number of DOF of the planar elipsograph mechanism in Fig. 9.1. Find the analytical expression of any point P on link 2.

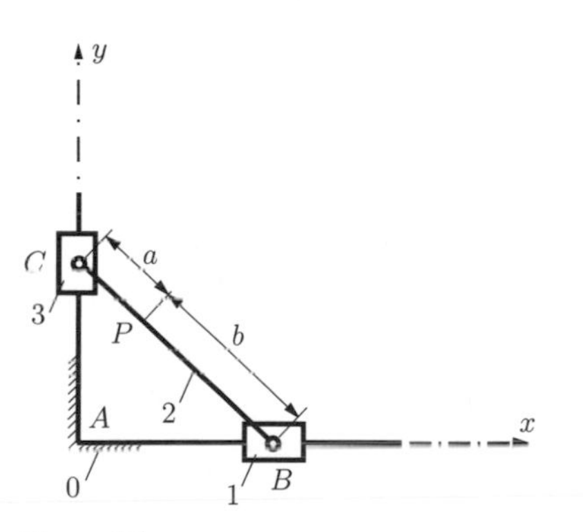

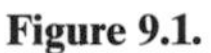

Figure 9.1.

2. Find the mobility of the planar mechanism represented in Fig. 9.2.

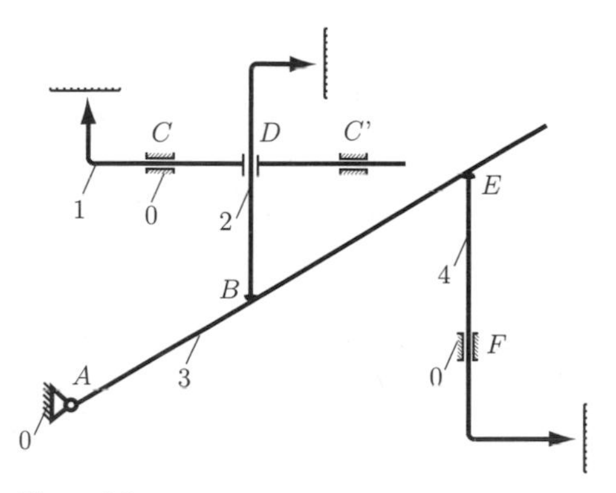

Figure 9.2.

3. Determine the family and the number of DOF for the mechanism depicted in Fig. 9.3.

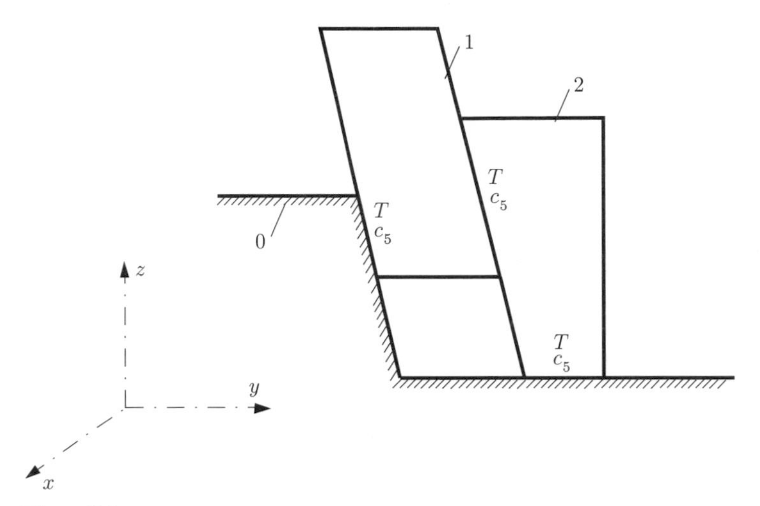

Figure 9.3.

4. Roller 2 of the mechanism in Fig. 9.4 undergoes an independent rotation about its axis that does not influence the motion of link 3. The purpose of roller 2 is to replace the sliding friction with a rolling friction. From a kinematical point of view, roller 2 is a passive element. Find the number of DOF of the mechanism.

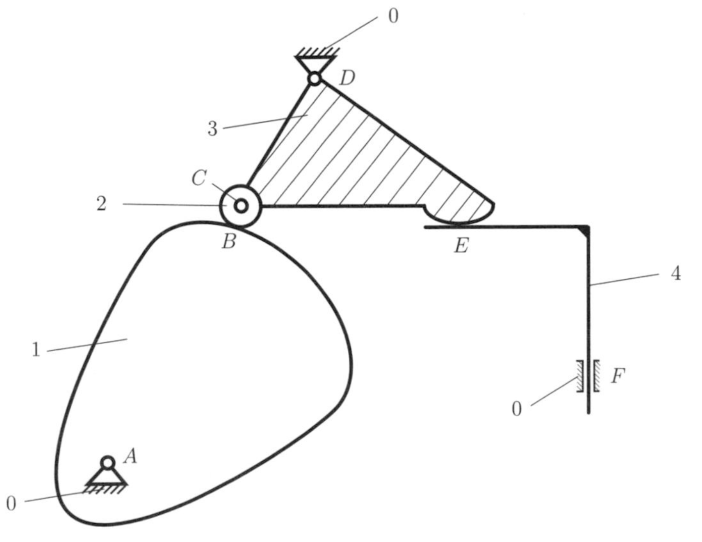

Figure 9.4.

5. Find the family and the number of DOF of the mechanism in Fig. 9.5.

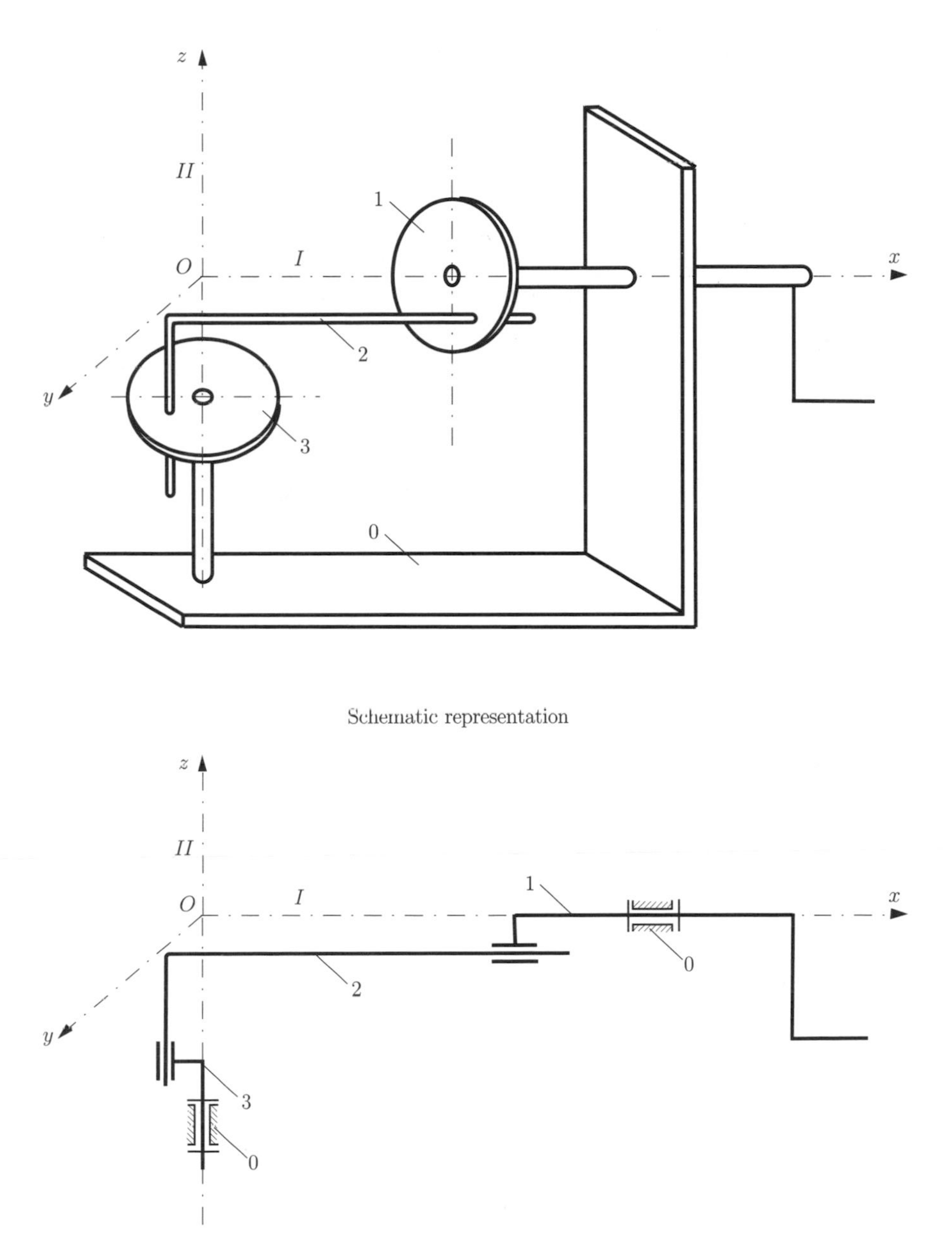

Figure 9.5.

9.2 Kinematics and Force Analysis

6. The following data are given for the mechanism shown in Fig. 9.6: $AB = CD = 0.04$ m and $AD = BC = 0.09$ m. Find the trajectory of point M located on link BC, for the case (a) $BM = MC$, and (b) $MC = 2BM$.

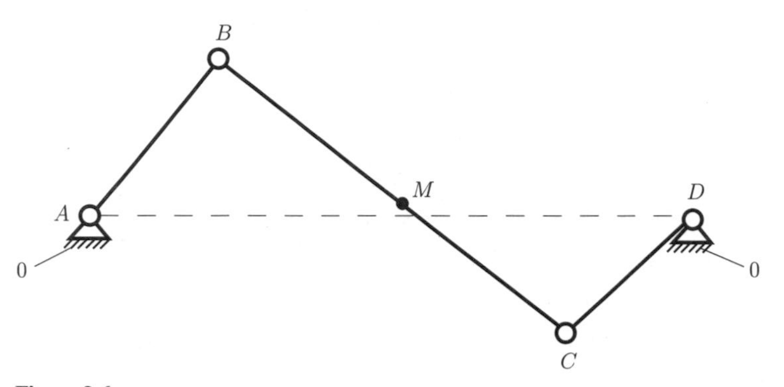

Figure 9.6.

7. The planar mechanism depicted in Fig. 9.7 has dimensions $AB = 0.03$ m, $BC = 0.065$ m, $CD = 0.05$ m, $BM = 0.09$ m, and $CM = 0.12$ m. Find the trajectory described by point M.

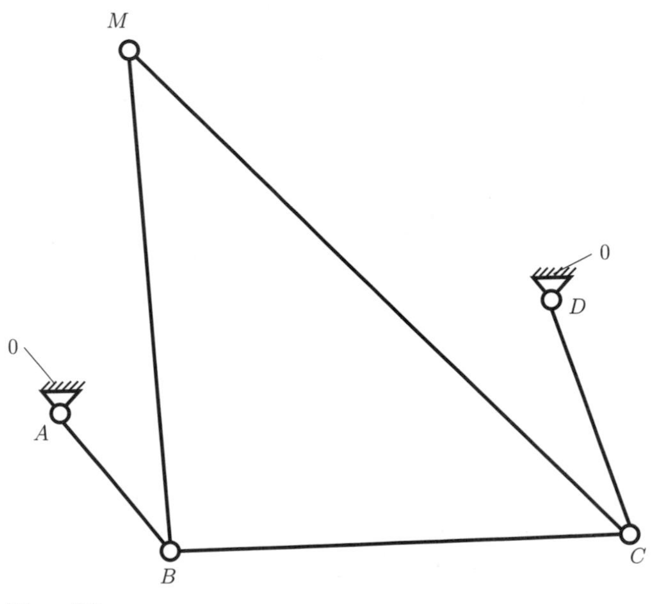

Figure 9.7.

8. The mechanism shown in Fig. 9.8 has dimensions $AB = 0.03\,\text{m}$, $BC = 0.12\,\text{m}$, $CD = 0.12$ m, $ED = 0.0$ m, $CF = 0.17$ m, $R_1 = 0.04$ m, $R_4 = 0.08$ m, $L_a = 0.025$ m, and $L_b = 0.105$ m. Find the trajectory of joint C.

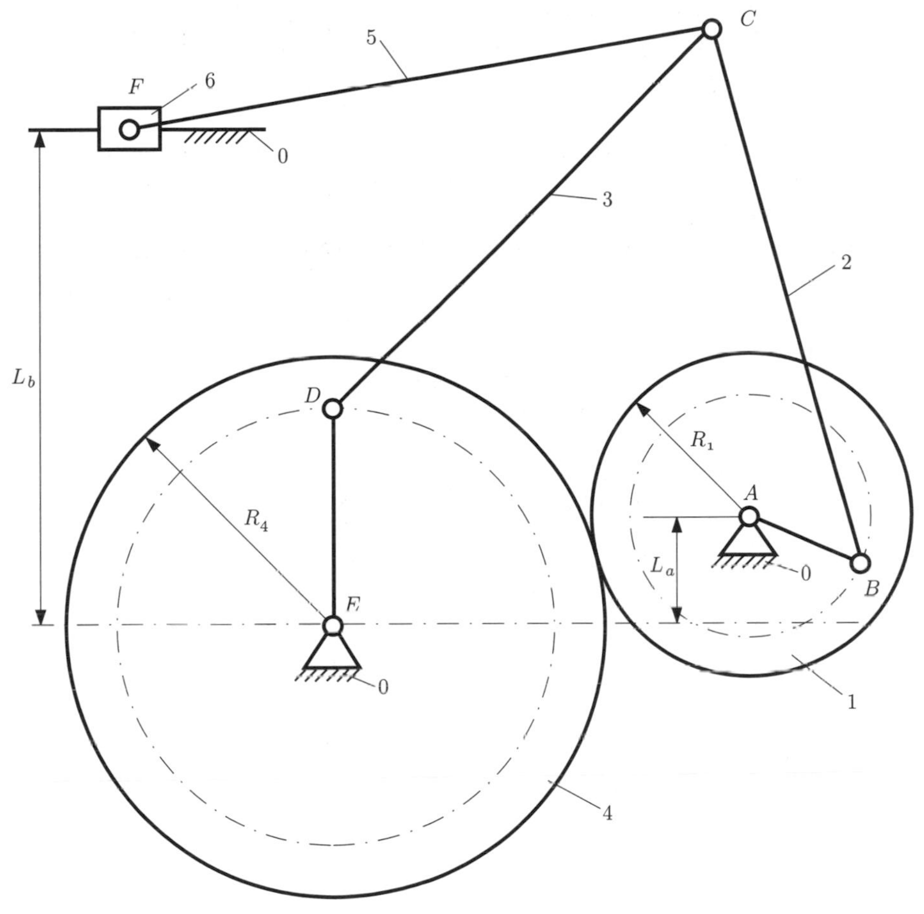

Figure 9.8.

9. **I.** The length of the links are known for the mechanism shown in Fig. 9.9. The angle of driver link 1 with the horizontal is $\phi = \phi_1$, and the angular speed of driver link 1 is $n = n_1 =$ constant. The input data for ten cases are given in the table in Fig. 9.9. Determine:

a) the family f of the mechanism and the number of DOF of the mechanism;
b) the positions of joints B, C, D, and E;
c) the linear velocities of joints B, C, D, E, and the angular velocities of links 2, 3, and 4;
d) the linear accelerations of joints B, C, D, E, and the angular accelerations of links 2, 3, and 4.

II. Links 1, 2, and 4 are homogeneous bars made of steel, each with a mass density $\rho = 7800$ kg m^3, and each with a constant cross-sectional area $A_s = 1$ cm^2. Sliders 3 and 5 have negligible dimensions and can be considered as material points with masses $m_3 = m_5 = 0.1$ kg. Find:

e) the forces and moments of inertia for each link;
f) the joint forces and the equilibrium moment if an external vertical force of magnitude $F_e = 250$ N acts on link 5 at point D.

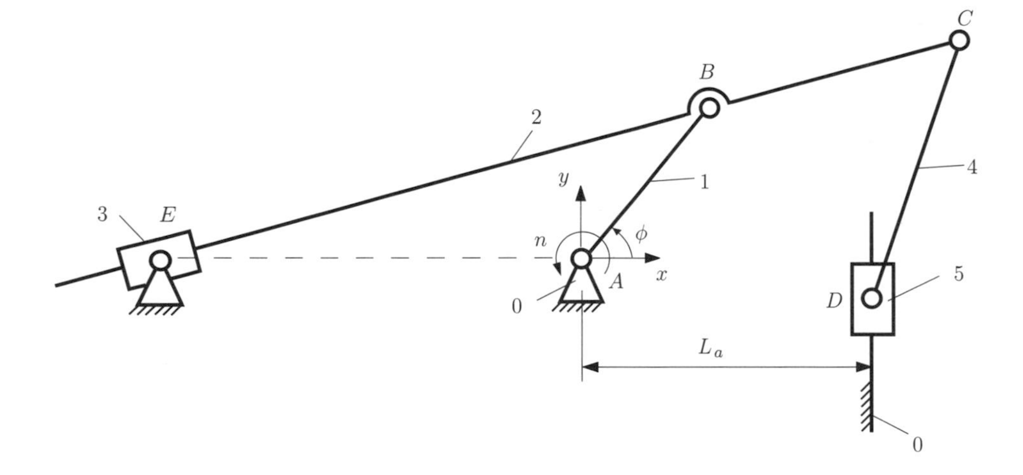

Index	AB [mm]	AE [mm]	BC [mm]	CD [mm]	L_a [mm]	ϕ [deg]	n [rpm]
1	20	50	30	60	20	30	60
2	30	70	50	80	25	60	90
3	40	100	70	120	35	120	100
4	50	120	80	150	45	145	120
5	60	150	100	170	50	150	130
6	80	200	140	220	70	210	200
7	100	250	170	300	90	240	230
8	120	300	200	350	100	330	300
9	150	350	250	400	120	15	350
10	200	500	300	600	180	45	400

Figure 9.9.

10. **I.** The mechanism shown in Fig. 9.10 has links of known length. The angle of driver link 1 with the horizontal is $\phi = \phi_1$, and the angular speed of driver link 1 is $n = n_1 =$ constant. The input data for ten cases are given in the table in Fig. 9.10. Determine:

a) the family f of the mechanism and the number of DOF of the mechanism;
b) the positions of joints B, D, E, F and the trajectory of C; the trajectory of C for a complete rotation of driver link 1, $\phi \in [0°, \ldots, 360°]$;
c) the linear velocities of the joints B, E, F, and the angular velocities of links 2, 3, and 4;
d) the linear accelerations of joints B, E, F, and the angular accelerations of links 2, 3, and 4.

II. Links 1, 2, and 4 are homogeneous bars made of steel having a mass density $\rho = 7800$ kg m^3, with a constant cross-sectional area $A_s = 2$ cm^2. Sliders 3 and 5 have negligible dimensions and can be considered as material points with masses $m_3 = m_5 = 0.2$ kg. Find:

e) the forces and moments of inertia for each link;
f) the joint forces and equilibrium moment if an external vertical force of magnitude $F_e = 500$ N acts on link 5 at point E.

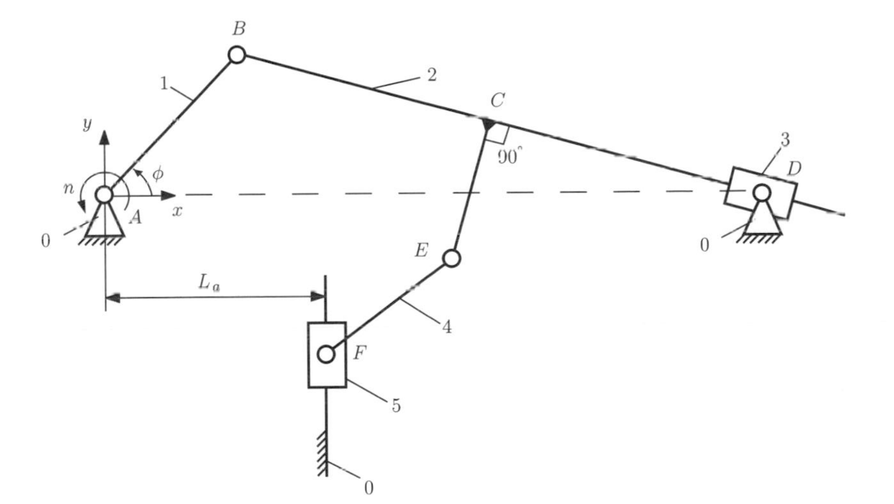

Index	AB [mm]	AD [mm]	BC [mm]	CE [mm]	EF [mm]	L_a [mm]	ϕ [deg]	n [rpm]
1	40	150	100	30	120	90	30	200
2	60	200	140	50	170	130	120	300
3	80	250	180	60	200	170	300	400
4	100	350	240	70	300	240	210	500
5	120	400	300	80	400	280	150	600
6	150	500	350	100	400	380	60	1100
7	180	600	400	100	500	430	240	1000
8	200	700	550	150	600	520	330	900
9	220	800	650	180	700	600	120	800
10	250	900	700	180	800	650	30	700

Figure 9.10.

11. **I.** The dimensions of the links are given for the mechanism shown in Fig. 9.11. The angle of driver link 1 with the horizontal is $\phi = \phi_1$, and the angular speed of driver link 1 is $n = n_1 =$ constant. The input data for ten cases are given in the table in Fig. 9.11. Determine:

a) the family f of the mechanism and the number of DOF;
b) the positions of joints B and D;
c) the linear velocities of joints B, D, and the angular velocities of links 2, 3, 4, and 5;
d) the linear accelerations of joints B, D, and the angular accelerations of links 2, 3, 4, and 5.

II. Links 1, 3, and 5 are homogeneous bars made of aluminum having a constant cross-sectional area $A_s = 1$ cm^2. Sliders 2 and 4 have negligible dimensions and can be considered as material points with masses $m_2 = m_4 = 0.1$ kg. Find:

e) the forces and moments of inertia for each link;
f) the joint forces and the equilibrium moment if an external torque of magnitude $M_e = 500$ N m acts on link 5 at point E.

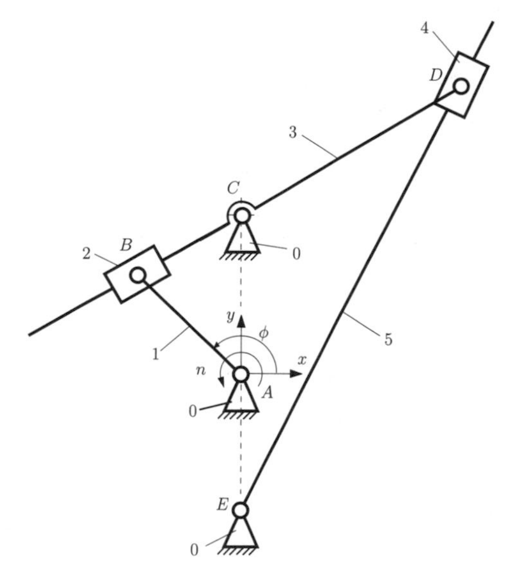

Index	AB [mm]	AC [mm]	AE [mm]	CD [mm]	ϕ [deg]	n [rpm]
1	140	60	250	150	30	500
2	180	100	300	200	120	600
3	220	100	400	250	210	700
4	250	150	400	300	150	800
5	50	20	100	80	30	900
6	80	40	150	100	120	1000
7	100	50	200	120	330	1100
8	120	60	250	140	150	1200
9	150	70	300	180	60	1300
10	200	100	350	250	120	1400

Figure 9.11.

12. **I.** The lengths of the links are known for the mechanism shown in Fig. 9.12. The dimensions of the links are given. The angle of driver link 1 with the horizontal is $\phi = \phi_1$, and the angular speed of driver link 1 is $n = n_1 =$ constant. The input data for ten cases are given in the table in Fig. 9.12. Find:

a) the family f of the mechanism and the number of DOF of the mechanism;
b) the positions of joints B, C, D, and F (Link 2 is a plate);
c) the linear velocities of joints B, C, D, F and the angular velocities of links 2, 3, 4, and 5;
d) the linear accelerations of joints B, C, D, F and the angular accelerations of links 2, 3, 4, and 5.

II. Links 1, 3, and 4 are homogeneous bars made of steel with a constant cross-sectional area $A_s = 1\ \text{cm}^2$. Link 2 is a homogeneous plate made of steel with a width of 1 cm. Slider 5 has negligible dimensions and can be considered as a material point with a mass $m_5 = 0.2$ kg. Find:

e) the forces and moments of inertia for each link;
f) the joint forces and equilibrium moment if an external force of magnitude $F_e =$ 500 N acts on link 5 at point F.

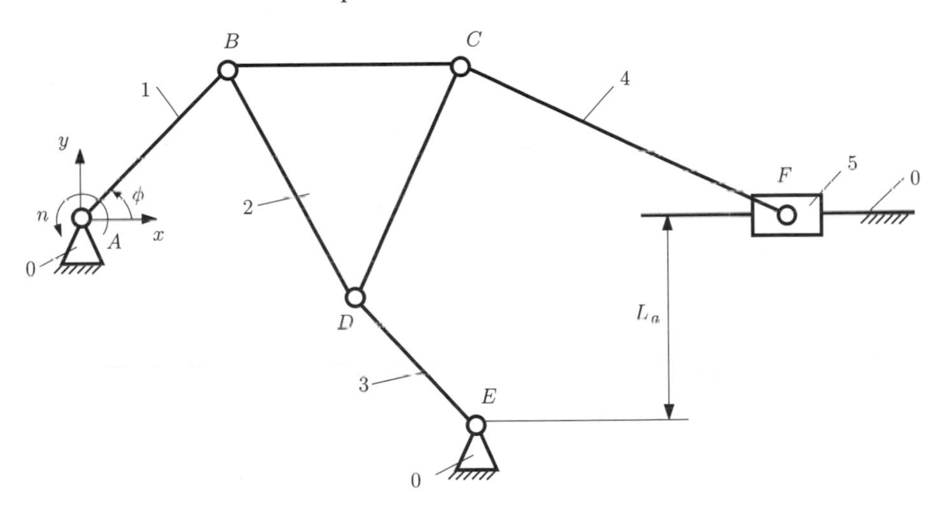

Index	AB [mm]	BD [mm]	DE [mm]	AE [mm]	BC [mm]	CD [mm]	CF [mm]	L_a [mm]	ϕ [deg]	n [rpm]
1	250	670	420	640	240	660	850	170	30	30
2	120	320	200	320	110	300	400	80	120	35
3	150	400	250	380	140	400	500	100	210	40
4	60	160	100	150	55	150	250	40	300	45
5	240	650	400	640	220	600	800	150	60	50
6	100	270	170	260	100	260	350	70	150	55
7	220	580	360	570	230	550	770	160	240	60
8	160	430	270	410	160	410	600	110	330	65
9	200	530	330	520	210	500	700	150	30	70
10	180	480	300	470	170	450	650	120	220	75

Figure 9.12.

13. **I.** The lengths of the links are known for the mechanism shown in Fig. 9.13. The angle of driver link 1 with the horizontal is $\phi = \phi_1$, and the angular speed of driver link 1 is $n = n_1 =$ constant. The input data for ten cases are given in the table in Fig. 9.13. Determine:

a) the family f and the number of DOF;
b) the positions of the joints; the trajectory of the center of mass of link 2 for a complete rotation of driver link 1, $\phi \in [0°, \ldots, 360°]$;
c) the linear velocities of the joints and the angular velocities of the links;
d) the linear accelerations of the joints and the angular accelerations of the links.

II. Links 1, 2, 4, and 5 are homogeneous bars made of aluminum having a constant cross-sectional area $A_s = 0.5$ cm^2. Slider 3 has negligible dimensions and can be considered as a material point with a mass $m_3 = 0.01$ kg. Find:

e) the forces and moments of inertia for each link;
f) the joint forces and the equilibrium moment if an external torque of magnitude $M_e = 500$ N m acts on link 5 at point F.

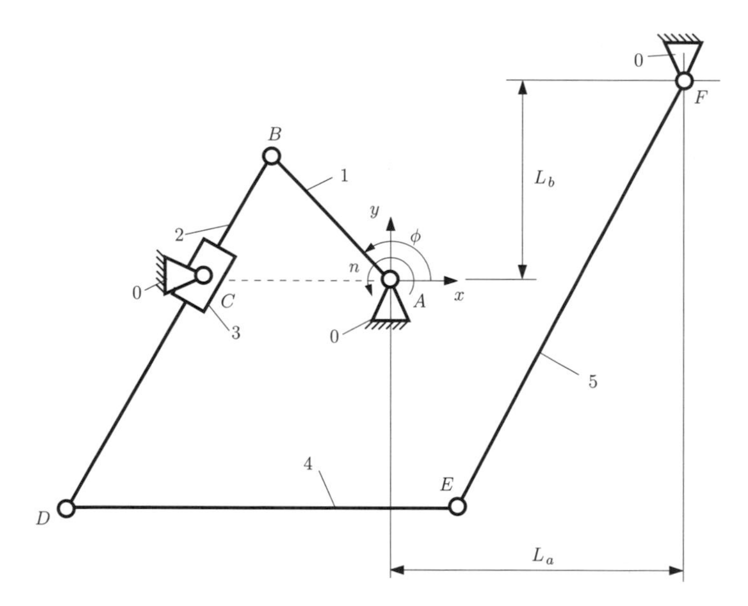

Index	AB [mm]	AC [mm]	BD [mm]	DE [mm]	EF [mm]	L_a [mm]	L_b [mm]	ϕ [deg]	n [rpm]
1	120	60	240	330	190	300	70	30	60
2	50	25	100	140	80	130	30	150	100
3	200	100	400	550	300	500	100	210	70
4	110	55	220	300	175	275	65	150	110
5	150	75	300	400	240	370	90	120	30
6	250	120	500	675	400	620	140	135	120
7	200	110	450	610	330	560	110	300	90
8	100	50	200	280	160	250	60	240	50
9	170	85	340	470	270	430	100	330	40
10	75	37	150	200	120	180	45	60	80

Figure 9.13.

14. **I.** The dimensions of the links are given for the mechanism shown in Fig. 9.14. The angle of driver link 1 with the horizontal is $\phi = \phi_1$, and the angular speed of driver link 1 is $n = n_1 =$ constant. The input data for ten cases are given in the table in Fig. 9.14. Determine:

   a) the family f and the number of DOF;
   b) the positions of the joints; the trajectory of the center of mass of link 2 for a complete rotation of driver link 1;
   c) the linear velocities of the joints and the angular velocities of the links;
   d) the linear accelerations of the joints and the angular accelerations of the links.

   II. Links 1, 2, 3, and 4 are homogeneous bars made of steel each having a constant cross-sectional area $A_s = 1$ cm^2. Slider 5 has negligible dimensions and can be considered as a material point with a mass $m_5 = 0.1$ kg. Find:

   e) the forces and moments of inertia for each link;
   f) the joint forces and equilibrium moment if an external force $F_e = 800$ N acts on link 5 at point F.

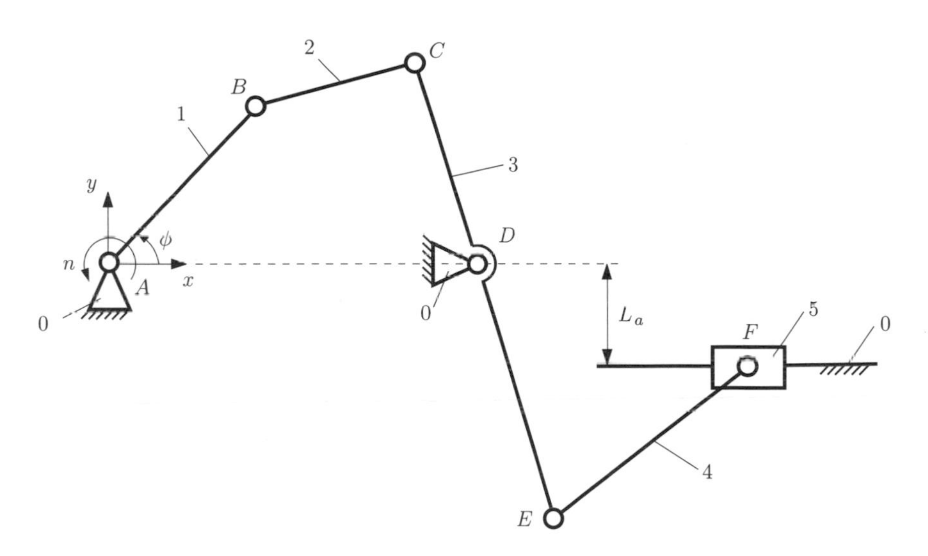

Index	AB [mm]	BC [mm]	AD [mm]	DE [mm]	EF [mm]	CD [mm]	L_a [mm]	ϕ [deg]	n [rpm]
1	100	260	240	80	250	140	20	240	200
2	180	470	430	180	400	270	70	120	220
3	150	400	360	130	400	210	40	30	250
4	250	650	600	200	600	350	100	330	100
5	80	210	190	60	180	120	20	210	120
6	160	420	380	130	350	240	20	150	140
7	200	500	460	160	500	280	50	60	150
8	90	240	220	130	350	140	30	300	160
9	50	130	120	45	150	70	20	240	180
10	220	570	530	180	500	300	80	120	100

Figure 9.14.

15. **I.** The lengths of the links are known for the mechanism shown in Fig. 9.15. The angle of driver link 1 with the horizontal is $\phi = \phi_1$, and the angular speed of driver link 1 is $n = n_1 =$ constant. The input data for ten cases are given in the table in Fig. 9.15. Determine:

a) the family f and the number of DOF;
b) the positions of the joints; the trajectory of the center of mass of link 2 for a complete rotation of driver link 1, $\phi \in [0°, \ldots, 360°]$;
c) the linear velocities of the joints and the angular velocities of the links;
d) the linear accelerations of the joints and the angular accelerations of the links.

II. Links 1, 2, and 5 are homogeneous bars made of steel each with a constant cross-sectional area $A_s = 0.5\ \text{cm}^2$. Sliders 3 and 4 have negligible dimensions and can be considered as material points each with masses $m_3 = m_4 = 0.1$ kg. Find:

e) the forces and moments of inertia for each link;
f) the joint forces and the equilibrium moment if an external torque of magnitude $M_e = 500$ N m acts on link 5 at point E.

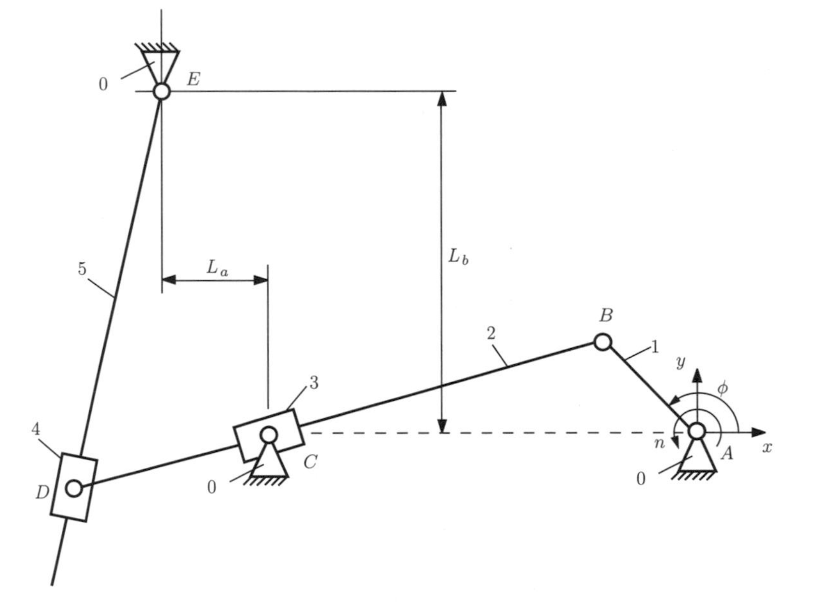

Index	AB [mm]	AC [mm]	BD [mm]	L_a [mm]	L_b [mm]	ϕ [deg]	n [rpm]
1	150	450	700	100	200	60	60
2	200	600	1000	150	250	120	55
3	250	800	1200	180	300	210	50
4	50	160	250	30	60	300	45
5	60	130	240	40	70	30	40
6	70	250	400	50	80	150	35
7	80	250	400	50	90	240	30
8	90	300	450	60	100	330	65
9	100	300	500	70	120	30	70
10	120	300	500	80	150	150	75

Figure 9.15.

16. **I.** The dimensions of the links are given for the mechanism shown in Fig. 9.16. The angle of driver link 1 with the horizontal is $\phi = \phi_1$, and the angular speed of driver link 1 is $n = n_1 =$ constant. The input data for ten cases are given in the table in Fig. 9.16. Find:

a) the family f of the mechanism and the number of DOF of the mechanism;
b) the positions of the joints;
c) the linear velocities of the joints and the angular velocities of the links;
d) the linear accelerations of the joints and the angular accelerations of the links.

II. Links 1, 2, and 4 are homogeneous bars made of aluminum each with a constant cross-sectional area $A_s = 1\ \text{cm}^2$. Sliders 3 and 5 have negligible dimensions and can be considered as material points each with masses $m_3 = m_5 = 0.1$ kg. Find:

e) the forces and moments of inertia for each link;
f) the joint forces and equilibrium moment if an external force of magnitude $F_e = 500$ N acts on link 5 at point D.

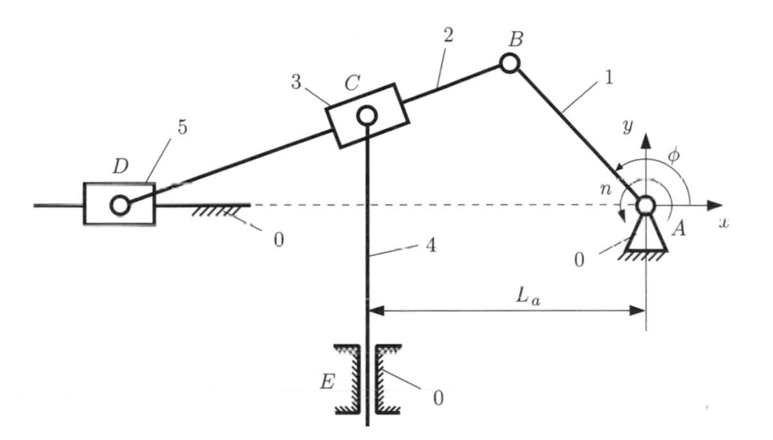

Index	AB [mm]	BD [mm]	L_a [mm]	ϕ [deg]	n [rpm]
1	180	700	210	210	400
2	250	900	300	30	500
3	120	400	150	240	600
4	150	500	180	60	700
5	170	600	200	210	800
6	220	800	250	30	450
7	200	650	230	240	550
8	240	850	270	30	650
9	80	300	100	210	750
10	100	350	120	60	850

Figure 9.16.

17. I. The lengths of the links are known for the mechanism shown in Fig. 9.17. The angle of driver link 1 with the horizontal is ϕ, and the angular speed of driver link 1 is $n =$ constant. The input data for ten cases are given in the table in Fig. 9.17. Determine:

a) the family f and the number of DOF;
b) the positions of the joints; the trajectory of the center of mass of link 4 for a complete rotation of driver link 1;
c) the linear velocities of the joints and the angular velocities of the links;
d) the linear accelerations of the joints and the angular accelerations of the links.

II. Links 1, 2, 4, and 5 are homogeneous bars made of steel each with a constant cross-sectional area $A_s = 1.5$ cm^2. Slider 3 has negligible dimensions and can be considered as a material point with mass $m_3 = 0.1$ kg. Find:

e) the forces and moments of inertia for each link;
f) the joint forces and equilibrium moment if an external torque of magnitude $M_e = 900$ N m acts on link 5 at point F.

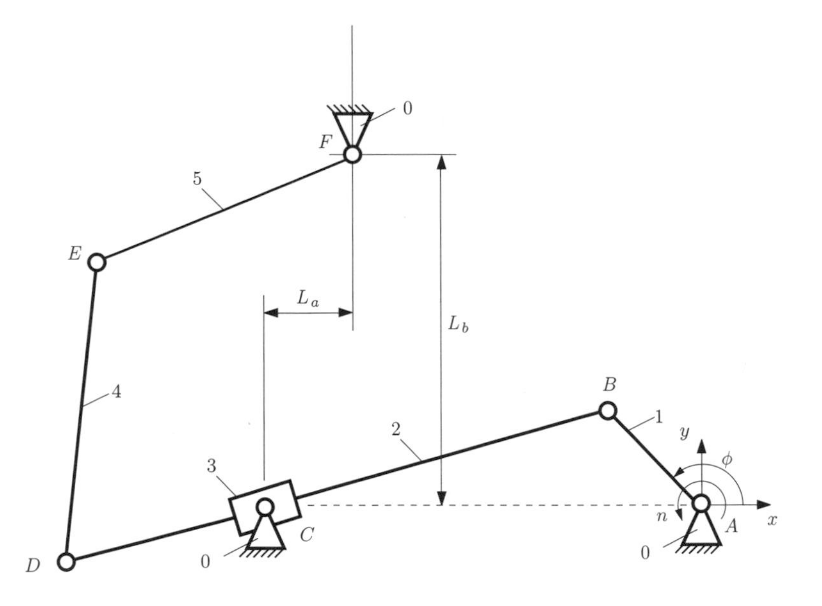

Index	AB [mm]	AC [mm]	BD [mm]	DE [mm]	EF [mm]	L_a [mm]	L_b [mm]	ϕ [deg]	n [rpm]
1	100	240	400	200	135	35	170	30	20
2	150	350	600	300	200	55	250	120	30
3	200	500	800	400	270	70	300	300	40
4	20	50	150	40	27	7	30	210	50
5	250	600	950	500	340	90	420	60	60
6	30	75	120	60	40	10	50	150	70
7	40	100	160	80	55	13	65	330	60
8	50	120	190	100	65	17	85	240	90
9	60	180	250	120	80	20	100	30	100
10	80	200	300	160	110	25	130	120	110

Figure 9.17.

18. **I.** The dimensions of the links are given for the mechanism in Fig. 9.18. The angle of driver link 1 with the horizontal is ϕ, and the angular speed of driver link 1 is $n =$ constant. The input data for ten cases are given in the table in Fig. 9.18. Find:

a) the family f of the mechanism and the number of DOF of the mechanism;
b) the positions of the joints;
c) the linear velocities of the joints and the angular velocities of the links;
d) the linear accelerations of the joints and the angular accelerations of the links.

II. Links 1, 2, 3, and 4 are homogeneous bars made of aluminum each with a constant cross-sectional area $A_s = 0.1$ cm^2. Slider 5 has negligible dimensions and can be considered as a material point with mass $m_5 = 0.1$ kg. Find:

e) the forces and moments of inertia for each link;
f) the joint forces and the equilibrium moment if an external force of magnitude $F_e = 600$ N acts on link 5 at F.

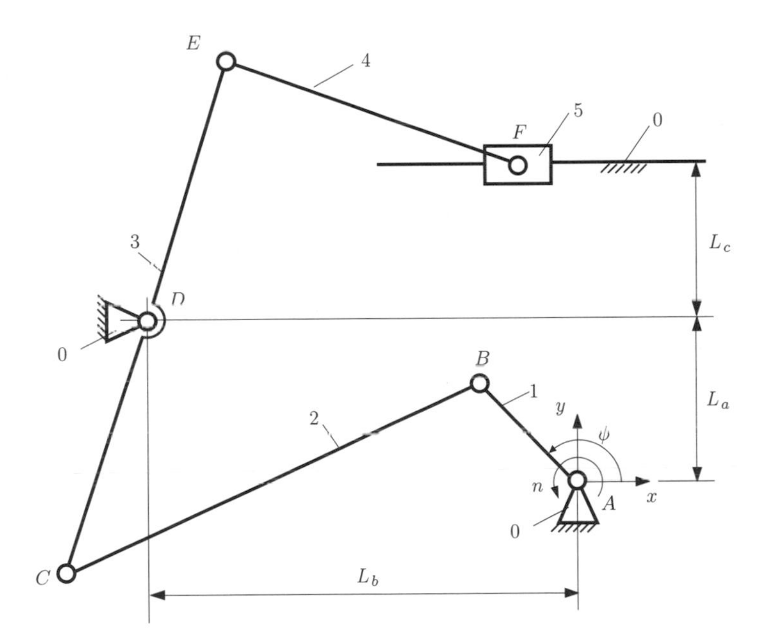

Index	AB [mm]	BC [mm]	$CD = DE$ [mm]	EF [mm]	L_a [mm]	L_b [mm]	L_c [mm]	ϕ [deg]	n [rpm]
1	120	450	180	300	150	450	140	300	900
2	150	550	220	400	180	530	180	210	1000
3	200	750	300	500	250	750	250	120	1100
4	250	940	380	700	310	930	300	30	1200
5	30	110	45	80	40	110	40	330	1300
6	40	150	60	120	50	145	50	240	1400
7	50	185	75	160	60	180	60	150	1500
8	60	240	90	180	75	225	70	60	1600
9	80	300	120	200	100	300	80	300	1700
10	100	370	150	250	120	360	90	210	1800

Figure 9.18.

19. **I.** The dimensions of the links are given for the mechanism in Fig. 9.19. The angle of driver link 1 with the horizontal is ϕ, and the angular speed of driver link 1 is $n =$ constant. The input data for ten cases are given in the table in Fig. 9.19. Find:

a) the family f of the mechanism and the number of DOF of the mechanism;
b) the positions of the joints;
c) the linear velocities of the joints and the angular velocities of the links;
d) the linear accelerations of the joints and the angular accelerations of the links.

II. Links 1, 2, 3, and 4 are homogeneous bars made of steel each with a constant cross-sectional area $A_s = 1$ cm^2. Slider 5 has negligible dimensions and can be considered as a material point with a mass $m_5 = 0.3$ kg. Find:

e) the forces and moments of inertia for each link;
f) the joint forces and the equilibrium moment if an external force of magnitude $F_e = 600$ N acts on link 5 at F.

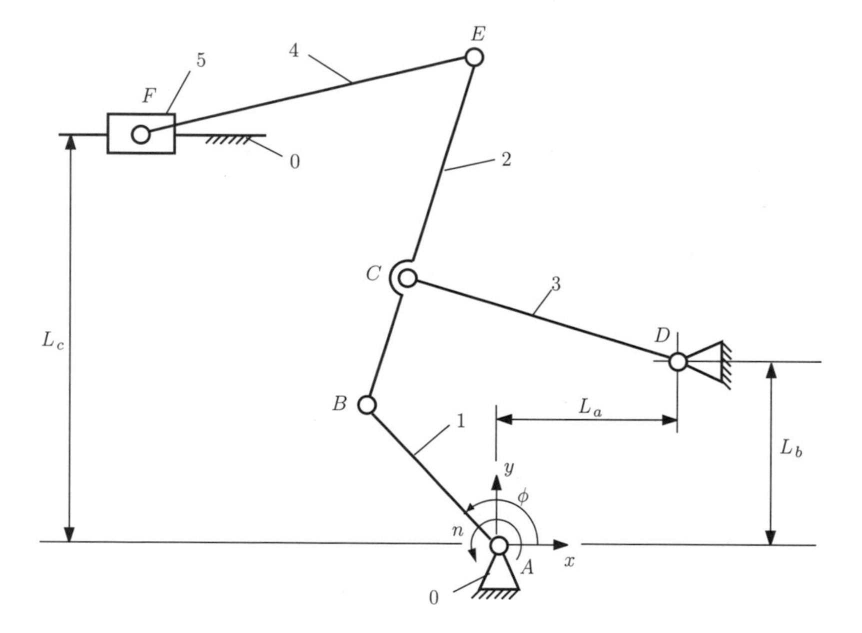

Index	AB [mm]	BC [mm]	CE [mm]	CD [mm]	EF [mm]	L_a [mm]	L_b [mm]	L_c [mm]	ϕ [deg]	n [rpm]
1	140	650	250	400	350	370	550	700	150	30
2	250	1200	400	800	700	650	1000	1200	120	70
3	120	550	180	350	300	320	480	600	30	100
4	200	900	300	600	600	500	800	1100	210	40
5	180	800	270	550	530	450	700	1000	240	60
6	100	450	150	300	250	250	400	500	60	80
7	160	700	250	500	480	400	600	800	150	50
8	220	1000	350	650	600	550	900	1200	30	120
9	150	670	220	450	320	330	600	750	210	110
10	90	400	120	250	210	220	350	460	240	90

Figure 9.19.

20. **I.** The dimensions of the links are given for the mechanism in Fig. 9.20. The angle of driver link 1 with the horizontal is ϕ, and the angular speed of driver link 1 is n = constant. The input data for ten cases are given in the table in Fig. 9.20. Find:

a) the family f and the number of DOF;
b) the positions of the joints;
c) the linear velocities of the joints and the angular velocities of the links;
d) the linear accelerations of the joints and the angular accelerations of the links.

II. Links 1, 2, 3, and 4 are homogeneous bars made of steel each with a constant cross-sectional area $A_s = 0.5$ cm^2. Slider 5 has negligible dimensions and can be considered as a material point with mass $m_5 = 0.1$ kg. Find:

e) the forces and moments of inertia for each link;
f) the joint forces and the equilibrium moment if an external force of magnitude $F_e = 500$ N acts on link 5 at E.

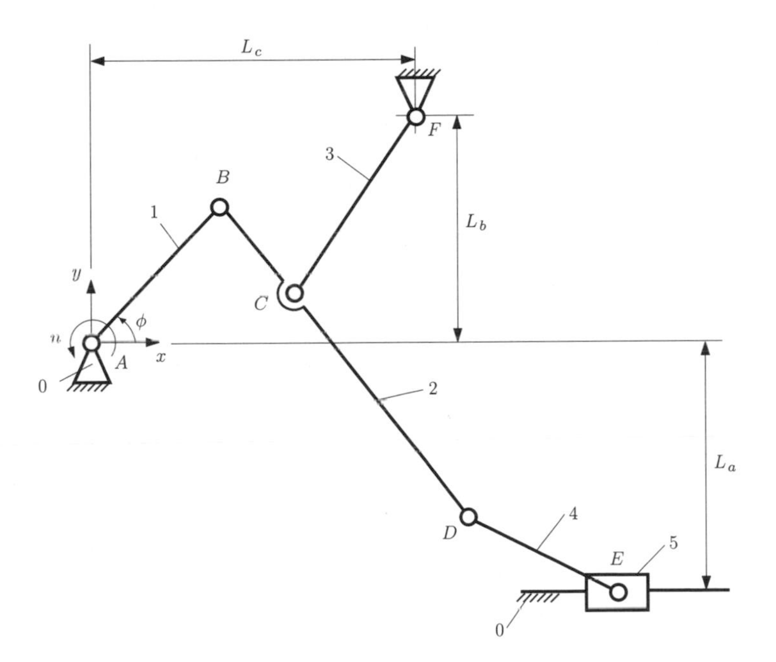

Index	AB [mm]	BC [mm]	CF [mm]	CD [mm]	DE [mm]	L_a [mm]	L_b [mm]	L_c [mm]	ϕ [deg]	n [rpm]
1	60	160	150	60	180	65	120	210	30	60
2	100	270	260	90	300	120	200	350	150	100
3	180	520	470	165	540	210	360	630	210	70
4	200	540	520	190	600	240	400	700	135	110
5	150	400	400	150	450	150	300	500	120	30
6	50	135	130	50	150	55	100	170	120	120
7	80	220	210	75	230	90	160	280	300	90
8	250	670	660	240	750	270	500	850	240	120
9	160	430	420	150	450	180	320	560	330	40
10	120	320	300	120	370	130	250	420	60	80

Figure 9.20.

21. **I.** The dimensions of the links are given for the mechanism in Fig. 9.21. The angle of driver link 1 with the horizontal is ϕ, and the angular speed of driver link 1 is $n =$ constant. The input data for ten cases are given in the table in Fig. 9.21. Find:

a) the family f of the mechanism and the number of DOF of the mechanism;
b) the positions of the joints;
c) the linear velocities of the joints and the angular velocities of the links;
d) the linear accelerations of the joints and the angular accelerations of the links.

II. Links 1, 2, and 4 are homogeneous bars made of aluminum each with a constant cross-sectional area $A_s = 1$ cm^2. Sliders 3 and 5 have negligible dimensions and can be considered as material points each with masses $m_3 = m_5 = 0.1$ kg. Find:

e) the forces and moments of inertia for each link;
f) the joint forces and the equilibrium moment if an external force of magnitude $F_e = 500$ N acts on link 3 at point C.

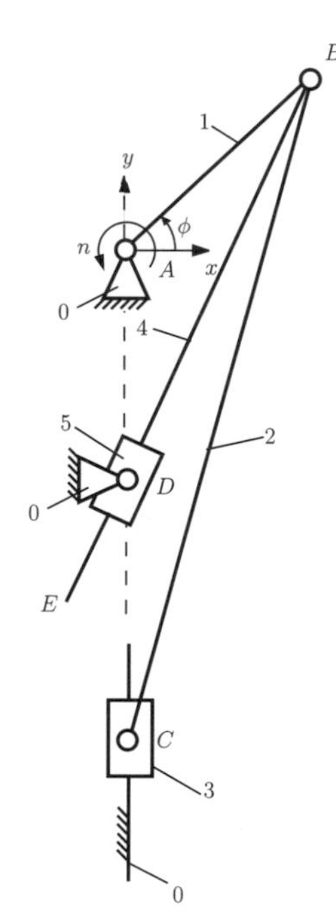

Index	AB [mm]	BC [mm]	AD [mm]	BE [mm]	ϕ [deg]	n [rpm]
1	20	50	25	60	30	200
2	40	100	50	110	150	250
3	60	150	70	170	210	300
4	80	200	90	220	150	350
5	100	250	110	300	120	400
6	250	630	280	650	120	450
7	200	550	230	550	300	500
8	180	450	200	500	240	550
9	150	380	170	400	330	600
10	120	300	130	350	60	650

Figure 9.21.

22. **I.** The dimensions of the links are given for the mechanism in Fig. 9.22. The angle of driver link 1 with the horizontal is $\phi = \phi_1$, and the angular speed of driver link 1 is $n = n_1$ = constant. The input data for ten cases are given in the table in Fig. 9.22. Find:

    a) the family f of the mechanism and the number of DOF of the mechanism;
    b) the positions of the joints;
    c) the linear velocities of the joints and the angular velocities of the links;
    d) the linear acceleration of the joints and the angular accelerations of the links.

    II. Links 1, 3, and 4 are homogeneous bars made of steel each having a constant cross-sectional area $A_s = 1\ \text{cm}^2$. Link 2 is a homogeneous plate made of steel with a width of 0.5 cm. Slider 5 has negligible dimensions and can be considered as a material point with mass $m_5 = 0.1$ kg. Find:

    e) the forces and moments of inertia for each link;
    f) the joint forces and the equilibrium moment if an external force of magnitude $F_e = 500$ N acts on link 5 at point F.

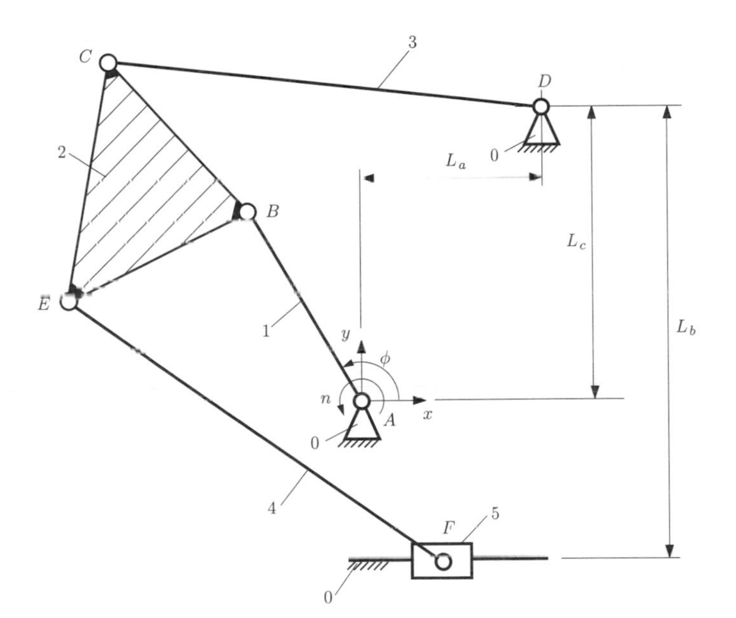

Index	AB [mm]	BC [mm]	BE [mm]	CE [mm]	CD [mm]	EF [mm]	L_a [mm]	L_b [mm]	L_c [mm]	ϕ [deg]	n [rpm]
1	150	300	600	850	330	1200	350	200	100	30	140
2	100	200	400	600	220	800	250	150	100	120	100
3	90	240	400	600	220	900	250	150	100	210	50
4	80	150	300	450	170	600	200	150	50	300	120
5	70	150	300	400	150	650	150	150	50	60	130
6	60	120	250	350	140	250	150	80	50	150	150
7	50	100	200	300	120	200	150	80	20	240	110
8	40	80	150	200	100	180	120	80	20	330	80
9	30	60	120	160	70	150	80	80	20	30	130
10	20	50	100	150	60	100	80	80	20	120	70

Figure 9.22.

23. **I.** The dimensions of the links are given for the mechanism in Fig. 9.23. The angle of driver link 1 with the horizontal is $\phi = \phi_1$, and the angular speed of driver link 1 is $n = n_1 =$ constant. The input data for ten cases are given in the table in Fig. 9.23. Find:

a) the family f and the number of DOF of the mechanism;
b) the positions of the joints;
c) the linear velocities of the joints and the angular velocities of the links;
d) the linear accelerations of the joints and the angular accelerations of the links.

II. Links 1, 3, and 4 are homogeneous bars made of aluminum with a constant cross-sectional area $A_s = 1$ cm^2. Sliders 2 and 5 have negligible dimensions and can be considered as material points with masses $m_2 = m_5 = 0.5$ kg. Find:

e) the forces and moments of inertia for each link;
f) the joint forces and the equilibrium moment if an external force of magnitude $F_e = 500$ N acts on link 5 at E.

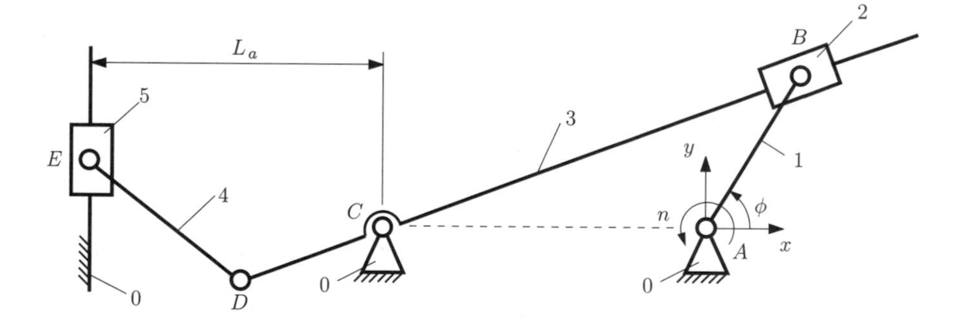

Index	AB [mm]	AC [mm]	CD [mm]	DE [mm]	L_a [mm]	ϕ [deg]	n [rpm]
1	150	220	280	200	230	60	100
2	200	300	500	250	400	120	40
3	180	300	400	200	360	30	90
4	140	200	350	180	300	150	60
5	80	120	250	100	200	60	110
6	250	350	600	360	500	150	30
7	160	250	400	150	350	210	90
8	100	150	220	120	200	30	120
9	180	260	400	220	300	120	70
10	120	180	250	200	180	240	50

Figure 9.23.

24. I. The dimensions of the links are given for the mechanism in Fig. 9.24. The angle of driver link 1 with the horizontal is $\phi = \phi_1$, and the angular speed of driver link 1 is $n = n_1$ = constant. The input data for ten cases are given in the table in Fig. 9.24. Find:

a) the family f and the number of DOF of the mechanism;
b) the positions of the joints;
c) the linear velocities of the joints and the angular velocities of the links;
d) the linear accelerations of the joints and the angular accelerations of the links.

II. Links 1, 3, and 4 are homogeneous bars made of steel each having a constant cross-sectional area $A_s = 0.5$ cm^2. Sliders 2 and 5 have negligible dimensions and can be considered as material points each having masses $m_2 = m_5 = 0.1$ kg. Find:

e) the forces and moments of inertia for each link;
f) the joint forces and the equilibrium moment if an external force of magnitude $F_e = 500$ N acts on link 5 at point E.

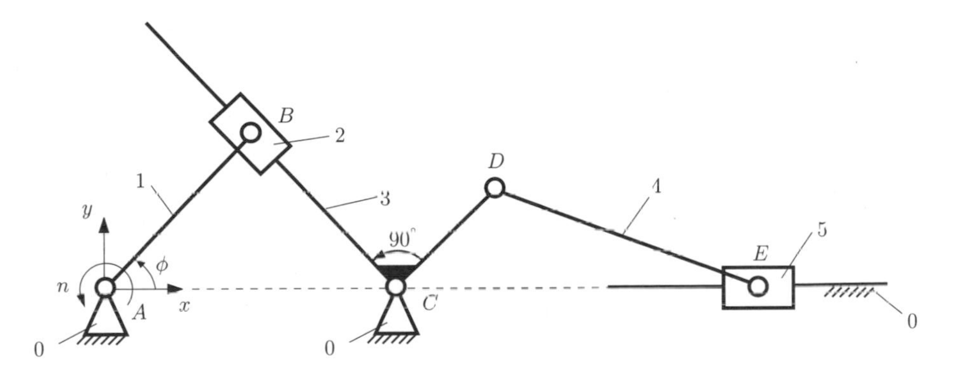

Index	AB [mm]	AC [mm]	CD [mm]	DE [mm]	ϕ [deg]	n [rpm]
1	200	60	200	500	30	80
2	160	90	150	400	120	70
3	80	40	100	300	210	60
4	250	100	280	800	150	50
5	100	50	150	400	240	40
6	240	70	200	550	120	30
7	60	30	70	200	60	90
8	150	50	180	450	150	100
9	220	80	200	600	60	110
10	180	80	200	500	150	120

Figure 9.24.

25. **I.** The dimensions of the links are given for the mechanism in Fig. 9.25. The angle of driver link 1 with the horizontal is $\phi = \phi_1$, and the angular speed of driver link 1 is $n = n_1 =$ constant. The input data for ten cases are given in the table in Fig. 9.25. Find:

a) the family f and the number of DOF of the mechanism;
b) the positions of the joints;
c) the linear velocities of the joints and the angular velocities of the links;
d) the linear accelerations of the joints and the angular accelerations of the links.

II. Links 1, 3, and 4 are homogeneous bars made of steel each with a constant cross-sectional area $A_s = 1$ cm^2. Sliders 2 and 5 have negligible dimensions and can be considered as material points each with masses $m_2 = m_5 = 1$ kg. Find:

e) the forces and moments of inertia for each link;
f) the joint forces and the equilibrium moment if an external force of magnitude $F_e = 700$ N acts on link 5 at point D.

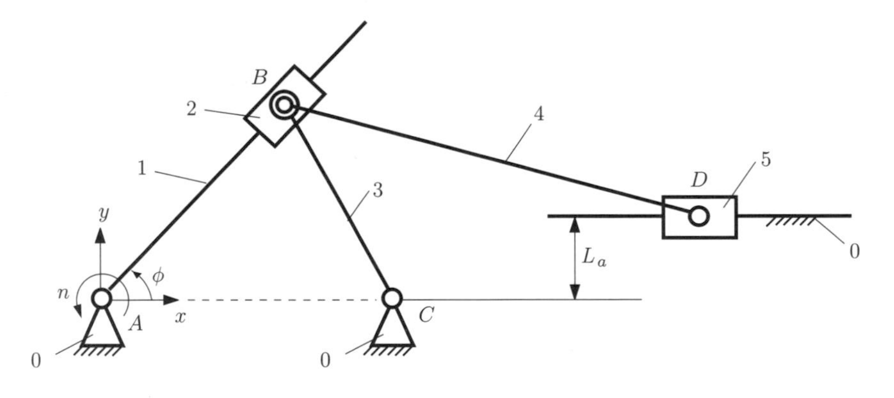

Index	AC [mm]	CB [mm]	BD [mm]	L_a [mm]	ϕ [deg]	n [rpm]
1	250	400	1500	50	60	600
2	80	150	400	20	330	1200
3	200	300	1000	50	30	800
4	150	300	900	100	210	1000
5	50	100	300	20	30	1500
6	180	300	1100	60	330	1100
7	100	150	600	30	150	1300
8	200	350	1000	120	120	900
9	100	180	600	40	60	1000
10	120	200	700	30	300	1400

Figure 9.25.

26. **I.** The dimensions of the links are given for the mechanism in Fig. 9.26. The angle of driver link 1 with the horizontal is $\phi = \phi_1$, and the angular speed of driver link 1 is $n = n_1 =$ constant. The input data for ten cases are given in the table in Fig. 9.26. Find:

a) the family f and the number of DOF of the mechanism;
b) the positions of the joints;
c) the linear velocities of the joints and the angular velocities of the links;
d) the linear accelerations of the joints and the angular accelerations of the links.

II. Links 1, 3, and 5 are homogeneous bars made of steel each with a constant cross-sectional area $A_s = 1$ cm^2. Sliders 2 and 4 have negligible dimensions and can be considered as material points each with masses $m_2 = m_4 = 0.5$ kg. Find:

e) the forces and moments of inertia for each link;
f) the joint forces and the equilibrium moment if an external torque of magnitude $M_e = 700$ N m acts on link 5 at point A.

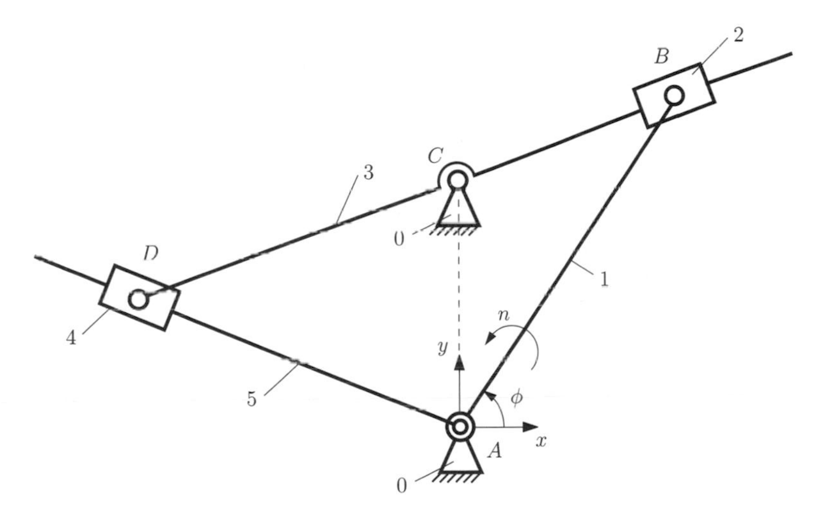

Index	AB [mm]	AC [mm]	CD [mm]	ϕ [deg]	n [rpm]
1	160	90	160	30	300
2	180	90	200	120	180
3	80	60	70	240	220
4	140	60	140	150	400
5	250	100	280	60	250
6	120	80	100	150	150
7	220	150	220	210	200
8	100	60	120	120	320
9	150	100	130	30	350
10	200	100	200	150	280

Figure 9.26.

27. **I.** The dimensions of the links are given for the mechanism in Fig. 9.27. The angle of driver link 1 with the horizontal is $\phi = \phi_1$, and the angular speed of driver link 1 is $n = n_1 =$ constant. The input data for ten cases are given in the table in Fig. 9.27. Find:

a) the family f and the number of DOF of the mechanism;
b) the positions of the joints;
c) the linear velocities of the joints and the angular velocities of the links;
d) the linear accelerations of the joints and the angular accelerations of the links.

II. Links 1, 2, and 4 are homogeneous bars made of steel with a constant cross-sectional area $A_s = 0.6\ \text{cm}^2$. Sliders 3 and 5 have negligible dimensions and can be considered as material points with masses $m_3 = m_5 = 0.5$ kg. Find:

e) the forces and moments of inertia for each link;
f) the joint forces and the equilibrium moment if an external force of magnitude $F_e = 500$ N acts on link 5 at E.

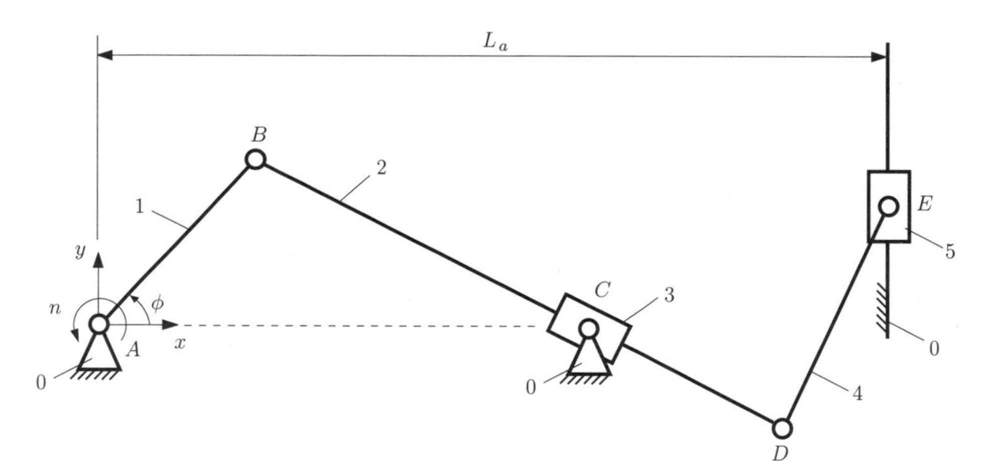

Index	AB [mm]	AC [mm]	$BD = L_a$ [mm]	DE [mm]	ϕ [deg]	n [rpm]
1	100	280	470	220	30	750
2	180	500	770	600	120	700
3	150	420	650	350	240	650
4	110	260	400	270	150	600
5	80	220	360	180	30	550
6	50	140	280	120	120	500
7	160	450	660	450	210	450
8	250	700	1000	550	150	400
9	200	550	850	400	60	350
10	220	600	900	450	120	300

Figure 9.27.

28. **I.** The dimensions of the links are given for the mechanism in Fig. 9.28. The angle of driver link 1 with the horizontal is $\phi = \phi_1$, and the angular speed of driver link 1 is $n = n_1 =$ constant. The input data for ten cases are given in the table in Fig. 9.28. Find:

a) the family f and the number of DOF of the mechanism;
b) the positions of the joints;
c) the linear velocities of the joints and the angular velocities of the links;
d) the linear accelerations of the joints and the angular accelerations of the links.

II. Links 1, 2, and 5 are homogeneous bars made of steel each having a constant cross-sectional area $A_s = 0.5$ cm^2. Sliders 3 and 4 have negligible dimensions and can be considered as material points with masses $m_3 = m_4 = 0.5$ kg. Find:

e) the forces and moments of inertia for each link;
f) the joint forces and the equilibrium moment if an external force of magnitude $F_e = 900$ N acts on link 5 at point G.

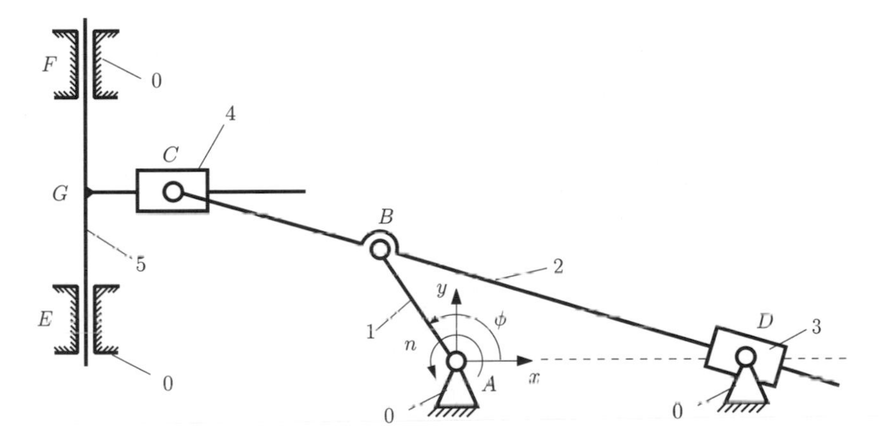

Index	AB [mm]	AD [mm]	BC [mm]	ϕ [deg]	n [rpm]
1	250	700	300	30	120
2	220	600	250	120	140
3	200	500	250	240	160
4	180	450	200	60	180
5	160	400	200	150	200
6	140	350	150	300	220
7	120	300	150	30	240
8	100	250	120	120	260
9	80	200	100	330	280
10	60	150	80	60	300

Figure 9.28.

29. **I.** The dimensions of the links are given for the mechanism shown in Fig. 9.29. The angle of driver link 1 with the horizontal is $\phi = \phi_1$, and the angular speed of driver link 1 is $n = n_1 =$ constant. The input data for ten cases are given in the table in Fig. 9.29. Find:

a) the family f and the number of DOF of the mechanism;
b) the positions of the joints;
c) the linear velocities of the joints and the angular velocities of the links;
d) the linear accelerations of the joints and the angular accelerations of the links.

II. Links 1, 3, and 5 are homogeneous bars made of steel with a constant cross-sectional area $A_s = 1$ cm^2. Sliders 2 and 4 have negligible dimensions and can be considered as material points with masses $m_2 = m_4 = 1$ kg. Find:

e) the forces and moments of inertia for each link;
f) the joint forces and the equilibrium moment if an external force of magnitude $F_e = 700$ N acts on link 5 at point G.

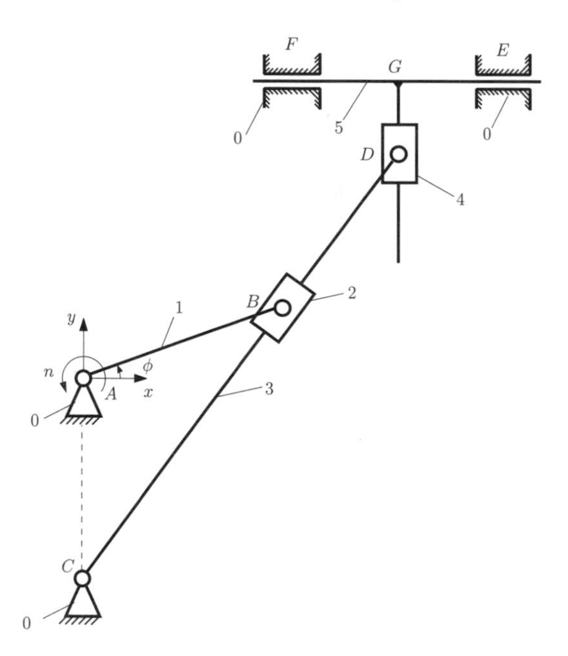

Index	AB [mm]	AC [mm]	CD [mm]	ϕ [deg]	n [rpm]
1	120	200	380	30	70
2	150	250	450	240	80
3	200	350	600	120	90
4	100	200	350	300	100
5	80	150	280	60	110
6	220	400	700	150	30
7	250	450	850	330	60
8	70	150	300	150	40
9	240	450	800	30	80
10	180	300	600	240	50

Figure 9.29.

9.3 Gears

30. The number of teeth of the gears in contact are given for the planetary gear train shown in Fig. 9.30. Gear 1 has N_1 external gear teeth, gear 2 has N_2 external gear teeth, and gear 2′ has $N_{2'}$ external gear teeth. Gears 2 and 2′ are fixed on the same shaft CC'. Gear 3 has N_3 external gear teeth, and gear 3′ has $N_{3'}$ external gear teeth. Gears 3 and 3′ are fixed on the same shaft EE'. Planet gear 4 has N_4 external gear teeth, and planet gear 5 has N_5 external gear teeth. Gear 1 rotates with a constant input angular speed n_1. The module of the gears is m. The input data for ten cases are given in the table in Fig. 9.30.

a) Find the absolute angular velocity of the output planet arm 6.
b) Find the motor torque if an external driven torque of magnitude $M_e = 500$ N m acts on planet arm 6 at point L.

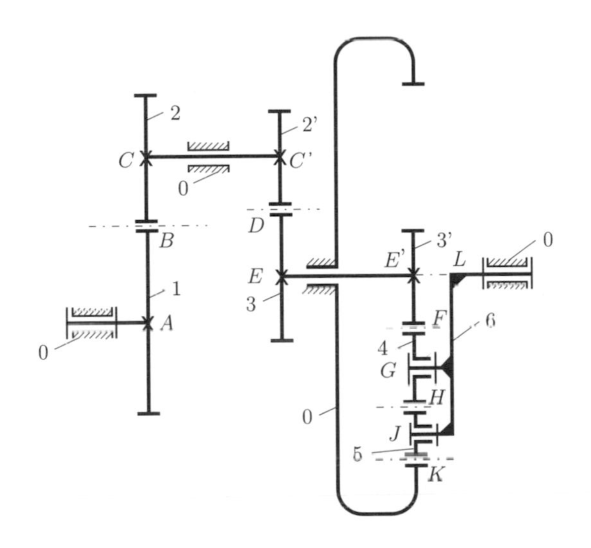

Index	N_1	N_2	$N_{2'}$	N_3	$N_{3'}$	N_4	N_5	n_1 [rpm]	m [mm]
1	36	40	21	30	24	18	14	320	30
2	21	35	18	25	23	17	27	440	22
3	27	38	19	42	21	15	12	320	36
4	24	38	17	30	19	13	23	360	28
5	32	40	19	36	18	23	15	320	26
6	24	32	18	27	20	17	18	420	28
7	29	44	20	40	22	17	12	400	33
8	23	42	15	25	17	12	18	420	33
9	28	42	14	21	16	21	14	630	28
10	22	40	13	26	15	22	11	480	30

Figure 9.30.

31. The planetary gear train considered in Fig. 9.31 has gears with the same module m. Sun gear 1 has N_1 external gear teeth, planet gear 2 has N_2 external gear teeth, gear 3 has N_3 external gear teeth, and gear 4 has N_4 external gear teeth. Gears 3 and 3′ are fixed on the same shaft. Sun gear 1 rotates with an input angular speed n_1 rpm, and arm 5 rotates with n_5 rpm. The input data for ten cases are given in the table in Fig. 9.31.

a) Find the number of DOF for the planetary gear train.
b) Find the absolute angular velocity of output gear 4.
c) Find the motor torques if an external driven torque of magnitude $M_e = 600$ N m acts on gear 4.

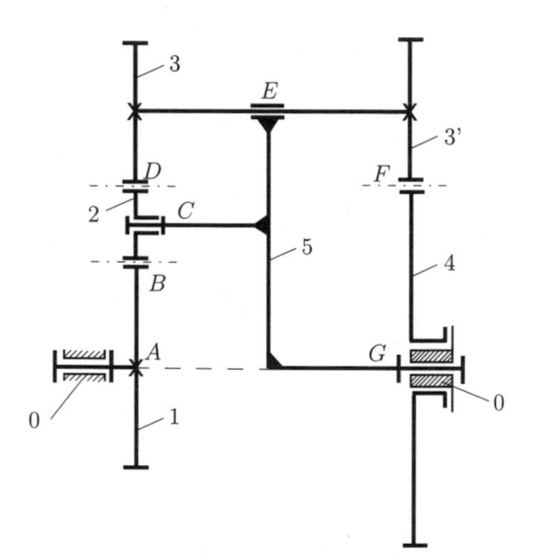

Index	N_1	N_2	N_3	N_4	n_1 [rpm]	n_5 [rpm]	m [mm]
1	22	18	20	54	290	110	24
2	18	15	22	45	94	50	30
3	22	12	18	36	320	77	28
4	25	20	15	60	380	200	22
5	22	15	15	40	299	99	26
6	32	16	18	70	370	160	20
7	14	12	16	30	150	70	39
8	25	18	12	48	432	625	22
9	25	21	20	70	102	51	22
10	28	14	21	60	205	340	24

Figure 9.31.

32. The number of teeth of the gears in contact are given for the planetary gear train considered in Fig. 9.32. Gear 1 has N_1 external gear teeth, gear 2 has N_2 external gear teeth, gear 2′ has $N_{2'}$ external gear teeth, gear 3 has N_3 external gear teeth, and gear 3′ has $N_{3'}$ external gear teeth. Planet gears 2 and 2′ are fixed on the same link, and planet gears 3 and 3′ are fixed on the same shaft. Gear 1 rotates with the input angular speed n_1 rpm, and arm 5 rotates at n_5 rpm. The module of the gears is m. The input data for ten cases are given in the table in Fig. 9.32.

a) Find the absolute angular velocity of output ring gear 4.
b) Find the motor torques if an external driven torque of magnitude $M_e = 400$ N m acts on ring gear 4.

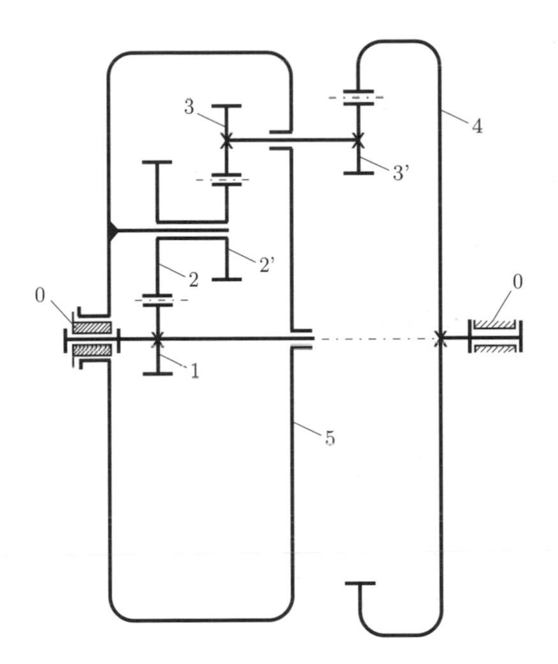

Index	N_1	N_2	$N_{2'}$	N_3	$N_{3'}$	n_1 [rpm]	n_5 [rpm]	m [mm]
1	17	25	25	23	12	125	10	24
2	15	27	18	24	16	440	80	24
3	16	23	18	28	23	606	60	26
4	18	29	21	29	29	350	60	20
5	20	23	24	30	23	750	75	22
6	22	28	24	30	28	235	100	20
7	24	24	30	40	20	785	200	20
8	16	21	33	44	21	650	50	26
9	13	22	15	30	11	775	75	28
10	12	16	14	21	14	540	100	36

Figure 9.32.

33. A planetary gear train is shown in Fig. 9.33. Gear 1 has N_1 external gear teeth, planet gear 2 has N_2 external gear teeth, gear 2′ has $N_{2'}$ external gear teeth, gear 3 has N_3 internal gear teeth, sun gear 3′ has $N_{3'}$ external gear teeth, and planet gear 4 has N_4 external gear teeth. Gears 2 and 2′ are fixed on the same shaft, and gears 3 and 3′ are fixed on the same shaft. Sun gear 1 rotates with an input angular speed of n_1 rpm. The module of the gears is m in mm. The input data for ten cases are given in the table in Fig. 9.33.

a) Find the absolute angular velocity of output arm 5.
b) Find the motor torque if an external driven torque of magnitude $M_e = 700$ N m acts on arm 5.

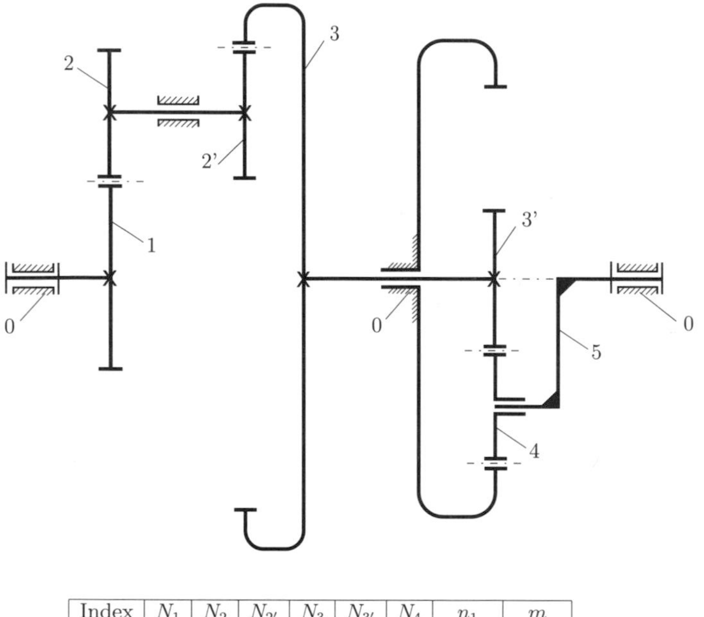

Index	N_1	N_2	$N_{2'}$	N_3	$N_{3'}$	N_4	n_1 [rpm]	m [mm]
1	11	22	17	51	12	32	550	26
2	17	31	16	51	21	28	434	20
3	11	18	20	75	18	26	450	26
4	16	20	14	70	16	24	300	28
5	22	24	18	66	17	34	600	22
6	17	21	20	68	15	25	280	24
7	21	25	20	63	20	30	450	22
8	17	32	16	68	20	25	360	20
9	19	30	15	46	23	30	530	22
10	23	24	20	50	19	19	480	22

Figure 9.33.

34. A planetary gear train is shown in Fig. 9.34. Ring gear 1 has N_1 internal gear teeth, planet gear 2 has N_2 external gear teeth, planet gear 2′ has $N_{2'}$ external gear teeth, planet gear 3 has N_3 internal gear teeth, and ring gear 4 has N_4 internal gear teeth. Gears 2 and 2′ are fixed on the same shaft and gears 3 and 3′ are fixed on the same shaft. Gear 1 rotates with the input angular speed n_1 rpm, and arm 5 rotates at n_5 rpm. The module of the gears is m. The input data for ten cases are given in the table in Fig. 9.34.

a) Find the absolute angular velocity of output ring gear 4.
b) Find the motor torques if an external driven torque of magnitude $M_e = 800$ N m acts on output ring gear 4.

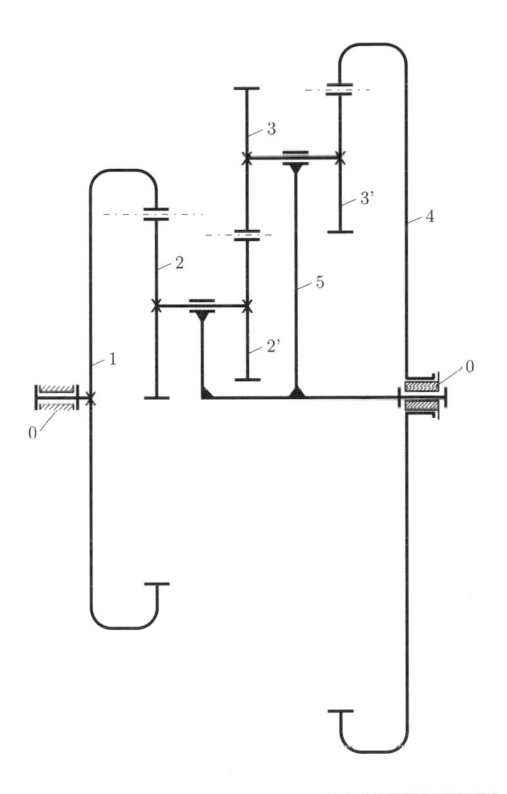

Index	N_1	N_2	$N_{2'}$	N_3	N_4	n_1 [rpm]	n_5 [rpm]	m [mm]
1	60	25	15	20	90	100	-150	28
2	54	18	15	21	94	76	-120	28
3	48	12	18	15	81	175	-320	30
4	45	14	21	18	94	194	-135	26
5	64	16	20	24	104	126	-160	22
6	75	15	20	26	120	164	-70	22
7	60	20	16	18	90	142	-128	28
8	84	21	15	24	120	140	-180	26
9	66	22	14	20	100	269	-231	28
10	60	24	16	26	92	159	-140	24

Figure 9.34.

35. The number of teeth of the gears in contact are given for the planetary gear train shown in Fig. 9.35. Sun gear 1 has N_1 external gear teeth, planet gear 2 has N_2 external gear teeth, gear 3 has N_3 internal gear teeth, and gear 3′ has $N_{3'}$ external gear teeth. Gear 4 has N_4 external gear teeth, and gear 5 has N_5 external gear teeth. Gear 1 rotates with a constant input angular speed n_1 rpm. The module of the gears is m. The input data for ten cases are given in the table in Fig. 9.35.

a) Find the absolute angular velocity of the output ring arm 6.

b) Find the motor torque if an external driven torque of magnitude $M_e = 900$ N m acts on ring gear 6.

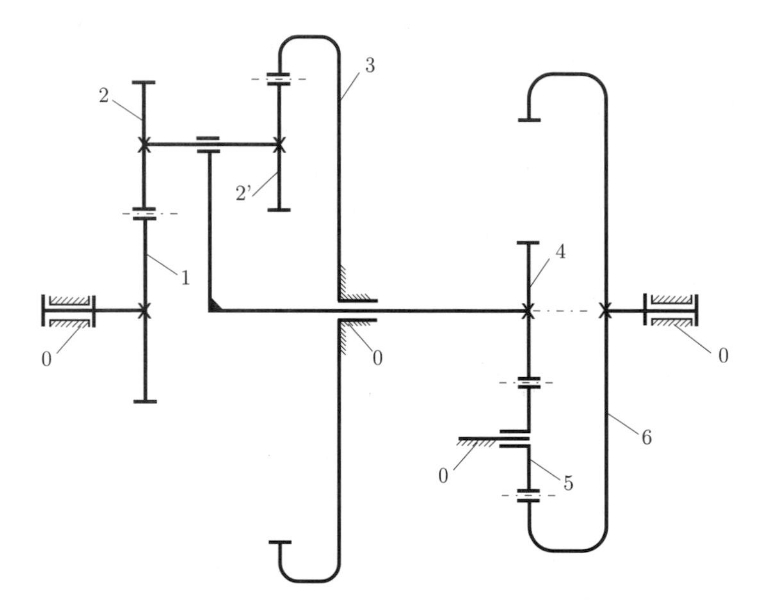

Index	N_1	N_2	N_3	N_4	N_5	n_1 [rpm]	m [mm]
1	11	19	40	29	24	110	210
2	14	18	45	12	22	50	512
3	17	23	51	14	25	77	128
4	17	29	60	19	33	200	989
5	13	15	39	15	12	99	910
6	17	20	58	13	19	160	309
7	14	28	55	16	18	70	246
8	17	17	51	14	21	625	160
9	19	21	54	17	22	51	610
10	18	22	52	19	15	340	490

Figure 9.35.

36. A planetary gear train is shown in Fig. 9.36. Sun ring gear 1 has N_1 internal gear teeth, planet gear 2 has N_2 external gear teeth, and planet gear $2'$ has $N_{2'}$ external gear teeth. Gears 2 and $2'$ are fixed on the same shaft. The gear 1 rotates with the input angular speed n_1 rpm. The module of the gears is m. The input data for ten cases are given in the table in Fig. 9.36.

a) Find the absolute angular velocity of output ring gear 4.
b) Find the motor torque if an external driven torque of magnitude $M_e = 800$ N m acts on output ring gear 4.

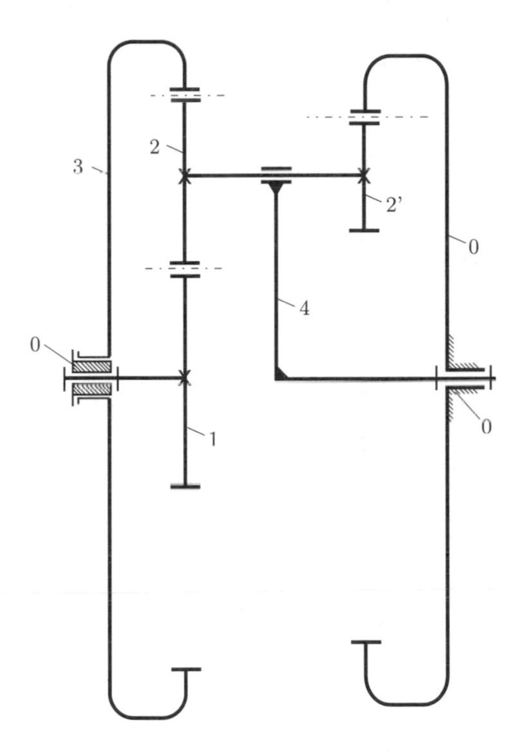

Index	N_1	N_2	$N_{2'}$	n_1 [rpm]	m [mm]
1	28	21	16	370	20
2	20	29	21	273	20
3	24	24	18	210	20
4	21	27	15	250	20
5	23	12	11	479	28
6	26	19	11	270	22
7	18	29	25	570	20
8	18	18	12	240	28
9	22	26	18	296	20
10	30	17	34	320	20

Figure 9.36.

37. A planetary gear train is shown in Fig. 9.37. Ring gear 1 has N_1 internal gear teeth, planet gear 2 has N_2 external gear teeth, planet gear 2′ has $N_{2'}$ external gear teeth, gear 3 has N_3 external gear teeth, gear 4 has N_4 external gear teeth, and ring gear 5 has N_5 internal gear teeth. Gears 2 and 2′ are fixed on the same shaft. Gear 1 rotates with the input angular speed n_1 rpm. The module of the gears is m. The input data for ten cases are given in the table in Fig. 9.37.

a) Find the absolute angular velocity of output ring gear 5.
b) Find the motor torque if an external driven torque of magnitude $M_e = 500$ N m acts on output ring gear 5.

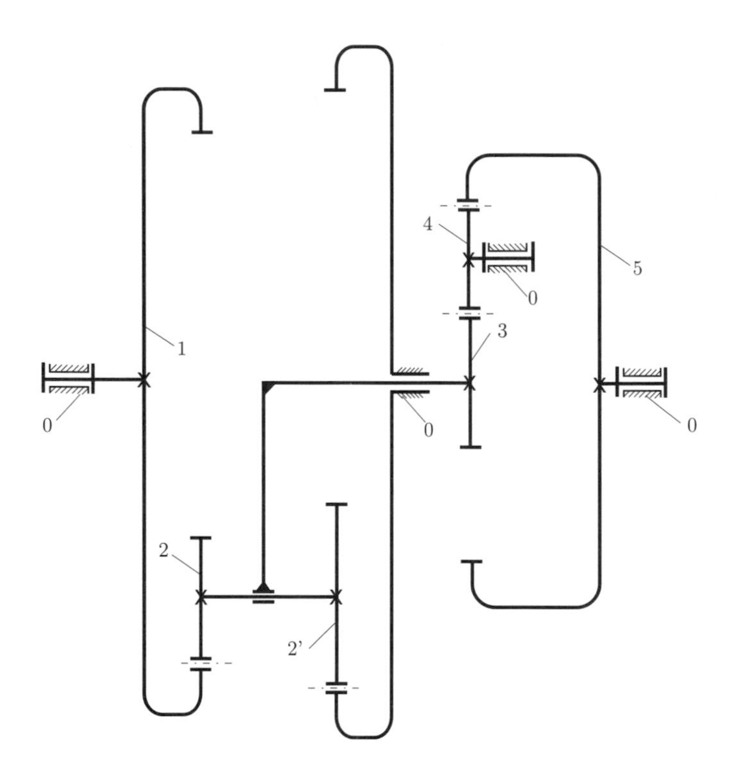

Index	N_1	N_2	$N_{2'}$	N_3	N_4	N_5	n_1 [rpm]	m [mm]
1	75	35	20	11	13	50	720	42
2	72	18	12	15	15	88	100	16
3	56	21	14	13	16	59	200	33
4	85	30	25	16	19	98	490	28
5	70	30	25	19	30	57	780	20
6	65	35	15	17	19	85	200	28
7	60	16	12	13	13	52	260	39
8	84	21	14	18	26	90	150	22
9	66	40	30	20	22	86	480	24
10	60	30	20	26	16	64	320	30

Figure 9.37.

38. A planetary gear train is shown in Fig. 9.38. Gear 1 has N_1 external gear teeth, planet gear 2 has N_2 external gear teeth, gear 3 has N_3 external gear teeth, gear 4 has N_4 external gear teeth, gear 4′ has $N_{4'}$ external gear teeth, and ring gear 5 has N_5 internal gear teeth. Gears 4 and 4′ are fixed on the same shaft. Gear 1 rotates with the input angular speed n_1. The module of the gears is m. The input data for ten cases are given in the table in Fig. 9.38.

a) Find the absolute angular velocity of output ring gear 5.
b) Find the motor torque if an external driven torque of magnitude $M_e = 500$ N m acts on output ring gear 5.

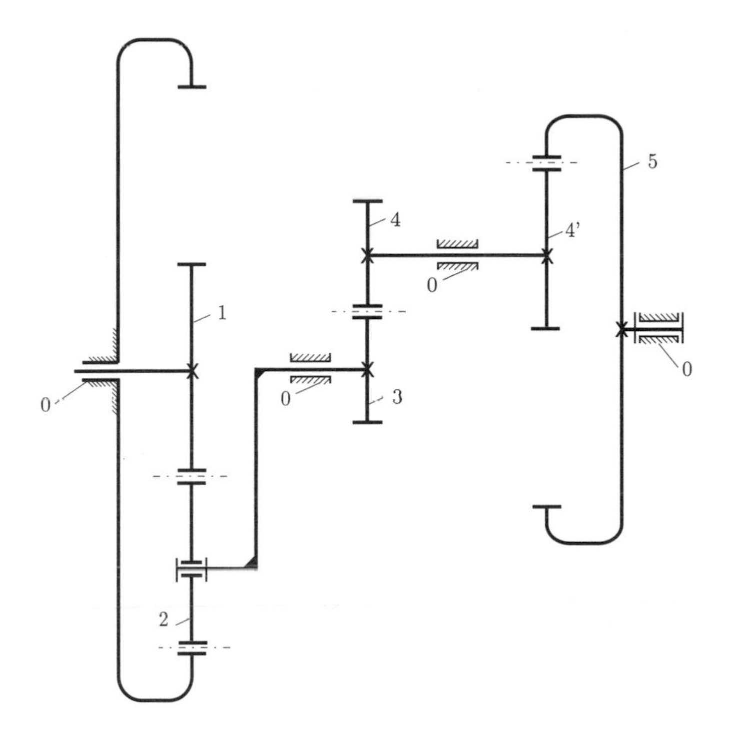

Index	N_1	N_2	N_3	N_4	$N_{4'}$	N_5	n_1 [rpm]	m [mm]
1	12	17	20	11	17	51	319	33
2	20	26	28	16	21	63	276	22
3	15	17	24	20	27	54	320	30
4	11	14	18	11	16	48	700	39
5	19	26	27	23	29	58	460	22
6	13	18	21	14	30	60	248	30
7	15	19	17	15	26	52	400	28
8	14	22	22	15	19	57	810	28
9	21	26	28	21	30	70	470	20
10	16	21	24	20	25	75	370	26

Figure 9.38.

39. The planetary gear train considered is shown in Fig. 9.39. Sun gear 1 has N_1 external gear teeth, planet gear 2 has N_2 external gear teeth, planet gear 2′ has $N_{2'}$ external gear teeth, sun gear 4 has N_4 external gear teeth, planet gear 5 has N_5 external gear teeth, and ring gear 6 has N_6 internal gear teeth. Gears 2 and 2′ are fixed on the same shaft.

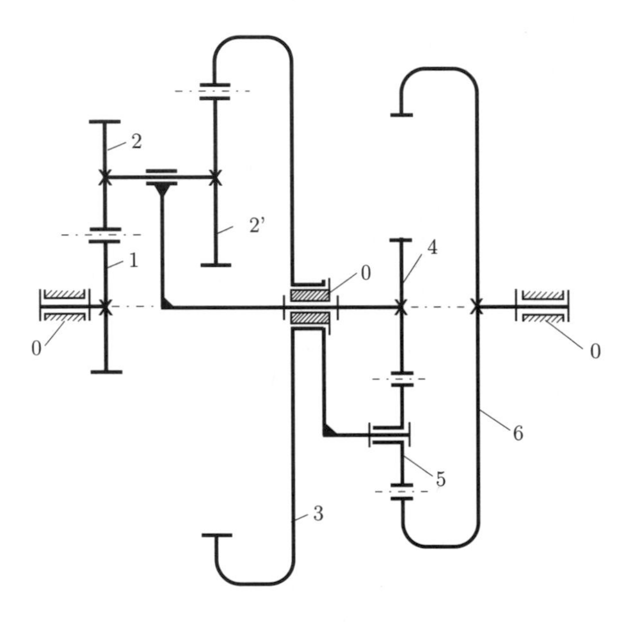

Index	N_1	N_2	$N_{2'}$	N_4	N_5	n_3 [rpm]	n_6 [rpm]	m [mm]
1	22	20	35	15	16	200	150	24
2	22	22	20	14	17	38	115	28
3	26	18	16	18	17	250	160	22
4	27	15	13	20	24	500	230	24
5	19	23	16	24	20	300	243	24
6	20	13	30	17	14	150	90	30
7	25	24	14	19	21	320	220	20
8	27	18	24	32	28	120	72	22
9	26	24	32	21	24	250	120	20
10	22	13	35	13	26	300	80	28

Figure 9.39.

Gear 1 rotates with the input angular speed n_1 rpm; the ring gear 6 rotates at the input angular speed n_6 rpm. The module of the gears is m. The input data for ten cases are given in the table in Fig. 9.39.

a) Find the absolute angular velocity of each gear.
b) Find the motor torques if an external driven torque of magnitude $M_e = 250$ N m acts on output ring gear 6.

9.4 Open Kinematic Chains

40. The dimensions of the links, for the planar manipulator in Fig. 9.40, are $AB = 3$ cm, $BC = 2$ cm, and $CD = 4$ cm. The input data are:

$$q_1 = 45^\circ, \qquad q_2 = 195^\circ, \qquad q_3 = 165^\circ,$$
$$\dot{q}_1 = 1 \text{ rad/s}, \qquad \dot{q}_2 = 2 \text{ rad/s}, \qquad \dot{q}_3 = -3 \text{ rad/s},$$
$$\ddot{q}_1 = 0.1 \text{ rad/s}^2, \qquad \ddot{q}_2 = -0.2 \text{ rad/s}^2, \qquad \ddot{q}_3 = 0.4 \text{ rad/s}^2.$$

Find the velocity and acceleration of point D.

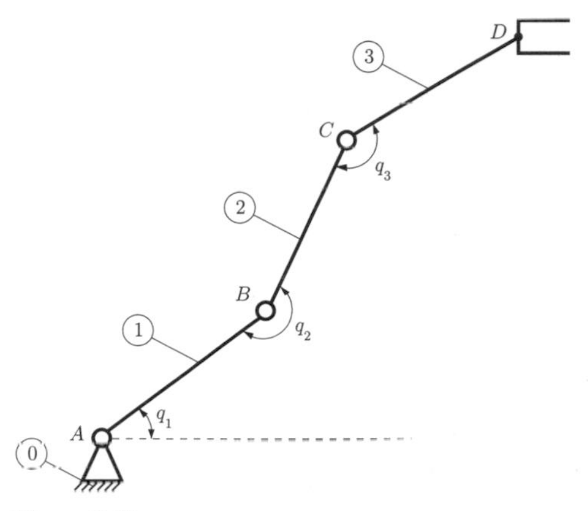

Figure 9.40.

41. Link AB, for the planar manipulator shown in Fig. 9.41, has a length $AB = 1$ m. The input data are:

$$q_1 = 35^\circ, \qquad q_2 = 95^\circ, \qquad q_3 = 1 \text{ m},$$
$$\dot{q}_1 = 0.1 \text{ rad/s}, \qquad \dot{q}_2 = -0.2 \text{ rad/s}, \qquad \dot{q}_3 = 0.5 \text{ m/s},$$
$$\ddot{q}_1 = 0.01 \text{ rad/s}^2, \qquad \ddot{q}_2 = 0.05 \text{ rad/s}^2, \qquad \ddot{q}_3 = 0.4 \text{ m/s}^2.$$

Find the velocity and acceleration of point C.

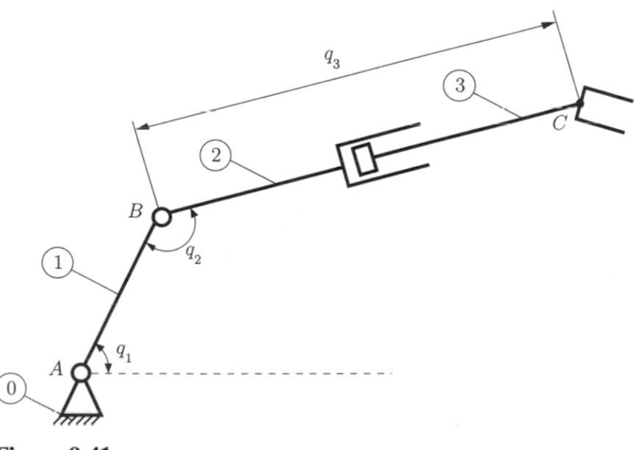

Figure 9.41.

42. Link BC, for the planar manipulator shown in Fig. 9.42, has a length $BC = 0.5$ m. The input data are:

$$q_1 = 45^\circ, \qquad q_2 = 1 \text{ m}, \qquad q_3 = 90^\circ,$$
$$\dot{q}_1 = 1 \text{ rad/s}, \qquad \dot{q}_2 = -2 \text{ m/s}, \qquad \dot{q}_3 = 2 \text{ rad/s},$$
$$\ddot{q}_1 = 0 \text{ rad/s}^2, \qquad \ddot{q}_2 = 0.5 \text{ m/s}^2, \qquad \ddot{q} = 0.4 \text{ rad/s}^2.$$

Find the velocity and acceleration of point C.

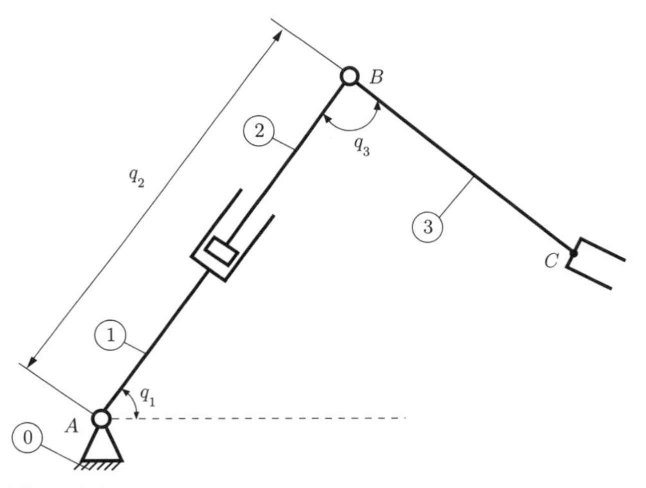

Figure 9.42.

43. The dimensions of the links, for the spatial manipulator shown in Fig. 9.43, are $AB = 3$ m and $BC = 4$ m. Links 1 and 2 are homogeneous bars made of steel with a constant cross-sectional area of 0.5 cm^2. Link 2 is perpendicular to link 1. Find and solve the equations of motion of the system for the following initial conditions:

$$q_1(0) = 45^\circ, \qquad q_2(0) = 30^\circ,$$
$$\dot{q}_1(0) = 1 \text{ rad/s}, \qquad \dot{q}_2(0) = 2 \text{ rad/s}.$$

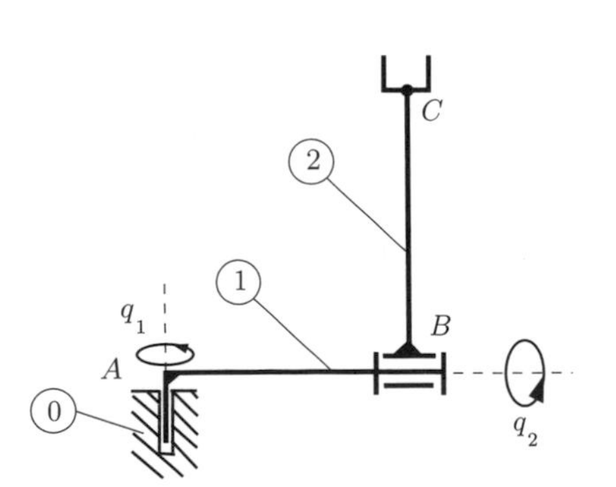

Figure 9.43.

44. The dimensions for the spatial manipulator shown in Fig. 9.44 are $BC = 0.5$ m, $CD = 0.5$ m, and the length of link 1 is 1.5 m. The links are homogeneous bars made of steel with a constant cross-sectional area of 1 cm^2. Link 2 is perpendicular to link 1, and link 3 is perpendicular to link 2. The initial conditions are

$$q_1(0) = 45°, \qquad q_2(0) = 0.1 \text{ m}, \qquad q_3(0) = 30°, \qquad q_4(0) = 60°,$$
$$\dot{q}_1(0) = 0.1 \text{ rad/s}, \quad \dot{q}_2(0) = 0.01 \text{ m/s}, \quad \dot{q}_3(0) = 0.1 \text{ rad/s}, \quad \dot{q}_4(0) = 0.3 \text{ rad/s}.$$

Find and solve the equations of motion of the system.

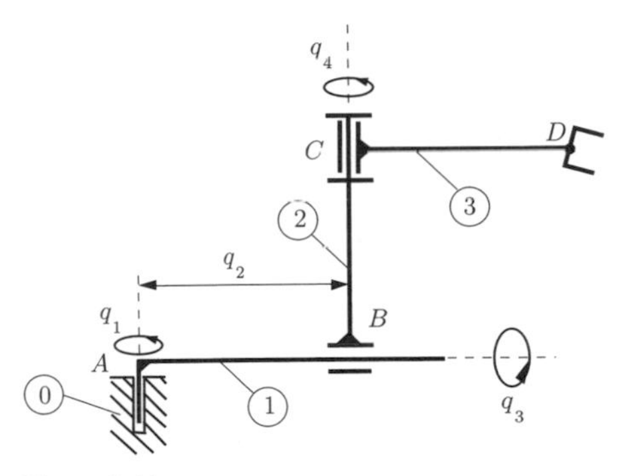

Figure 9.44.

45. Figure 9.45 is a schematic representation of a robot (Kane), with four links 1, 2, 3, and 4. Link 1 rotates in a "fixed" reference frame (0) about a vertical axis fixed in both (0) and 1. Link 1 supports link 2, and link 2 rotates relative to 1 about a horizontal axis fixed in both 1 and 2. Link 2 supports link 3, and link 3 performs translational motion relative to 2 while carrying 4. Link 4 rotates relative to 3 about an axis fixed in both 3 and 4. To characterize the instantaneous configuration of the arm, the generalized coordinates q_1, q_2, q_3, q_4 are employed.

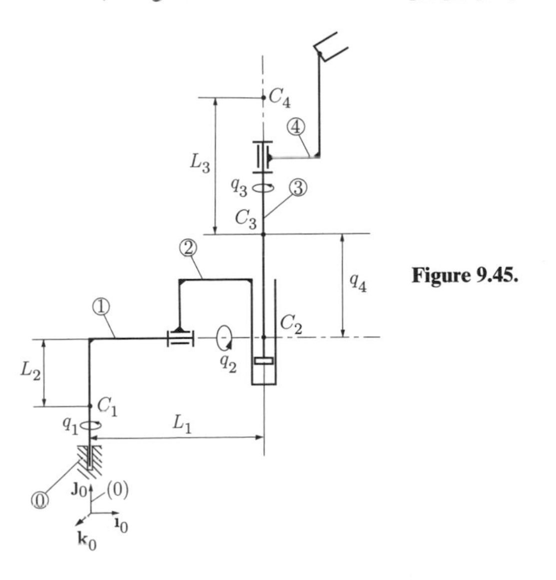

Figure 9.45.

The first three generalized coordinates are the radian measures of rotation angles, while q_4 is the distance between the mass centers, C_2 and C_3, of 2 and 3, respectively. The dimensions of interest are designated L_1, L_2 in Fig. 9.45, where C_1, C_2, C_3, and C_4 are the mass centers of links 1, 2, 3, and 4, respectively. Let $\mathbf{i}_0, \mathbf{j}_0, \mathbf{k}_0$ form a set of Cartesian perpendicular unit vectors fixed in the reference frame (0) as shown in Fig. 9.45. Let $\mathbf{i}_4, \mathbf{j}_4, \mathbf{k}_4$ be a set of unit vectors fixed in 4 in such a way that $\mathbf{i}_0 = \mathbf{i}_4$, $\mathbf{j}_0 = \mathbf{j}_4$, $\mathbf{k}_4 = \mathbf{k}_0$ when $q_1 = q_2 = q_3 = 0$. One can introduce the generalized speeds u_1, u_2, u_3, u_4 as

$$\begin{aligned} u_1 &= \boldsymbol{\omega}_{40} \cdot \mathbf{i}_4. \\ u_2 &= \boldsymbol{\omega}_{40} \cdot \mathbf{j}_4, \\ u_3 &= \boldsymbol{\omega}_{40} \cdot \mathbf{k}_4, \\ u_4 &= \dot{q}_4, \end{aligned} \tag{9.1}$$

where $\boldsymbol{\omega}_{40}$ denotes the angular velocity of 4 in reference frame (0). One may first verify that

$$\boldsymbol{\omega}_{40} = (\dot{q}_1 s_2 s_3 + \dot{q}_2 c_3)\mathbf{i}_4 + (\dot{q}_1 c_2 + \dot{q}_3)\mathbf{j}_4 + (-\dot{q}_1 s_2 c_3 + \dot{q}_2 s_3)\mathbf{k}_4, \tag{9.2}$$

where s_i and c_i denote sin q_i and cos q_i, $i = 1, 2, 3$, respectively. Substitution into Eq. (9.1) then yields

$$\begin{aligned} u_1 &= \dot{q}_1 s_2 s_3 + \dot{q}_2 c_3, \\ u_2 &= \dot{q}_1 c_2 + \dot{q}_3, \\ u_3 &= -\dot{q}_1 s_2 c_3 + \dot{q}_2 s_3, \\ u_4 &= \dot{q}_4. \end{aligned} \tag{9.3}$$

Equation (9.3) can be solved uniquely for $\dot{q}_1, \dot{q}_2, \dot{q}_3, \dot{q}_4$. Specifically,

$$\begin{aligned} \dot{q}_1 &= (u_1 s_3 - u_3 c_3)/s_2, \\ \dot{q}_2 &= (u_1 c_3 + u_3 s_3), \\ \dot{q}_3 &= u_2 + (u_3 c_3 - u_1 s_3)c_2/s_2, \\ \dot{q}_4 &= u_4, \end{aligned} \tag{9.4}$$

with singularities at $q_2 = 0^\circ$ and $q_2 = 180^\circ$ posing no problem. Thus, u_r as defined in Eq. (9.1) form a set of generalized speeds for the robot. The mass of link r is m_r, $r = 1, 2, 3, 4$. The central inertia dyadic of link r can be expressed as

$$\bar{I}_r = I_{xr}\mathbf{i}_r\mathbf{i}_r + I_{yr}\mathbf{j}_r\mathbf{j}_r + I_{zr}\mathbf{k}_r\mathbf{k}_r, \tag{9.5}$$

where I_{xr}, I_{yr}, I_{zr} denote the central principal moments of inertia for links $r = 1, 2, 3, 4$.

In the case of the robot arm, there are two kinds of forces that contribute to the generalized active forces F_r, $r = 1, 2, 3, 4$, namely, contact forces applied in order to drive links 1, 2, 3, 4, and gravitational forces exerted on the links by the Earth.

Consider first the contact forces. The set of contact forces transmitted from the fixed frame to link 1 (through the bearings and by means of the motor) can be replaced with a couple of torque $\mathbf{T}_{01}$ together with a force $\mathbf{F}_{01}$ applied to link 1 at C_1. Similarly, the set of contact forces transmitted from 2 to 1 can be replaced with a couple of torque $\mathbf{T}_{21}$ together with a force $\mathbf{F}_{21}$ applied to link 1 at C_2 (which is a point fixed in 1). The law of action and reaction then guarantees that the set of contact forces transmitted from link 1 to link 2 is equivalent to a couple of torque $-\mathbf{T}_{21}$ together with the force $-\mathbf{F}_{21}$ applied to link 2 at C_2. Next, the set of contact forces exerted on link 2 by link 3 can be replaced with a couple of torque $\mathbf{T}_{32}$ together with a force $\mathbf{F}_{32}$ applied to link 2 at $(C_3)_2$, the point of link 2 which instantaneously coincides with the point C_3. The set of forces exerted by link 2 on link 3 is, therefore, equivalent to a couple torque $-\mathbf{T}_{32}$ together with the force $-\mathbf{F}_{32}$ applied to link 3 at C_3. Similarly, torque $\mathbf{T}_{43}$ and forces $\mathbf{F}_{43}$ come into play in connection with the interactions of link 3 and link 4 at G_4. The gravitational forces exerted on the links of the robot by the Earth are denoted by $\mathbf{G}_r$, $r = 1,\ 2,\ 3,\ 4$, respectively.

The following notations are introduced:

$$\begin{aligned}
\tau_1 &= \mathbf{T}_{01} \cdot \mathbf{J}_1 = k_1(q_1^* - q_1) - k_2\dot{q}_1, \\
\tau_2 &= \mathbf{T}_{21} \cdot \mathbf{I}_2 = k_3(q_2 - q_2^*) + k_4\dot{q}_2 + g[(m_3 + m_4)q_4 + m_4 L_3]s_2, \\
\tau_3 &= \mathbf{T}_{43} \cdot \mathbf{J}_3 = k_5(q_3 - q_3^*) + k_6\dot{q}_3, \\
\sigma &= \mathbf{F}_{32} \cdot \mathbf{J}_3 = k_7(q_4 - q_4^*) + k_8\dot{q}_4 - g(m_3 + m_4)c_2,
\end{aligned} \tag{9.6}$$

where $k_1, \ldots, k_8$ are constant "gains" $k_1 = 3.0$ N m, $k_2 = 5.0$ N m s, $k_3 = 1.0$ N m, $k_4 = 3.0$ N m s, $k_5 = 0.3$ N m, $k_6 = 0.6$ N m s, $k_7 = 30$ N m, $k_8 = 41$ N s, and $q_r^* = \pi/3$ rad $r = 1, 2, 3$, while $q_4^* = 0.1$ m.

The initial numerical data are $L_1 = 0.1$ m, $L_2 = 0.1$ m, $L_3 = 0.7$ m, $m_1 = 9$ kg, $m_2 = 6$ kg, $m_3 = 4$ kg, $m_4 = 1$ kg, $I_{x1} = 0.01$ kg m^2, $I_{y1} = 0.02$ kg m^2, $I_{z1} = 0.01$ kg m^2, $I_{x2} = 0.06$ kg m^2, $I_{y2} = 0.01$ kg m^2, $I_{z2} = 0.05$ kg m^2, $I_{x3} = 0.4$ kg m^2, $I_{y3} = 0.01$ kg m^2, $I_{z3} = 0.4$ kg m^2, $I_{x4} = 0.0005$ kg m^2, $I_{y4} = 0.001$ kg m^2, and $I_{z4} = 0.001$ kg m^2.

Find and solve the equations of motion of the system.

APPENDIX 1

R–RRR–RRT Mechanism – *Mathematica* Program for Position Analysis

```
(* R-RRR-RRT Mechanism: Position analysis *)

Off[General::spell];
Off[General::spell1];

(* Input data *)
AB = 0.15 ;     (* m *)
BC = 0.40 ;     (* m *)
CD = 0.37 ;     (* m *)
Lc = CD ;
CE = 0.23 ;     (* m *)
EF = CE ;
La = 0.30 ;     (* m *)
Lb = 0.45 ;     (* m *)

xD = La ;
yD = Lb ;
xF = - Lc ;

pi = N [ Pi ] ;
increment = 0 ;

For [ fi = 0 , fi <= 2 pi , fi += pi / 12 ,

(* Position of joint B *)

xB = AB Cos [ fi ] ;
yB = AB Sin [ fi ] ;
```

```
(* Position of joint C *)

eq23a = ( xc - xB )^2 + ( yc - yB )^2 - BC^2 ;
eq23b = ( xc - xD )^2 + ( yc - yD )^2 - CD^2 ;
solution = Solve [ { eq23a == 0 , eq23b == 0 },
                     { xc , yc } ] ;
(* Two solutions for C *)
xc1 = xc /. solution [ [ 1 ] ] ;
yc1 = yc /. solution [ [ 1 ] ] ;
xc2 = xc /. solution [ [ 2 ] ] ;
yc2 = yc /. solution [ [ 2 ] ] ;
(* Select the correct position for C *)
If [ xc1 < xD  , xC = xc1 ; yC = yc1 ,
                 xC = xc2 ; yC = yc2 ] ;

(* Position of joint E *)

(* Parameters m and n of line CD: y=n x + n *)
m = ( yC - yD ) / ( xC - xD ) ;
n = yC - m xC ;
eq24 = ( xe - xC )^2 + ( ye - yC )^2 - CE^2 ;
eq27 = ye - m xe - n ;
solutionE = Solve [ { eq24 == 0 , eq27 == 0 } ,
                      { xe , ye } ] ;
(* Two solutions for E *)
xe1 = xe /. solutionE [ [ 1 ] ] ;
ye1 = ye /. solutionE [ [ 1 ] ] ;
xe2 = xe /. solutionE [ [ 2 ] ] ;
ye2 = ye /. solutionE [ [ 2 ] ] ;
(* Select the correct position for E *)
If [ xe1 <= xC , xE = xe1 ; yE = ye1 ,
                 xE = xe2 ; yE = ye2  ];

(* Position of joint F *)

eq28 = ( xE - xF )^2 + ( yE - yf )^2 - EF^2 ;
solutionF = Solve [ { eq28 == 0 } , { yf } ] ;
(* Two solutions for yF *)
yf1 = yf /. solutionF [ [ 1 ] ] ;
yf2 = yf /. solutionF [ [ 2 ] ] ;
(* Select the correct position for yF *)
If [ yf1 < yE , yF = yf1 ,
                yF = yf2 ] ;
```

```
markers = Table [ {
               Point [ { 0 , 0 } ] ,
               Point [ { xB , yB } ] ,
               Point [ { xC , yC } ] ,
               Point [ { xD , yD } ] ,
               Point [ { xE , yE } ] ,
               Point [ { xF , yF } ] ,
               Point [ { (xF+xE)/2 , (yF+yE)/2 } ],
               Point [ { (xB+xC)/2 , (yB+yC)/2 } ]
                    } ] ;

name = Table [ {
               Text [ "A" ,{0  , 0  },{ 3 , 0 } ] ,
               Text [ "B" ,{xB , yB },{-2 , 0 } ] ,
               Text [ "C" ,{xC , yC },{-2 ,-2 } ] ,
               Text [ "D" ,{xD , yD },{-3 , 0 } ] ,
               Text [ "E" ,{xE , yE },{3  , 0 } ] ,
               Text [ "F" ,{xF , yF },{3  , 0 } ] ,
               Text [ "C4",{(xF+xE)/2,(yF+yE)/2 } ,
                                           {-2 , 0 } ] ,
               Text [ "C2" ,{(xB+xC)/2,(yB+yC)/2} ,
                                           { 2 , 0 } ]
                } ] ;


graph [ increment ] = Graphics [
          { { RGBColor [ 1 , 0 , 0 ] ,
                Line [ { {0,0},{xB,yB} } ] } ,
            { RGBColor [ 0 , 0 , 1 ] ,
                Line [ { {xB,yB} , {xC,yC}} ] } ,
            { RGBColor [ 1 , 0, 1 ] ,
              Line [ { {xD,yD},  {xE,yE}} ] } ,
            { RGBColor [ 0 , 0 , 0 ] ,
                Line [ { {xE,yE},{xF,yF} } ] } ,
            { RGBColor [ 0 , 1 , 1 ] ,
                PointSize [ 0.01 ] , markers } ,
            { name } } ] ;

Show [ Graphics [ graph [ increment ] ] ,
         PlotRange -> { { -.5 , .5 } ,
                        { -.2 , .7 } } ,
        Frame -> True,
        AxesOrigin -> {xF,0},
        FrameLabel->{"x","y"},
        Axes -> {False,True},
       AspectRatio -> Automatic ] ;
```

```
increment++ ;
C2x [ increment ] = ( xB + xC ) / 2 ;
C2y [ increment ] = ( yB + yC ) / 2 ;

] ;  (* End of FOR loop *)

(* All positions on the same graphic *)

Show [Table[graph [i] ,{ i , increment } ] ,
        PlotRange -> { { -.5 , .5 } ,
                       { -.2 , .7 } } ,
        Frame -> True,
        AxesOrigin -> {xF,0},
        FrameLabel->{"x","y"},
        Axes -> {False,True},
       AspectRatio -> Automatic ] ;


(* All positions on the same graphic *)

Show [Table[graph [i] ,{ i , increment } ] ,
        PlotRange -> { { -.5 , .5 } ,
                       { -.2 , .7 } } ,
        Frame -> True,
        AxesOrigin -> {xF,0},
        FrameLabel->{"x","y"},
        Axes -> {False,True},
       AspectRatio -> Automatic ] ;
```

APPENDIX 2

R–TRR–RRT Mechanism – *Mathematica* Program for Position Analysis

```
(* Main program for position analysis R-TRR-RRT mechanism *)

Apply [ Clear , Names [ "Global`*" ] ] ;
Off[General::spell]; Off[General::spell1];

<<PosVec.m   (* package position vectors *)

AC = 0.10 ;               (* m *)
BC = 0.30 ;               (* m *)
BD = 0.90 ;               (* m *)
La = 0.10 ;               (* m *)
pi = N [ Pi ] ;

increment=0;
For [fi=0  ,fi<= 2 pi, fi += pi/12,

(* procedure position vectors *)
vecPOS=POS[{AC,BC,BD,La,fi}];

rB=vecPOS[[1]];rC=vecPOS[[2]]; rD=vecPOS[[3]];
rF=vecPOS[[4]]; rC4=vecPOS[[8]];
xB=rB[[1]]; yB=rB[[2]]; xC=rC[[1]]; yC=rC[[2]];
xF=rF[[1]]; yF=rF[[2]]; xD=rD[[1]]; yD=rD[[2]];
xC4=rC4[[1]]; yC4=rC4[[2]];

(* Mechanism drawing *)
markers = Table [ { Point [ { 0   , 0   } ] , Point [ { xF , yF } ] ,
Point [ { xB , yB } ], Point [ { xC , yC } ], Point [{xD , yD } ] ,
Point [ { xD , yD } ] , Point [ {xC4 , yC4} ] } ] ;

name = Table [{ Text [ "A" , {  0, 0  }, {  3 ,  0 } ] ,
Text [ "F" , { xF, yF }, { -2 , -2 } ] ,
Text [ "B" , { xB, yB }, {  3 ,  0 } ] ,
Text [ "C" , { xC, yC }, { -3 ,  0 } ] ,
Text [ "D" , { xD, yD }, { -2 , -2 } ] } ] ;
increment++ ;
```

```
vecx [ increment ] =(xB + xD)/2 ; vecy [ increment ] = (yB + yD)/2;
graph [ increment ] = Graphics [ {
{ RGBColor [ 1 , 0 , 0 ] , Line [ { {0,0},{xF,yF} } ] } ,
{ RGBColor [ 0 , 0 , 1 ] , Line [ { {xB,yB},{xD,yD} } ] } ,
{ RGBColor [ 0 , 0 , 1 ] , Line [ { {xB,yB},{xC,yC} } ] } ,
{ RGBColor [ 0 , 0 , 1 ] , PointSize [0.02],markers },{name}}];
Show [ Graphics [ graph [ increment ] ] ,
PlotRange -> { { -1 , 2 } , { -1 , 1 } },AspectRatio -> Automatic ] ;
] ;  (* end of the loop *)
(* end of the main program *)

(* PosVec.m *)
(* Package position analysis for R-TRR-RRT mechanism *)

BeginPackage["Mechanism`PosVec`","LinearAlgebra`CrossProduct`"]

POS::usage ="POS[{AC_,BC_,BD_,La_,fi_}] Computes the position vectors"

Begin["`Private`"]

POS[{AC_,BC_,BD_,La_,fi_}]:= Block[{xC,yC,AB,ABs,xb,yb,xb1,yb1,xb2,yb2,
AB1,AB2,AF,xF,yF,xd,yd,xd1,xd2,yd1,yd2,xD,yD,xC1,yC1,xC3,yC3,xC4,yC4},

pi=N[Pi];

(* Position of joint C *)
xC = AC  ;
yC = 0.  ;
rC={xC,yC,0};

(* Position of joint B *)
eq214a = xb - ABs Cos [ fi ] ;
eq214b = yb - ABs Sin [ fi ] ;
eq214c = yb^2 + ( xC - xb )^2 - BC^2 ;
solution= Solve [ { eq214a == 0 , eq214b == 0 , eq214c == 0 } ,
                  { xb , yb , ABs } ] ;

xb1 = xb  /. solution [ [ 1 ] ] ; yb1 = yb  /. solution [ [ 1 ] ] ;
xb2 = xb  /. solution [ [ 2 ] ] ; yb2 = yb  /. solution [ [ 2 ] ] ;
AB1 = ABs /. solution [ [ 1 ] ] ; AB2 = ABs /. solution [ [ 2 ] ] ;

If [ 0 <= fi < pi ,
     If [ yb1 > yb2 ,xB = xb1 ; yB = yb1 ; AB = AB1 ,
                     xB = xb2 ; yB = yb2 ; AB = AB2 ]] ;

If [ pi <= fi <= 2 pi ,
     If [ yb1 < yb2 , xB = xb1 ; yB = yb1 ; AB = AB1 ,
                     xB = xb2 ; yB = yb2 ; AB = AB2 ]] ;

AF = La + BC + 0.2 ;
xF = AF Cos [ fi ] ;
yF = AF Sin [ fi ] ;

(* Center of mass of link 1 (AF) *)
xC1 = xF / 2 ;
yC1 = yF / 2 ;
```

```
(* Position analysis of joint D *)
yD = La ;
eq215 = ( xB - xd )^2 + ( yB - yD )^2 - BD^2 ;

solution1 = Solve [ { eq215 == 0 } , { xd } ] ;

xd1 = xd /. solution1 [ [ 1 ] ] ;
xd2 = xd /. solution1 [ [ 2 ] ] ;

If [ xd1 > xd2 , xD = xd1 , xD = xd2 ] ;


(* Center of mass of link 3 (BC) *)

xC3 = ( xB + xC ) / 2 ;
yC3 = yB / 2 ;

(* Center of mass of link 4 (BD) *)

xC4 = ( xB + xD ) / 2 ;
yC4 = ( yB + yD ) / 2 ;

(* Position vectors *)

rB={xB,yB,0};
rC={xC,yC,0};
rD={xD,yD,0};
rF={xF,yF,0};
rC1={xC1,yC1,0};
rC2=rB;
rC3={xC3,yC3,0};
rC4={xC4,yC4,0};
rC5=rD;

Return[Chop[{rB,rC,rD,rF,rC1,rC2,rC3,rC4,rC5}]];
]


End[]

(*Protect[PosVec]*)

EndPackage[]

Null
```

APPENDIX 3

R–RTR–RRT Mechanism – *Mathematica* Program for Position Analysis

```
(* Position analysis for R-RTR-RRT mechanism *)

Apply [ Clear, Names["Global`*"]];

Off[General::spell];
Off[General::spell1];

Dist[xA_,yA_,xB_,yB_] := Sqrt[(xA-xB)^2 + (yA-yB)^2];

AB = 0.10 ;      (* m *)
CD = 0.075 ;     (* m *)
AC = 0.15 ;      (* m *)
DE = 0.20 ;      (* m *)
pi = N [ Pi ] ;  (* numerical value of PI *)

finitial = pi/2;
step = pi/8;
fifinal =  finitial + 2 pi;
increment = 0;

(* position of joint A *)
xA = 0;
yA = AC;
(* position of joint C *)
xC = 0;
yC = 0;

yE = 0;

For [ fi=finitial, fi<=fifinal, fi += step,

(* position of joint B *)
xB = xA + AB*Cos[fi];
yB = yA + AB*Sin[fi];
```

```
(* Position of joint D *)
If [xB == xC , fi3=fi ,
                   fi3=ArcTan[(yB-yC)/(xB-xC)]];

eq2 = (xd - xC)^2 + (yd - yC)^2 - CD^2;
eq3 = (xd - xC)*Sin[fi3] - (yd - yC)*Cos[fi3];

sol4 = Solve [ {eq2 == 0 ,eq3 == 0}, {xd,yd}];
xD1 = Simplify[xd/.sol4[[1]]];
yD1 = Simplify[yd/.sol4[[1]]];
xD2 = Simplify[xd/.sol4[[2]]];
yD2 = Simplify[yd/.sol4[[2]]];

(* constraints to select point D *)

If [ increment == 0,
  If [yD1 < yC , xD = xD1 ; yD = yD1,
                  xD = xD2 ; yD = yD2 ],

   dist1 = Dist[xD1,yD1,xDold,yDold];
   dist2 = Dist[xD2,yD2,xDold,yDold];
   If [dist1 < dist2 , xD = xD1 ; yD = yD1,
                       xD = xD2 ; yD = yD2]];

xDold = xD;
yDold = yD;

(* Position of joint E *)
eq4 = (xe - xD)^2+(yE - yD)^2 - DE^2;
sol7 = Solve [ eq4 == 0, xe];
xE1 = xe/.sol7[[1]];
xE2 = xe/.sol7[[2]];

(* constraints to select point E *)
If [ increment == 0,
  If [xE1 > xC , xE = xE1 , xE = xE2 ],

    dist1 = Dist[xE1,yE,xEold,yE];
    dist2 = Dist[xE2,yE,xEold,yE];
    If [dist1 < dist2 , xE = xE1 , xE = xE2 ]];

xEold = xE;

markers = Table [ { Point [ { xA , yA } ] ,
Point [ { xB , yB } ] , Point [ { xC , yC } ] ,
Point [ { xD , yD } ] , Point [ { xE , yE } ] } ] ;
```

```
name = Table [ {
               Text [" A " ,{ xA , yA }, { -1 , 0 } ] ,
               Text [" B " ,{ xB , yB }, { -1 , 0 } ] ,
               Text [" C " ,{ xC , yC }, { -1 ,-1 } ] ,
               Text [" D " ,{ xD , yD }, { -2 , 0 } ] ,
               Text [" E " ,{ xE , yE }, {  0 ,-2 } ]
                } ] ;

increment++ ;

veC4x [ increment ] = (xD+xE)/2. ;
veC4y [ increment ] = (yD+yE)/2. ;

graph [ increment ] = Graphics [
{{ RGBColor [ 1 , 0 , 0 ] , Line [ { {xA,yA},{xB,yB} } ] } ,
 { RGBColor [ 0 , 0 , 1 ] , Line [ { {xB,yB},{xC,yC},{xD,yD}}]} ,
 { RGBColor [ 0 , 0 , 0 ] , Line [ { {xD,yD},{xE,yE} } ] } ,
 { RGBColor [ 0 , 0 , 1 ] , PointSize [ 0.02 ] , markers } ,
 { name } } ] ;

Show [ Graphics [ graph [ increment ] ] ,
PlotRange -> {{-0.15 , 0.3 },{ -.1 , .275 } } ,
Frame -> True, AxesOrigin -> {0,0},
FrameLabel->{"x[m]","y[m]"}, Axes -> {True,True},
AspectRatio -> Automatic] ;

   ] ; (* end of FOR loop *)


(*  Show all the positions on the same graphic    *)

Show [ Table [ graph [ i ] , { i , increment } ] ,
PlotRange -> {{-0.15 ,0.3 },{ -.1 , .275 } } ,
Frame -> True, AxesOrigin -> {0,0},
FrameLabel->{"x[m]","y[m]"}, Axes -> {True,True},
AspectRatio -> Automatic ] ;
```

APPENDIX 4

R–RTR–RRT Mechanism – *Mathematica* Program for Velocity and Acceleration Analysis for $\phi = 45°$

```
(* R-RTR-RRT mechanism:  velocity and acceleration analysis *)

Apply [ Clear, Names["Global`*"]];
Off[General::spell];
Off[General::spell1];
pi = N[Pi];

inidata={AB->0.1, CD ->0.075, DE->0.2, AC->0.15,
fi[t]->pi/4, fi'[t]->4.712, fi''[t]->0};

xC = 0;
yC = 0;

(* Driver link AB *)

(* Position analysis *)
xA = 0;
yA = AC;
xB = xA + AB*Cos[fi[t]];
yB = yA + AB*Sin[fi[t]];
Print["xB = ", xB," = ", xB/.inidata," m" ];
Print["yB = ", yB," = ", yB/.inidata," m" ];

(* Velocity analysis *)
dxB = D[xB,t];
dyB = D[yB,t];
vB = Simplify[Sqrt[dxB^2 + dyB^2]];
Print["VxB = ", dxB," = ", dxB/.inidata," m/s " ];
Print["VyB = ", dyB," = ", dyB/.inidata," m/s "];
Print["VB = ", vB, " = ", vB/.inidata," m/s "];

(* Acceleration analysis *)
ddxB = D[dxB,t];
ddyB = D[dyB,t];
aB = Simplify[Sqrt[axB^2 + ayB^2]];
Print["axB = ", ddxB," = ", ddxB/.inidata," m/s^2" ];
Print["ayB = ", ddyB," = ", ddyB/.inidata," m/s^2"];
Print["aB = ", aB," = ", aB/.inidata," m/s^2"];
```

```
(* Link 3 *)
(* Position analysis*)
eq1 = yB - yC - (xB - xC)*Tan[fi3[t]];
sol1 = Solve [ eq1 == 0 , fi3[t]];
fi3s = fi3[t]/.sol1[[1]];
fi3n = fi3s/.inidata;
Print["Fi3 = ",fi3s];
Print[" "];
Print["Fi3 = ",fi3n," rad = ",fi3n*180/pi," degrees"];

(* Velocity analysis *)
deq1 = D[eq1,t];
sol2 = Solve [ deq1 ==0 , fi3'[t]];
omega3 = fi3'[t]/.sol2[[1]];
ruleF3 = {fi3[t]->fi3n};
omega3n = omega3/.ruleF3/.inidata;
Print["omega3 = ", omega3/.fi''[t]->0];
Print[" "];
Print["omega3 = ", omega3n," rad/s"];

(* Acceleration analysis *)
ddeq1 = D[deq1,t];
sol3 = Solve [ ddeq1 == 0 , fi3''[t]];
alpha3 = fi3''[t]/.sol3[[1]];
ruleDF3 = {fi3'[t]->omega3n};
alpha3n = alpha3/.ruleF3/.ruleDF3/.inidata;
Print["alpha3 = ", alpha3/.fi''[t]->0];
Print[" "];
Print["alpha3 = ", alpha3n," rad/s^2"];
ruleDDF3 = {fi3''[t]->alpha3n};

(* Joint D *)

(* Position analysis *)
eq21 = (xD[t] - xC)^2 + (yD[t] - yC)^2 - CD^2;
eq31 = (xD[t] - xC)*Sin[fi3[t]] -
       (yD[t] - yC)*Cos[fi3[t]];
sol4 = Solve[{eq21==0 ,eq31 ==0}, {xD[t],yD[t]}];
xDs = Simplify[xD[t]/.sol4[[1]]];
yDs = Simplify[yD[t]/.sol4[[1]]];
xDn = xDs/.inidata/.ruleF3;
yDn = yDs/.inidata/.ruleF3;
Print["xD = ", xDs," = ", xDn," m"];
Print["yD = ", yDs," = ", yDn," m"];

(* Velocity analysis *)
deq21 = D[eq21,t];
deq31 = D[eq31,t];
sol5 = Solve [ {deq21==0 ,deq31==0},
                {xD'[t],yD'[t]}];
vxDs = xD'[t]/.sol5[[1]];
vyDs = yD'[t]/.sol5[[1]];
ruleD = {xD[t]->xDn,yD[t]->yDn};
vxDn = vxDs/.ruleD/.ruleF3/.ruleDF3;
vyDn = vyDs/.ruleD/.ruleF3/.ruleDF3;
vD = Sqrt[vxDs^2 + vyDs^2];
vDn =  vD/.ruleD/.ruleF3/.ruleDF3;
```

```
Print["vxD = ", vxDs];
Print[" "];
Print["vxD = ", vxDn," m/s"];
Print[" "];
Print["vyD = ", vyDs];
Print[" "];
Print["vyD = ", vyDn," m/s"];
Print[" "];
Print["vD = ", vDn," m/s"];

(* Acceleration analysis *)
ddeq21 = D[deq21,t];
ddeq31 = D[deq31,t];
sol6 = Solve [ {ddeq21 == 0 ,ddeq31 == 0}, {xD''[t],yD''[t]}];
axDs = xD''[t]/.sol6[[1]];
ayDs = yD''[t]/.sol6[[1]];
ruleDD = {xD'[t]->vxDn,yD'[t]->vyDn};
axDn = axDs/.ruleDD/.ruleD /.ruleF3/.ruleDF3/.ruleDDF3;
ayDn = ayDs/.ruleDD/.ruleD /.ruleF3/.ruleDF3/.ruleDDF3;
aD = ExpandAll[Sqrt[axDn^2 + ayDn^2]];
ruleDDD = {xD''[t]->axDn,yD''[t]->ayDn};
Print["axD = ", axDs];
Print[" "];
Print["axD = ", axDn," m/s^2"];
Print[" "];
Print["ayD = ", ayDs];
Print[" "];
Print["ayD = ", ayDn," m/s^2"];
Print[" "];
Print["aD = ", aD," m/s^2"];

(* Joint E *)

(* Position analysis *)
yE = 0;
eq4 = (xE[t] - xD[t])^2+(yE - yD[t])^2 - DE^2;
sol7 = Solve [ eq4 == 0, xE[t]];
xEs = xE[t]/.sol7[[2]];
xEn = xEs/.ruleD/.inidata;
Print["xE = ", xEs];
Print[" "];
Print["xE = ", xEn," m"];

(* Velocity analysis *)
deq4 = D[eq4,t];
sol8 = Solve [ deq4 == 0 , xE'[t]];
vxEs = xE'[t]/.sol8[[1]];
ruleE = {xE[t]->xEn};
vxEn = vxEs/.ruleD/.ruleDD/.ruleE;

Print["vxE = ", vxEs];
Print["vxE = ", vxEn," m/s"];
Print["vyE = ", 0," m/s"];
Print["vE = vxE = ", vxEn," m/s"];
```

```
(* Acceleration analysis *)
ddeq4 = D[deq4,t];
sol9 = Solve [ ddeq4 == 0 , xE''[t]];
axEs = xE''[t]/.sol9[[1]];
ruleDE = {xE'[t]->vxDn};
axEn = axEs/.ruleD/.ruleDD/.ruleDDD/.ruleE/.ruleDE;
ruleDDE = {xE''[t]->axDn};
Print["axE = ", axEs];
Print[" "];
Print["aE = axE = ", axEn," m/s^2"];

(* Link 4 *)

(* Position analysis *)
eq5 = yE - yD[t] - (xE[t] - xD[t])*Tan[fi4[t]];
sol10 = Solve [ eq5 == 0 , fi4[t]];
fi4s = fi4[t]/.sol10[[1]];
fi4n = fi4s/.ruleD/.ruleE;
Print["Fi4 = ", fi4s," = ", fi4n," rad = ",
       fi4n*180/pi," degrees"];

(* Velocity analysis *)
deq5 = D[eq5,t];
sol11= Solve [ deq5 == 0 , fi4'[t]];
omega4s = fi4'[t]/.sol11[[1]];
ruleF4={fi4[t]->fi4n};
omega4n =omega4s/.ruleD/.ruleE/.ruleDD/.ruleDE/.ruleF4;
Print["Omega4 = ", omega4s];
Print["Omega4 = ",omega4n," rad/s "];

(* Acceleration analysis *)
ddeq5 = D[deq5,t];
sol12= Solve [ ddeq5 == 0 , fi4''[t]];
alpha4s = fi4''[t]/.sol12[[1]];
ruleDF4={fi4'[t]->omega4n};
alpha4n=
alpha4s/.ruleD/.ruleE/.ruleDD/.ruleDE/.ruleDDD/.ruleDDE/.ruleF4/.ruleDF4;
Print["alpha4 = ", alpha4s];
Print[" "];
Print["alpha4 = ",alpha4n,"rad/s^2 "];
```

APPENDIX 5

R–RRR–RRT Mechanism – *Mathematica* Program for Velocity and Acceleration Analysis for $\phi = 45°$

```
(* R-RRR-RRT Mechanism: velocity and acceleration analysis *)

Off[General::spell];
Off[General::spell1];

Apply [ Clear , Names [ "Global`*" ] ] ;

AB = 0.15 ;          (* m *)
BC = 0.40 ;          (* m *)
CD = 0.37 ;          (* m *)
Lc = CD   ;
CE = 0.23 ;          (* m *)
EF = CE   ;
La = 0.30 ;          (* m *)
Lb = 0.45 ;          (* m *)

pi = N [ Pi ] ;      (* numerical value of PI *)
fi = pi / 4 ;         (* driver link angle *)

(* Position  analysis *)
Print [ "Position analysis" ] ;
Print [ "" ] ;

(* position of joint D *)
xD = La ;
yD = Lb ;

(* position of joint B *)
xB = AB Cos [ fi ] ;
yB = AB Sin [ fi ] ;
Print [ " xB = " , xB," m"] ;
Print [ " yB = " , yB," m"] ;
```

```
(* position of joint C *)
eq23a = ( xc - xB )^2 + ( yc - yB )^2 - BC^2 ;
eq23b = ( xc - xD )^2 + ( yc - yD )^2 - CD^2 ;
solutionC = Solve [ { eq23a == 0 , eq23b == 0 } ,  { xc , yc } ] ;
xc1 = xc /. solutionC [ [ 1 ] ] ;
yc1 = yc /. solutionC [ [ 1 ] ] ;
xc2 = xc /. solutionC [ [ 2 ] ] ;
yc2 = yc /. solutionC [ [ 2 ] ] ;
If [ xc1 < xc2 , xC = xc1 ; yC = yc1 , xC = xc2 ; yC = yc2 ] ;
Print [ " xC = " , xC," m"] ;
Print [ " yC = " , yC," m"] ;

(* position of joint E *)

(* parameters m and n *)

m = ( yC - yD ) / ( xC - xD ) ;
n = yC - m xC ;

eq24 = ( xe - xC )^2 + ( ye - yC )^2 - CE^2 ;
eq27 = ye - m xe - n ;

solutionE = Solve [ { eq24 == 0 , eq27 == 0 } ,
                    { xe , ye } ] ;

xe1 = xe /. solutionE [ [ 1 ] ] ;
ye1 = ye /. solutionE [ [ 1 ] ] ;
xe2 = xe /. solutionE [ [ 2 ] ] ;
ye2 = ye /. solutionE [ [ 2 ] ] ;

If [ xe1 <= xC , xE = xe1 ; yE = ye1 ,
                 xE = xe2 ; yE = ye2 ] ;

Print [ " xE = " , xE," m" ] ;
Print [ " yE = " , yE," m" ] ;

(* position of joint F *)

xF = - Lc ;
eq28 = ( xE - xF )^2 + ( yE - yf )^2 - EF^2 ;

solutionF = Solve [ { eq28 == 0 } , { yf } ] ;
yf1 = yf /. solutionF [ [ 1 ] ] ;
yf2 = yf /. solutionF [ [ 2 ] ] ;
If [ yf1 < yE , yF = yf1 ,
                yF = yf2 ] ;

Print [ " yF = " , yF," m" ] ;

ruleP={xb[t] -> xB, yb[t] -> yB,
       xc[t] -> xC,yc[t] -> yC,
       xe[t] -> xE, ye[t] -> yE,
       yf[t] -> yF,
       mf[t] -> m, nf[t] -> n};
```

```
(* Linear and angular velocity analysis *)

(* angular velocity of the driver element is 100 rpm *)
omega = 2 pi 100 / 60 ;

Print [ "" ] ;
Print [ "Velocity analysis" ] ;
Print [ "" ] ;

(* vxB - velocity in the x direction of the joint B *)
(* vyB - velocity in the y direction of the joint B *)

(* linear velocity of joint B *)

dxB = D [ AB Cos [ fif [ t ] ] , t ] ;
dyB = D [ AB Sin [ fif [ t ] ] , t ] ;

rulefi = {fif[t] -> fi, fif'[t] -> omega, fif''[t] -> 0} ;

vxB = dxB /. rulefi ;
vyB = dyB /. rulefi ;

Print [ " vxB = " , vxB," m/s" ] ;
Print [ " vyB = " , vyB," m/s" ] ;

(* linear velocity of joint C *)

eq35aa = ( xc[t] - xb[t] )^2 + ( yc[t] - yb[t] )^2 - BC^2 ;
eq35bb = ( xc[t] - xD )^2 + ( yc[t] - yD )^2 - CD^2 ;

deq35a = D [ eq35aa , t ] ;
deq35b = D [ eq35bb , t ] ;

rulevB = {xb'[t]-> vxB, yb'[t] -> vyB} ;

deq35as = deq35a /.ruleP/.rulevB;
deq35bs = deq35b /.ruleP/.rulevB;

solutionvC = Solve [ { deq35as == 0 , deq35bs == 0 } ,
                        { xc'[t] , yc'[t] } ] ;

vxC = xc'[t] /. solutionvC [ [ 1 ] ] ;
vyC = yc'[t] /. solutionvC [ [ 1 ] ] ;

Print [ " vxC = " , vxC," m/s" ] ;
Print [ " vyC = " , vyC," m/s" ] ;

(* linear velocity of joint E *)

eq37aa = yD - mf[t] xD - nf[t] ;
eq37ba = yc[t] - mf[t] xc[t] - nf[t] ;
eq37ca = ye[t] - mf[t] xe[t] - nf[t] ;
eq37da = ( xe[t] - xc[t] )^2 + ( ye[t] - yc[t] )^2 - CE^2 ;
```

```
deq37a = D [ eq37aa , t ] ;
deq37b = D [ eq37ba , t ] ;
deq37c = D [ eq37ca , t ] ;
deq37d = D [ eq37da , t ] ;

rulevC = {xc'[t] -> vxC, yc'[t] -> vyC } ;

deq37as = deq37a /.ruleP/.rulevC;
deq37bs = deq37b /.ruleP/.rulevC;
deq37cs = deq37c /.ruleP/.rulevC;
deq37ds = deq37d /.ruleP/.rulevC;

solutionvE = Solve [ { deq37as == 0 , deq37bs == 0 ,
                          deq37cs == 0 , deq37ds == 0 } ,
                     { xe'[t] , ye'[t] , mf'[t] , nf'[t] } ] ;

vxE = xe'[t] /. solutionvE [ [ 1 ] ] ;
vyE = ye'[t] /. solutionvE [ [ 1 ] ] ;
mprime = mf'[t] /. solutionvE [ [ 1 ] ] ;
nprime = nf'[t] /. solutionvE [ [ 1 ] ] ;

Print [ " vxE = " , vxE," m/s" ] ;
Print [ " vyE = " , vyE," m/s" ] ;

(* linear velocity of joint F *)

eq38 = ( xe[t] - xF )^2 + ( ye[t] - yf[t] )^2 - EF^2 ;
deq38 = D [ eq38 , t ] ;
rulevE = {xe'[t] -> vxE, ye'[t] -> vyE} ;
deq38s = deq38 /.ruleP/.rulevE;
solutionvF = Solve [ { deq38s == 0 } , { yf'[t] } ] ;
vyF = yf'[t] /. solutionvF [ [ 1 ] ] ;
Print [ " vyF = " , vyF," m/s" ] ;
ruleV={xb'[t] -> vxB, yb'[t] -> vyB,
       xc'[t] -> vxC, yc'[t] -> vyC,
       xe'[t] -> vxE, ye'[t] -> vyE,
       yf'[t] -> vyF,
       mf'[t] -> mprime, nf[t] -> nprime};

(* Angular velocity analysis *)

(* angular velocity of link BC (2) *)

fi2[t]=ArcTan [( yc[t] - yb[t] ) / ( xc[t] - xb[t] ) ] ;
dfi2 = D [ fi2[t] , t ] ;
w2 = dfi2 /.ruleP/.ruleV ;
Print [ " angular velocity of link BC : w2 = " ,w2," rad/s" ];

(* angular velocity of link DE (3) *)

fi3[t] = ArcTan [ ( ye[t] - yD ) / ( xe[t] - xD ) ] ;
dfi3 = D [ fi3[t] , t ] ;
w3 = dfi3 /.ruleP/.ruleV ;
Print [ " angular velocity of link DE : w3 = " ,w3," rad/s" ];
```

```
(* angular velocity of the link EF (4) *)

fi4[t] = ArcTan [ ( yf[t] - ye[t] ) / ( xF - xe[t] ) ] ;
dfi4 = D [ fi4[t] , t ] ;
w4 = dfi4 /.ruleP/.ruleV ;
Print [ " angular velocity of link EF : w4 = " ,w4," rad/s" ];

(* Acceleration analysis *)

Print [ "" ] ;
Print [ "Acceleration analysis" ] ;
Print [ "" ] ;

(* linear acceleration of joint B *)

ddxb = D [ dxB , t ] ;
ddyb = D [ dyB , t ] ;
axB = ddxb /.rulefi ;
ayB = ddyb /.rulefi ;

(* linear acceleration of center of mass of link AB *)

axC1 = axB/2 ; ayC1 = ayB/2 ;
Print [ " axB = " , axB," m/s^2" ] ;
Print [ " ayB = " , ayB," m/s^2" ] ;
Print [ " axC1 = " , axC1," m/s^2" ] ;
Print [ " ayC1 = " , ayC1," m/s^2" ] ;

(* linear acceleration of joint C *)

ddeq35a = D [ deq35a , t ] ;
ddeq35b = D [ deq35b , t ] ;
ruleaB = { xb''[t] -> axB, yb''[t] -> ayB} ;
ddeq35as = ExpandAll [ ddeq35a ]/.ruleP/.ruleV/.ruleaB;
ddeq35bs = ExpandAll [ ddeq35b ]/.ruleP/.ruleV/.ruleaB;

solutionaC = Solve [ { ddeq35as == 0 , ddeq35bs == 0 } ,
                     { xc''[t] , yc''[t] } ] ;
axC = xc''[t] /. solutionaC [ [ 1 ] ] ;
ayC = yc''[t] /. solutionaC [ [ 1 ] ] ;

(* linear acceleration of center of mass of the link BC *)

axC2 = ( axB + axC ) / 2 ;    ayC2 = ( ayB + ayC ) / 2 ;
Print [ " axC = " , axC," m/s^2" ] ;
Print [ " ayC = " , ayC," m/s^2" ] ;
Print [ " axC2 = " , axC2," m/s^2" ] ;
Print [ " ayC2 = " , ayC2," m/s^2" ] ;

(* linear acceleration of joint E *)
```

```
ddeq37a = D [ deq37a , t ] ;
ddeq37b = D [ deq37b , t ] ;
ddeq37c = D [ deq37c , t ] ;
ddeq37d = D [ deq37d , t ] ;
ruleaC = { xc''[t] -> axC, yc''[t] -> ayC } ;
ddeq37as = ddeq37a /.ruleP/.ruleV/.ruleaC;
ddeq37bs = ddeq37b /.ruleP/.ruleV/.ruleaC;
ddeq37cs = ddeq37c /.ruleP/.ruleV/.ruleaC;
ddeq37ds = ddeq37d /.ruleP/.ruleV/.ruleaC;
solutionaE = Solve [ { ddeq37as == 0 , ddeq37bs == 0 ,
                         ddeq37cs == 0 , ddeq37ds == 0 } ,
               { xe''[t] , ye''[t] , mf''[t] , nf''[t] } ] ;
axE = xe''[t] /. solutionaE [ [ 1 ] ] ;
ayE = ye''[t] /. solutionaE [ [ 1 ] ] ;

(* linear acceleration of center of mass of link DE *)

axC3 = axE/2; ayC3 = ayE/2 ;
Print [ " axE = " , axE," m/s^2" ] ;
Print [ " ayE = " , ayE," m/s^2" ] ;
Print [ " axC3 = " , axC3," m/s^2" ] ;
Print [ " ayC3 = " , ayC3," m/s^2" ] ;

(* linear acceleration of joint F *)

axF = 0 ;
ddeq38 = D [ deq38 , t ] ;
ruleaE = { xe''[t] -> axE, ye''[t] -> ayE } ;
ddeq38s = ddeq38 /.ruleP/.ruleV/.ruleaE;
solutionaF = Solve [ { ddeq38s == 0 } , { yf''[t] } ] ;
ayF = yf''[t] /.solutionaF [ [ 1 ] ] ;

(* linear acceleration of center of mass of link EF *)

axC4 = ( axE + axF ) / 2 ;    ayC4 = ( ayE + ayF ) / 2 ;
Print [ " axF = " , axF," m/s^2" ] ;
Print [ " ayF = " , ayF," m/s^2" ] ;
Print [ " axC4 = " , axC4," m/s^2" ] ;
Print [ " ayC4 = " , ayC4," m/s^2" ] ;
ruleA={xb''[t] -> axB, yb''[t] -> ayB,
       xc''[t] -> axC, yc''[t] -> ayC,
       xe''[t] -> axE, ye''[t] -> ayE,
       yf''[t] -> ayF};
(* angular acceleration of the link BC *)

ddfi2 = D [ dfi2 , t ] ;
alpha2 = ddfi2 /.ruleP/.ruleV/.ruleA;
Print [" angular acceleration of link BC : alpha2 = ",alpha2,
" rad/s^2" ] ;

(* angular acceleration of the link DE *)
```

```
ddfi3 = D [ dfi3 , t ] ;
alpha3 = ddfi3 /.ruleP/.ruleV/.ruleA;
Print [ " angular acceleration of link DE : alpha3 = ",
alpha3," rad/s^2" ] ;

(* angular acceleration of the link EF *)
ddfi4 = D [ dfi4 , t ] ;
alpha4 = ddfi4 /.ruleP/.ruleV/.ruleA;
Print [ " angular acceleration of link EF : alpha4 = ",alpha4,
" rad/s^2" ] ;

Position analysis

 xB = 0.106066 m
 yB = 0.106066 m
 xC = -0.0696798 m
 yC = 0.46539 m
 xE = -0.299481 m
 yE = 0.474956 m
 yF = 0.256034 m

Velocity analysis

 vxB = -1.11072 m/s
 vyB = 1.11072 m/s
 vxC = 0.0702851 m/s
 vyC = 1.68835 m/s
 vxE = 0.113976 m/s
 vyE = 2.73787 m/s
 vyF = 2.77458 m/s
 angular velocity of link BC : w2 = -3.28675 rad/s
 angular velocity of link DE : w3 = -4.56707 rad/s
 angular velocity of link EF : w4 = -0.520622 rad/s

Acceleration analysis

 axB = -11.6314 m/s^2
 ayB = -11.6314 m/s^2
 axC1 = -5.81572 m/s^2
 ayC1 = -5.81572 m/s^2
 axC = 7.42779 m/s^2
 ayC = -7.11978 m/s^2
 axC2 = -2.10182 m/s^2
 ayC2 = -9.37561 m/s^2
 axE = 12.0451 m/s^2
 ayE = -11.5456 m/s^2
 axC3 = 6.02254 m/s^2
 ayC3 = -5.77279 m/s^2
 axF = 0 m/s^2
 ayF = -7.60013 m/s^2
 axC4 = 6.02254 m/s^2
 ayC4 = -9.57286 m/s^2
 angular acceleration of link BC : alpha2 = -47.7584 rad/s^2
 angular acceleration of link DE : alpha3 = 18.391 rad/s^2
 angular acceleration of link EF : alpha4 = -55.1071 rad/s^2
```

APPENDIX 6

R–RRR–RRT Mechanism – *Mathematica* Program for Velocity and Acceleration Analysis for $\phi = 45^\circ$ by Using the Contour Method

```
(* R-RRR-RRT Mechanism: velocity and acceleration analysis *)
(* contour method *)

Off[General::spell];
Off[General::spell1];
Apply [ Clear , Names [ "Global`*" ] ] ;

pi = N [ Pi ] ;

(* initial data *)
AB = 0.15 ; BC = 0.40 ;  CD = 0.37 ;
CE = 0.23 ;    EF = CE   ; La = 0.30 ;
Lb = 0.45 ; Lc = CD   ;   fi = pi / 4 ;

(* position  analysis *)

xD = La ; yD = Lb ;
xB = AB Cos [ fi ] ; yB = AB Sin [ fi ] ;
eq23a = ( xc - xB )^2 + ( yc - yB )^2 - BC^2 ;
eq23b = ( xc - xD )^2 + ( yc - yD )^2 - CD^2 ;
solutionC = Solve [ { eq23a == 0 , eq23b == 0 } , { xc , yc } ] ;
xc1 = xc /. solutionC [ [ 1 ] ] ; yc1 = yc /. solutionC [ [ 1 ] ] ;
xc2 = xc /. solutionC [ [ 2 ] ] ; yc2 = yc /. solutionC [ [ 2 ] ] ;
If [ xc1 < xc2 , xC = xc1 ; yC = yc1 , xC = xc2 ; yC = yc2 ] ;
m = ( yC - yD ) / ( xC - xD ) ; n = yC - m xC ;
eq24 = ( xe - xC )^2 + ( ye - yC )^2 - CE^2 ;
eq27 = ye - m xe - n ;
solutionE = Solve [ { eq24 == 0 , eq27 == 0 } , { xe , ye } ] ;
xe1 = xe /. solutionE [ [ 1 ] ] ; ye1 = ye /. solutionE [ [ 1 ] ] ;
xe2 = xe /. solutionE [ [ 2 ] ] ; ye2 = ye /. solutionE [ [ 2 ] ] ;
If [ xe1 <= xC , xE = xe1 ; yE = ye1 , xE = xe2 ; yE = ye2 ] ;
xF = - Lc ;
eq28 = ( xE - xF )^2 + ( yE - yf )^2 - EF^2 ;
solutionF = Solve [ { eq28 == 0 } , { yf } ] ;
yf1 = yf /. solutionF [ [ 1 ] ] ; yf2 = yf /. solutionF [ [ 2 ] ] ;
If [ yf1 < yE , yF = yf1 , yF = yf2 ] ;

rAB = { xB , yB , 0 } ;
rAC = { xC , yC , 0 } ;
rAD = { xD , yD , 0 } ;
rAE = { xE , yE , 0 } ;
rAF = { xF , yF , 0 } ;
```

```
Cross [ xx_ , yy_ ] := { xx[[2]] yy[[3]] - xx[[3]] yy[[2]] ,
                         xx[[3]] yy[[1]] - xx[[1]] yy[[3]] ,
                         xx[[1]] yy[[2]] - xx[[2]] yy[[1]] } ;

(* angular velocity of the driver element is 100 rpm *)
w10 = 2 pi 100 / 60 ;

Print [ "" ] ;
Print [ "Velocity analysis" ] ;
Print [ "" ] ;

(* contour 1 *)

omega10  = { 0 , 0 , w10  } ; (* driver element *)
omega21u = { 0 , 0 , w21u } ;
omega32u = { 0 , 0 , w32u } ;
omega03u = { 0 , 0 , w03u } ;

eq323a = omega10 + omega21u + omega32u + omega03u ;
eq323b = Cross [ rAB , omega21u ] + Cross [ rAC , omega32u ] +
         Cross [ rAD , omega03u ] ;

soluC1 = Solve [ { eq323a [ [ 3 ] ] == 0 ,
                   eq323b [ [ 1 ] ] == 0 ,
                   eq323b [ [ 2 ] ] == 0 } ,
                 { w21u , w32u , w03u } ] ;

w21 = w21u /. soluC1 [ [ 1 ] ] ;
w32 = w32u /. soluC1 [ [ 1 ] ] ;
w03 = w03u /. soluC1 [ [ 1 ] ] ;

w20 = w10 + w21 ;
w30 = - w03 ; (* w30 = w20 + w32; *)

omega21 = { 0 , 0 , w21 } ;
omega32 = { 0 , 0 , w32 } ;
omega03 = { 0 , 0 , w03 } ;
vB = Cross [ omega10 , rAB ] ;
rBC = rAC - rAB ;
omega20 = omega10 + omega21 ;
vC = vB + Cross [ omega20 , rBC ] ;

Print [ " angular velocity of link 2, w20 = " , w20," rad/s" ] ;
Print [ " angular velocity of link 3, w30 = " , w30," rad/s" ] ;
Print [ " linear velocity of joint B, vxB = " , vB [ [ 1 ] ], " m/s"] ;
Print [ "                             vyB = " , vB [ [ 2 ] ], " m/s"] ;
Print [ " linear velocity of joint C, vxC = " , vC [ [ 1 ] ], " m/s"] ;
Print [ "                             vyC = " , vC [ [ 2 ] ], " m/s"] ;

(* contour 2 *)
w30 = - w03 ;
omega30 = - omega03 ; (* known from the previous contour *)
omega43u = { 0 , 0 , w43u } ;
omega54u = { 0 , 0 , w54u } ;
V05u = { 0 , v05u , 0 } ;
```

```
eq324a = omega30 + omega43u + omega54u ;
eq324b = Cross [ rAD , omega30 ] + Cross [ rAE , omega43u ] +
           Cross    [ rAF , omega54u ] + V05u ;

soluC2 = Solve [ { eq324a [ [ 3 ] ] == 0 ,
                          eq324b [ [ 1 ] ] == 0 ,
                          eq324b [ [ 2 ] ] == 0 } ,
                       { w43u , w54u , v05u } ] ;
w43 = w43u /. soluC2 [ [ 1 ] ] ;
w54 = w54u /. soluC2 [ [ 1 ] ] ;
v05 = v05u /. soluC2 [ [ 1 ] ] ;

omega43 = { 0 , 0 , w43 } ;
omega54 = { 0 , 0 , w54 } ;
V05 = { 0 , v05 , 0 } ;

w40 = w30 + w43 ;
rDE = rAE - rAD ;
vE = Cross [ omega30 , rDE ] ;
omega40 = omega30 + omega43 ;
rEF = rAF - rAE ;
vF = Chop[vE + Cross [ omega40 , rEF ]] ;

Print [ " angular velocity of link 4, w40 = " , w40," rad/s" ] ;
Print [ " linear velocity of joint E, vxE = " , vE [ [ 1 ] ], " m/s" ] ;
Print [ "                             vyE = " , vE [ [ 2 ] ], " m/s" ] ;
Print [ " linear velocity of joint F, vxF = " , vF[ [ 1 ] ], " m/s" ] ;
Print [ "                             vyF = " , vF[ [ 2 ] ], " m/s" ] ;


Print [ "" ] ;
Print [ "Acceleration analysis" ] ;
Print [ "" ] ;

(* contour 1 *)
Alpha21u = { 0 , 0 , alpha21u } ;
Alpha32u = { 0 , 0 , alpha32u } ;
Alpha03u = { 0 , 0 , alpha03u } ;
rCD = rAD - rAC ;

eq410a = Alpha21u + Alpha32u + Alpha03u ;
eq410b = Cross [ rAB, Alpha21u ] + Cross [ rAC, Alpha32u ] +
          Cross [ rAD, Alpha03u ] - w10^2 rAB - w20^2 rBC -
          w03^2 rCD ;
solutC21 = Solve [ { eq410a [ [ 3 ] ] == 0 ,
                     eq410b [ [ 1 ] ] == 0 ,
                     eq410b [ [ 2 ] ] == 0 } ,
                   { alpha21u, alpha32u, alpha03u } ] ;
alpha21 = alpha21u /. solutC21 [ [ 1 ] ] ;
alpha32 = alpha32u /. solutC21 [ [ 1 ] ] ;
alpha03 = alpha03u /. solutC21 [ [ 1 ] ] ;
```

```
Alpha21 = { 0 , 0 , alpha21 } ;
Alpha32 = { 0 , 0 , alpha32 } ;
Alpha03 = { 0 , 0 , alpha03 } ;
alpha20 = alpha21 ;
alpha30 = - alpha03 ; (* =  alpha20 + alpha32 *)
aB = - w10^2 rAB;
Alpha20 = Alpha21;
aC = aB + Cross [ Alpha20, rBC ] - w20^2 rBC ;

Print [ " angular acc. of link 2, alpha20 = ", alpha20, " rad/s^2" ];
Print [ " angular acc. of link 3, alpha30 = ", alpha30, " rad/s^2" ];
Print [ " linear acc. of joint B, axB = ", aB [[1]], " m/s^2" ] ;
Print [ "                         ayB = ", aB [[2]], " m/s^2" ] ;
Print [ " linear acc. of joint C, axC = ", aC[ [ 1 ] ], " m/s^2" ] ;
Print [ "                          ayC = ", aC[ [ 2 ] ], " m/s^2" ] ;

(* contour 2 *)

Alpha30 =  { 0, 0, alpha30 } ;
Alpha43u = { 0, 0, alpha43u } ;
Alpha54u = { 0, 0, alpha54u } ;
A05u = { 0, a05u , 0 } ;
eq411a = Alpha30 + Alpha43u + Alpha54u ;
eq411b = Cross [ rAD, Alpha30 ] + Cross [ rAE, Alpha43u ] +
         Cross [ rAF, Alpha54u ] + A05u - w30^2 rDE -
          w40^2 rEF ;

solutC22 = Solve [ { eq411a [ [ 3 ] ] == 0 ,
                     eq411b [ [ 1 ] ] == 0 ,
                     eq411b [ [ 2 ] ] == 0 } ,
                   { alpha43u , alpha54u , a05u } ] ;

alpha43 = alpha43u /. solutC22  [ [ 1 ] ] ;
alpha54 = alpha54u /. solutC22   [ [ 1 ] ] ;
a05 = a05u /. solutC22  [ [ 1 ] ] ;
alpha40 = alpha30 + alpha43 ;
Alpha43 = { 0, 0, alpha43 } ;
aE = Cross [ Alpha30 , rDE ] - w30^2 rDE ;
Alpha40 = Alpha30 + Alpha43 ;
aF =Chop[ aE + Cross [ Alpha40, rEF ] - w40^2 rEF];

Print [ " angular acc. of link 4, alpha40 = ", alpha40, " rad/s^2" ] ;
Print [ " linear acc. of joint E, axE = ", aE[ [ 1 ] ], " m/s^2" ] ;
Print [ "                         ayE = ", aE[ [ 2 ] ], " m/s^2" ] ;
Print [ " linear acc. of joint F, axF = " , aF [ [ 1 ] ]," m/s^2" ] ;
Print [ "                         ayF = " , aF [ [ 2 ] ]," m/s^2" ] ;
```

```
Velocity analysis

 angular velocity of link 2, w20 = -3.28675 rad/s
 angular velocity of link 3, w30 = -4.56707 rad/s
 linear velocity of joint B, vxB = -1.11072 m/s
                             vyB = 1.11072 m/s
 linear velocity of joint C, vxC = 0.0702851 m/s
                             vyC = 1.68835 m/s
 angular velocity of link 4, w40 = -0.520622 rad/s
 linear velocity of joint E, vxE = 0.113976 m/s
                             vyE = 2.73787 m/s
 linear velocity of joint F, vxF = 0 m/s
                             vyF = 2.77458 m/s

Acceleration analysis

 angular acc. of link 2, alpha20 = -47.7584 rad/s^2
 angular acc. of link 3, alpha30 = 18.391 rad/s^2
 linear acc. of joint B, axB = -11.6314 m/s^2
                         ayB = -11.6314 m/s^2
 linear acc. of joint C, axC = 7.42779 m/s^2
                         ayC = -7.11978 m/s^2
 angular acc. of link 4, alpha40 = -55.1071 rad/s^2
 linear acc. of joint E, axE = 12.0451 m/s^2
                         ayE = -11.5456 m/s^2
 linear acc. of joint F, axF = 0 m/s^2
                         ayF = -7.60013 m/s^2
```

APPENDIX 7

R–TRR–RRT Mechanism – *Mathematica* Program for Velocity and Acceleration Analysis for $\phi = 45°$ by Using the Contour Method

```
(* Main program for vel. and acc. analysis R-TRR-RRT mechanism *)

Apply [ Clear , Names [ "Global`*" ] ] ;

Off[General::spell];
Off[General::spell1];

<<PosVec.m   (* package position vectors          *)
<<VelAcc.m   (* package velocity & acceleration   *)

(* geometrical  dimensions of the mechanism in m  *)
AC = 0.10 ;
BC = 0.30 ;
BD = 0.90 ;
La = 0.10 ;
pi = N [ Pi ] ;

fi=pi/4;        (* crank angle [rad]              *)

w10=100 pi/30;  (* crank angular speed [rad/s]    *)

(* procedure position vectors *)

vecPOS=POS[{AC,BC,BD,La,fi}];
rB=vecPOS[[1]];
rC=vecPOS[[2]];
rD=vecPOS[[3]];
rF=vecPOS[[4]];
rC1=vecPOS[[5]];
rC2=vecPOS[[6]];
rC3=vecPOS[[7]];
rC4=vecPOS[[8]];
rC5=vecPOS[[9]];

(* procedure velocity & acceleration  *)
```

```
vel=VELA[{fi,w10,rB,rC,rD,rC1,rC3,rC4}];
aC1=vel[[1]];
aC2=vel[[2]];
aC3=vel[[3]];
aC4=vel[[4]];
aC5=vel[[5]];
alpha30=vel[[6]];
alpha40=vel[[7]];
vB=vel[[8]];
vD=vel[[9]];
omega30=vel[[10]];
omega40=vel[[11]];

Print[" acc. CG link 1 = ",aC1," m/s^2 "]
Print[" acc. CG link 2 = ",aC2," m/s^2 "]
Print[" acc. CG link 3 = ",aC3," m/s^2 "]
Print[" acc. CG link 4 = ",aC4," m/s^2 "]
Print[" acc. CG link 5 = ",aC5," m/s^2 "]
Print[" ang. acc. link 3 = ",alpha30," rad/s^2 "]
Print[" ang. acc. link 4 = ",alpha40," rad/s^2 "]
Print[" velocity of B = ",vB," m/s "]
Print[" velocity of D = ",vD," m/s "]
Print[" ang. vel. link 3 = ",omega30," rad/s "]
Print[" ang. vel. link 4 = ",omega40," rad/s "]

acc. CG link 1 = {-23.2629, -23.2629, 0} m/s^2
acc. CG link 2 = {-20.026, -47.2778, 0} m/s^2
acc. CG link 3 = {-10.013, -23.6389, 0} m/s^2
acc. CG link 4 = {-18.2625, -23.6389, 0} m/s^2
acc. CG link 5 = {-16.4991, 0, 0} m/s^2
ang. acc. link 3 = {0, 0, -25.0325} rad/s^2
ang. acc. link 4 = {0, 0, 52.4141} rad/s^2
velocity of B = {-3.33304, 2.03186, 0} m/s
velocity of D = {-3.69101, 0, 0} m/s
ang. vel. link 3 = {0, 0, 13.0118} rad/s
ang. vel. link 4 = {0, 0, -2.29239} rad/s

(* VelAcc.m *)
(* Package vel. and acc. analysis for R-TRR-RRT mechanism *)

BeginPackage["Mechanism`Velacc`","LinearAlgebra`CrossProduct`"]
VELA::usage =
     "VELA[{fi_,w10_,rB_,rC_,rD_,rC1_,rC3_,rC4_}]
Computes the velocity and acceleration vectors"
Begin["`Private`"]
VELA[{fi_,w10_,rB_,rC_,rD_,rC1_,rC3_,rC4_}]:=
Block[
{omega10,omega32s,w32s,omega03s,w03s,w32,w03,v21,v21s,w30,
vB21,w20,omega43s,w43s,omega54s,w54s,vD05s,v05s,w43,w54,w40,
alpha32s,alp32s,alpha03s,alp03s,a21x,a21y,aB21s,a21cor,rBC,
alp32,alp03,a21,rCB,alpha43s,alp43s,alpha54s,aD05s,
rBD,alp43,alp54,a05,a05s},
```

```
(* Contour 1: 0-1-2-3-0  *)
omega10  = { 0 , 0 , w10} ; (* driver element *)
omega32s = { 0 , 0 , w32s } ;
omega03s = { 0 , 0 , w03s } ;
v21x = v21s Cos [ fi ] ; v21y = v21s Sin [ fi ] ;
vB21s = { v21x , v21y , 0 } ;
eq325a = omega10 + omega32s + omega03s ;
eq325b = Cross[ rB , omega32s ] + Cross[ rC , omega03s ] + vB21s;
solution4 = Solve [ { eq325a [ [ 3 ] ] == 0 , eq325b [ [ 1 ] ] == 0 ,
             eq325b [ [ 2 ] ] == 0 } , { w32s , w03s , v21s } ] ;
w32 = w32s /.solution4 [ [ 1 ] ] ;
w03 = w03s /.solution4 [ [ 1 ] ] ;
v21 = v21s /.solution4 [ [ 1 ] ] ;
w30 = - w03 ;
vB21 ={v21 Sin[fi], v21 Sin[fi], 0 } ;
w20 = w10;
omega30={0, 0, w30};
vB = Cross[omega30 , rB-rC] ;

(* Contour 2: 0-3-4-5 *)
omega43s = { 0 , 0 , w43s } ;
omega54s = { 0 , 0 , w54s } ;
vD05s = { v05s , 0 , 0 } ;
eq326a = omega30 + omega43s + omega54s ;
eq326b = Cross [ rC , omega30 ]  + Cross [ rB , omega43s ] +
          Cross [ rD , omega54s ] + vD05s ;
solution5 = Solve [ { eq326a [ [ 3 ] ] == 0 , eq326b [ [ 1 ] ] == 0 ,
             eq326b [ [ 2 ] ] == 0 },  { w43s , w54s , v05s } ] ;
w43 = w43s /. solution5 [ [ 1 ] ] ;
w54 = w54s /. solution5 [ [ 1 ] ] ;
v05 = v05s /. solution5 [ [ 1 ] ] ;
w40 = w30 + w43 ;
omega40 = { 0 , 0, w40 } ;
vD = vB + Cross [omega40 , rD-rB] ;

(* Contour 1: 0-1-2-3-0 *)
alpha32s = { 0 , 0 , alp32s } ; alpha03s = { 0 , 0 , alp03s } ;
a21x = a21s Cos [ fi ] ; a21y = a21s Sin [ fi ] ;
aB21s = { a21x , a21y , 0 } ; a21cor= 2 Cross [ omega10 , vB21 ] ;
rBC=rC-rB;
eq413a = alpha32s + alpha03s ;
eq413b = aB21s + a21cor +  Cross [ rB, alpha32s ] +
          Cross [ rC, alpha03s ] - w10^2 rB - w30^2 rBC ;
solution8 = Solve [ { eq413a [ [ 3 ] ] == 0 , eq413b [ [ 1 ] ] == 0 ,
                      eq413b [ [ 2 ] ] == 0 } ,
                     { alp32s , alp03s , a21s }];
alp32 = alp32s /. solution8 [ [ 1 ] ] ;
alp03 = alp03s /. solution8 [ [ 1 ] ] ;
a21   = a21s   /. solution8 [ [ 1 ] ] ;
alp30 = - alp03 ;
rCB = - rBC;
alpha30 = { 0 , 0 , alp30 } ;
aB = Cross[alpha30, rCB] - w30^2 rCB ;
```

```
(* Contour 2: 0-3-4-5-0 *)
alpha43s = { 0 , 0 , alp43s } ;
alpha54s = { 0 , 0 , alp54s } ;
aD05s   = { a05s , 0 , 0 } ;
rBD = rD - rB ;
eq415a = alpha30 + alpha43s + alpha54s ;
eq415b = Cross[rC,alpha30]+Cross[rB,alpha43s]+Cross[rD,alpha54s]+aD05s-
         w30^2 rCB-w40^2 rBD;
solution9 = Solve [ { eq415a [ [ 3 ] ] == 0 ,eq415b [ [ 1 ] ] == 0 ,
            eq415b [ [ 2 ] ] == 0 },
           { alp43s , alp54s , a05s } ];
alp43 = alp43s /. solution9 [ [ 1 ] ] ;
alp54 = alp54s /. solution9 [ [ 1 ] ] ;
a05  = a05s    /. solution9 [ [ 1 ] ] ;
alp40 = alp30 + alp43 ;
alpha40={ 0 , 0 , alp40 } ;
aD =Chop[aB+Cross[alpha40, rBD] - w40^2 rBD];
aC1=-w10^2 rC1;
aC2=aB;
aC3=Cross[alpha30,rC3-rC]-w30^2 (rC3-rC);
aC4=aB+Cross[alpha40,rC4-rB]-w40^2 (rC4-rB);
aC5=aD;
Return[Chop[{aC1,aC2,aC3,aC4,aC5,alpha30,alpha40,
            vB,vD,omega30,omega40}]];
]
End[]
(*Protect[Velacc]*)
EndPackage[]
Null
```

APPENDIX 8

R–TRR–RRT Mechanism – *Mathematica* Program for Joint Reactions

```
(* Main program for joint forces analysis R-TRR-RRT mechanism *)

Apply [ Clear , Names [ "Global`*" ] ] ;

Off[General::spell];
Off[General::spell1];

<<PosVec.m    (* package position vectors           *)
<<VelAcc.m    (* package velocity & acceleration    *)
<<Torsor.m    (* package inertia & external loads *)
<<Reaction.m (* package joint forces                *)

(* geometrical  dimensions of the mechanism         *)

AC = 0.10 ;                 (* m *)
BC = 0.30 ;                 (* m *)
BD = 0.90 ;                 (* m *)
La = 0.10 ;                 (* m *)
pi = N [ Pi ] ;

fi=pi/4;          (* crank angle [rad]             *)
w10=100 pi/30;  (* crank angular speed [rad/s]    *)
Fe={100,0,0};   (* driven force on link 5 [N]     *)

(* procedure position vectors *)
vecPOS=POS[{AC,BC,BD,La,fi}];
```

```
rB=vecPOS[[1]];
rC=vecPOS[[2]];
rD=vecPOS[[3]];
rF=vecPOS[[4]];
rC1=vecPOS[[5]];
rC2=vecPOS[[6]];
rC3=vecPOS[[7]];
rC4=vecPOS[[8]];
rC5=vecPOS[[9]];

xB=rB[[1]]; yB=rB[[2]];
xC=rC[[1]]; yC=rC[[2]];
xF=rF[[1]]; yF=rF[[2]];
xD=rD[[1]]; yD=rD[[2]];
xC1=rC1[[1]]; yC1=rC1[[2]];
xC3=rC3[[1]]; yC3=rC3[[2]];
xC4=rC4[[1]]; yC4=rC4[[2]];

(* procedure velocity & acceleration  *)
vel=VELA[{fi,w10,rB,rC,rD,rC1,rC3,rC4}];
aC1=vel[[1]];
aC2=vel[[2]];
aC3=vel[[3]];
aC4=vel[[4]];
aC5=vel[[5]];
alpha30=vel[[6]];
alpha40=vel[[7]];


(* procedure inertia & external loads *)
to=
TOR[{aC1,aC2,aC3,aC4,aC5,alpha30,alpha40}];

Fin1=to[[1]];
Min1=to[[2]];
G1=to[[3]];

Fin2=to[[4]];
Min2=to[[5]];
G2=to[[6]];

Fin3=to[[7]];
Min3=to[[8]];
G3=to[[9]];

Fin4=to[[10]];
Min4=to[[11]];
G4=to[[12]];
```

```
Fin5=to[[13]];
Min5=to[[14]];
G5=to[[15]];

F1=Fin1+G1;
F2=Fin2+G2;
F3=Fin3+G3;
F4=Fin4+G4;
F5=Fin5+G5;
M1=Min1;
M2=Min2;
M3=Min3;
M4=Min4;
M5=Min5;

(* procedure joint forces *)
fo=
REA[{fi,rB,rC,rD,rC1,rC2,rC3,rC4,
     F1,M1,F2,M2,F3,M3,F4,M4,F5,M5,Fe}];

Me= fo[[1]];
F01=fo[[2]];
F12=fo[[3]];
F23=fo[[4]];
F03=fo[[5]];
F34=fo[[6]];
F45=fo[[7]];
F05=fo[[8]];
Print[" Equilibrium moment Me =",Me," Nm "]
Print[" Reaction force F01 =",F01," N "]
Print[" Reaction force F12 =",F12," N "]
Print[" Reaction force F23 =",F23," N "]
Print[" Reaction force F03 =",F03," N "]
Print[" Reaction force F34 =",F34," N "]
Print[" Reaction force F45 =",F45," N "]
Print[" Reaction force F05 =",F05," N "]
```

```
Equilibrium moment Me ={0, 0, 37.347} Nm
Reaction force F01 ={-77.451, 68.747, 0} N
Reaction force F12 ={-71.9362, 71.9362, 0} N
Reaction force F23 ={-71.1552, 73.3975, 0} N
Reaction force F03 ={-37.1565, -60.6433, 0} N
Reaction force F34 ={-107.11, 14.4145, 0} N
Reaction force F45 ={-100.643, 19.3109, 0} N
Reaction force F05 ={0, -18.9283, 0} N
```

```
(* Torsor.m *)
(* Package inertia & external loads *)

BeginPackage["Mechanism`Torsor`",
            "LinearAlgebra`CrossProduct`"]

TOR::usage =
"TOR[{aC1_,aC2_,aC3_,aC4_,aC5_,alpha30_,alpha40_}]
 Computes the torsor vectors"

Begin["`Private`"]

TOR[{aC1_,aC2_,aC3_,aC4_,aC5_,alpha30_,alpha40_}]:=

Block[
{L1e,IC1,m1,AF,
 Lslider,innermass,outermass,m2,
 L3e,IC3,m3,L4e,IC4,m4,m5
},

(* ========================== *)
(* Inertia forces and moments *)
(* ========================== *)
pi=N[Pi];
BC = 0.30 ;
La = 0.10 ;
AF = La + BC + 0.2 ;
BD = 0.90 ;

d = 0.005 ;     (* dimensions of the cross section *)
h = 0.010 ;     (* of the links                      *)
ro = 7800 ; (* the density in kg/m^3             *)

 (* link 1 *)

L1e=AF + pi h /4.;  (* equivalent length          *)
m1=ro L1e d h ;     (* mass of link 1             *)
IC1=m1 L1e^2/ 12 ;  (* mass moment of inertia     *)

Fin1=-m1 aC1;       (* inertia force                *)
Min1={0 ,0 ,0 } ;   (* angular acceleration = 0   *)
```

```
(* link 2  *)

do = d + 0.01 ;
ho = h + 0.01 ;
Lslider = 0.020 ;
innermass= ro Lslider d h ;     (* inner mass        *)
outermass= ro Lslider do ho ;   (* outer mass        *)
m2= outermass - innermass ;     (* slider mass      *)
Fin2 = - m2 aC2;
Min2 = { 0 , 0 , 0 } ;
g = 9.81 ; (* acceleration gravity *)
G1 = { 0 , - m1 g , 0 };
G2 = { 0 , - m2 g , 0 };

(* link 3 *)

L3e= BC + pi h / 4 ;(* equivalent length       *)
m3= ro L3e d h ;     (* mass of link 3         *)
IC3 = m3 L3e^2 / 12 ;
Fin3 = - m3 aC3 ;          (* inertia force     *)
Min3 = - IC3 alpha30;      (* moment of inertia *)
G3 = { 0 , - m3 g , 0 };


(* link 4 *)

L4e= BD + pi h / 4 ;      (* equivalent length    *)
m4= ro L4e d h ;           (* mass                 *)
IC4 = m4 L4e^2 / 12 ;
Fin4 = - m4 aC4 ;             (* inertia force       *)
Min4 = - IC4 alpha40 ;        (* moment of inertia   *)
G4 = { 0 , - m4 g , 0 };

(* link 5 *)

Fin5 = - m2 aC5 ;
Min5 = { 0 , 0 , 0 } ;
G5 = { 0 , - m2 g , 0 };

Return[Chop[{Fin1,Min1,G1,Fin2,Min2,G2,Fin3,Min3,G3,
             Fin4,Min4,G4,Fin5,Min5,G5}]];
]
End[]
(*Protect[Torsor]*)
EndPackage[]
Null

(* Reaction.m *)
(* Package joint forces  *)
```

```
BeginPackage["Mechanism`Reaction`",
            "LinearAlgebra`CrossProduct`"]
REA::usage =
     "REA[{fi_,rB_,rC_,rD_,rC1_,rC2_,rC3_,rC4_,
          F1_,M1_,F2_,M2_,F3_,M3_,F4_,M4_,F5_,M5_,Fe_}]
       Computes the joint reactions."

Begin["`Private`"]

REA[{fi_,rB_,rC_,rD_,
     rC1_,rC2_,rC3_,rC4_,
     F1_,M1_,F2_,M2_,F3_,M3_,
     F4_,M4_,F5_,M5_,Fe_}]:=

Block[
{R05,R05y,eqMB45,solution101,
 R45,R45x,R45y,R54,eqMB4,eqF5x,solution102,
 eqMD4,eqF45x,solution103,
 R23,R23x,R23y,R32,
 eqMC3,eqF2del,del,solution104,
 eqMB3,eqF23del,solution105,
 eqMC23,solution106,R12,R12x,R12y,
 M,Ms,R01,R01x,R01y,eqF1del,eqMB12,
 eqMC123,solution107,F21,F32,F43},

pi=N[Pi];

(* ----------------------- *)
   R05 = { 0 , R05y , 0 };
(* ----------------------- *)

(* Σ M_B for 5&4 *)
eqMB45=Cross[rD -rB,R05+F5+Fe]+
       Cross[rC4-rB,F4]+M4;

solution101=Solve[eqMB45[[3]]==0,R05y];

F05={0, R05y/.solution101[[1]], 0};

(* ------------------------ *)
   R45 = { R45x , R45y , 0 };
(* ------------------------ *)

   R54 = - R45 ;

(* Σ M_B for 4 *)
eqMB4=Cross[rD-rB,R54]+
      Cross[rC4-rB,F4]+M4;
```

```
(* Σ F for 5 *)
eqF5x=(F5 + Fe + R45)[[1]];

solution102=Solve[{eqMB4[[3]]==0,eqF5x==0},
            {R45x,R45y}];

F45 = { R45x/.solution102[[1]] ,
        R45y/.solution102[[1]] , 0 };

F54 = - F45;

(* ------------------------ *)
   R34 = { R34x , R34y , 0 } ;
(* ------------------------ *)

(* Σ M_D for 4 *)
eqMD4=Cross[rB-rD,R34]+
      Cross[rC4-rD,F4]+M4;

(* Σ F for 4&5 *)
eqF45x=(F4+R34+F5+Fe)[[1]];

solution103=Solve[{eqMD4[[3]]==0,eqF45x==0},
            {R34x,R34y}];

F34 = { R34x/.solution103[[1]] ,
        R34y/.solution103[[1]] , 0 };

F43 = - F34;

(* ------------------------- *)
   R23 = { R23x , R23y , 0 };
(* ------------------------- *)

R32 = - R23 ;

del={Cos[fi],Sin[fi],0};

(* Σ M_C for 3 *)
eqMC3=Cross[rB-rC,R23+F43]+
      Cross[rC3-rC,F3]+M3;

(* Σ F_del for 2 *)
eqF2del=(F2+R32).del;

solution104=Solve[{eqMC3[[3]]==0,eqF2del==0},
            {R23x,R23y}];

F23 = { R23x/.solution104[[1]] ,
        R23y/.solution104[[1]] , 0 };
```

```
F32 = - F23 ;

R12 = { R12x , R12y , 0 } ;
R32 = - R23 ;

(* --------------------- *)
   R03 = {R03x, R03y, 0};
(* --------------------- *)

(* Σ M_B for 3 *)
eqMB3=Cross[rC-rB,R03]+
      Cross[rC3-rB,F3]+M3;

(* Σ F_del for 2&3 *)
eqF23del=(R03+F3+F2+F43).del;

solution105=Solve[{eqMB3[[3]]==0,eqF23del==0},
           {R03x,R03y}];

F03 = { R03x/.solution105[[1]] ,
        R03y/.solution105[[1]] , 0 };

(* --------------------- *)
   R12 = {R12x, R12y, 0};
(* --------------------- *)

(* Σ M_C for 2&3 *)
eqMC23=Cross[rB-rC,R12+F2+F43]+M2+
       Cross[rC3-rC,F3]+M3;

solution106=Solve[{eqMC23[[3]]==0,R12.del==0},
           {R12x,R12y}];

F12 = { R12x/.solution106[[1]] ,
        R12y/.solution106[[1]] , 0 };

F21 = -F12;

(* --------------------- *)
   M  = { 0 , 0 , Ms};
   R01 = {R01x, R01y, 0};
(* --------------------- *)

(* Σ F_del for 1 *)
eqF1del=(F1+R01).del;

(* Σ M_B for 1&2 *)
eqMB12=Cross[-rB,R01]+Cross[rC1-rB,F1]+M1+M+M2;
```

```
(* Σ M_C for 1&2&3 *)
eqMC123=Cross[-rC,R01]+
        Cross[rC1-rC,F1]+M1+M+
        Cross[rB-rC,F2+F43]+M2+
        Cross[rC3-rC,F3]+M3;

solution107=Solve[{eqMB12[[3]]==0,
                   eqMC123[[3]]==0,
                   eqF1del==0},
                 {R01x,R01y,Ms}];

F01 = { R01x/.solution107[[1]] , R01y/.solution107[[1]] , 0 };
Me  = { 0 , 0 , Ms/.solution107[[1]]};

Return[Chop[{Me,F01,F12,F23,F03,F34,F45,F05}]];
]
End[]
(*Protect[Reaction]*)
EndPackage[]
Null
```

APPENDIX 9

Laws of Motion

Consider the motion of a system $\{S\}$ of ν particles $P_1, \ldots, P_\nu$ ($\{S\} = \{P_1, \ldots, P_\nu\}$) in an inertial reference frame (0). The equation of motion for the ith particle is

$$\mathbf{F}_i = m_i \mathbf{a}_i, \tag{1}$$

where $\mathbf{F}_i$ is the resultant of all contact and distance forces acting on P_i; m_i is the mass of P_i; and $\mathbf{a}_i$ is the acceleration of P_i in (0). Equation (1) is the expression of *Newton's second law*.

If the *inertia force* $\mathbf{F}_{\text{in}i}$ for P_i in (0) is defined as

$$\mathbf{F}_{\text{in}i} = -m_i \mathbf{a}_i, \tag{2}$$

then Eq. (1) may be written as

$$\mathbf{F}_i + \mathbf{F}_{\text{in}i} = \mathbf{0}. \tag{3}$$

Equation (3) is the expression of *D'Alembert's principle*.

If $\{S\}$ is a *holonomic* system possessing n degrees of freedom, then the position vector $\mathbf{r}_i$ of P_i relative to a point O fixed in reference frame (0) may be expressed as a vector function of n generalized coordinates $q_1, \ldots, q_n$ and time t:

$$\mathbf{r}_i = \mathbf{r}_i(q_1, \ldots, q_n, t). \tag{4}$$

The velocity $\mathbf{v}_i$ of P_i in (0) may now be written as

$$\mathbf{v}_i = \sum_{r=1}^{n} \frac{\partial \mathbf{r}_i}{\partial q_r} \frac{\partial q_r}{\partial t} + \frac{\partial \mathbf{r}_i}{\partial t} = \sum_{r=1}^{n} \frac{\partial \mathbf{r}_i}{\partial q_r} \dot{q}_r + \frac{\partial \mathbf{r}_i}{\partial t}, \tag{5}$$

or as

$$\mathbf{v}_i = \sum_{r=1}^{n} (\mathbf{v}_i)_r \dot{q}_r + \frac{\partial \mathbf{r}_i}{\partial t}, \tag{6}$$

where $(\mathbf{v}_i)_r$ is called the rth *partial velocity* of P_i in (0) and is defined as

$$(\mathbf{v}_i)_r = \frac{\partial \mathbf{r}_i}{\partial q_r} = \frac{\partial \mathbf{v}_i}{\partial \dot{q}_r}. \tag{7}$$

Next, replace Eq. (3) with

$$\sum_{i=1}^{\nu}[\mathbf{F}_i + \mathbf{F}_{\text{in}i}]\cdot(\mathbf{v}_i)_r = 0. \tag{8}$$

If a *generalized active force* F_r and a *generalized inertia force* F_r^* are defined as

$$F_r = \sum_{i=1}^{\nu}(\mathbf{v}_i)_r\cdot\mathbf{F}_i = \sum_{i=1}^{\nu}\frac{\partial \mathbf{r}_i}{\partial q_r}\cdot\mathbf{F}_i = \sum_{i=1}^{\nu}\frac{\partial \mathbf{v}_i}{\partial \dot{q}_r}\cdot\mathbf{F}_i, \tag{9}$$

$$F_r^* = \sum_{i=1}^{\nu}(\mathbf{v}_i)_r\cdot\mathbf{F}_{\text{in}i} = \sum_{i=1}^{\nu}\frac{\partial \mathbf{r}_i}{\partial q_r}\cdot\mathbf{F}_{\text{in}i} = \sum_{i=1}^{\nu}\frac{\partial \mathbf{v}_i}{\partial \dot{q}_r}\cdot\mathbf{F}_{\text{in}i}, \tag{10}$$

then Eq. (8) may be written as

$$F_r + F_r^* = 0, \qquad r = 1,\ldots,n. \tag{11}$$

Equations (11) are *Kane's dynamical equations*.

Consider the generalized inertia force F_r^*:

$$\begin{aligned} F_r^* &= \sum_{i=1}^{\nu}\mathbf{F}_{\text{in}i}\cdot(\mathbf{v}_i)_r = -\sum_{i=1}^{\nu}m_i\mathbf{a}_i\cdot(\mathbf{v}_i)_r = -\sum_{i=1}^{\nu}m_i\ddot{\mathbf{r}}_i\cdot\frac{\partial \mathbf{r}_i}{\partial q_r} \\ &= -\sum_{i=1}^{\nu}\left[\frac{d}{dt}\left(m_i\dot{\mathbf{r}}_i\cdot\frac{\partial \mathbf{r}_i}{\partial q_r}\right) - m_i\dot{\mathbf{r}}_i\cdot\frac{d}{dt}\left(\frac{\partial \mathbf{r}_i}{\partial q_r}\right)\right]. \end{aligned} \tag{12}$$

Now

$$\frac{d}{dt}\left(\frac{\partial \mathbf{r}_i}{\partial q_r}\right) = \sum_{k=1}^{n}\frac{\partial^2 \mathbf{r}_i}{\partial q_r \partial q_k}\dot{q}_k + \frac{\partial^2 \mathbf{r}_i}{\partial q_r \partial t} = \frac{\partial \mathbf{v}_i}{\partial q_r}, \tag{13}$$

and furthermore, using Eq. (5) gives

$$\frac{\partial \mathbf{v}_i}{\partial \dot{q}_r} = \frac{\partial \mathbf{r}_i}{\partial q_r}. \tag{14}$$

Substitution of Eq. (13) and Eq. (14) in Eq. (12) leads to

$$\begin{aligned} F_r^* &= -\sum_{i=1}^{\nu}\left[\frac{d}{dt}\left(m_i\mathbf{v}_i\cdot\frac{\partial \mathbf{v}_i}{\partial \dot{q}_r}\right) - m_i\mathbf{v}_i\cdot\frac{\partial \mathbf{v}_i}{\partial q_r}\right] \\ &= -\left[\frac{d}{dt}\frac{\partial}{\partial \dot{q}_r}\left(\sum_{i=1}^{\nu}\frac{1}{2}m_i\mathbf{v}_i\cdot\mathbf{v}_i\right) - \frac{\partial}{\partial q_r}\left(\sum_{i=1}^{\nu}\frac{1}{2}m_i\mathbf{v}_i\cdot\mathbf{v}_i\right)\right]. \end{aligned} \tag{15}$$

The *kinetic energy* T of $\{S\}$ in reference frame (0) is defined as

$$T = \frac{1}{2}\sum_{i=1}^{\nu} m_i \mathbf{v}_i \cdot \mathbf{v}_i. \tag{16}$$

Therefore, the generalized inertia forces F_r^* can be written as

$$F_r^* = -\frac{d}{dt}\left(\frac{\partial T}{\partial \dot{q}_r}\right) + \frac{\partial T}{\partial q_r}, \tag{17}$$

and Kane's dynamical equations can be written as

$$F_r - \frac{d}{dt}\left(\frac{\partial T}{\partial \dot{q}_r}\right) + \frac{\partial T}{\partial q_r} = 0, \tag{18}$$

or

$$\frac{d}{dt}\left(\frac{\partial T}{\partial \dot{q}_r}\right) - \frac{\partial T}{\partial q_r} = F_r, \tag{19}$$

and

$$\frac{d}{dt}\left(\frac{\partial T}{\partial \dot{q}_r}\right) - \frac{\partial T}{\partial q_r} = \sum_{i=1}^{\nu} \frac{\partial \mathbf{r}_i}{\partial q_r} \cdot \mathbf{F}_i, \qquad r = 1, \ldots, n. \tag{20}$$

Equations (20) are known as *Lagrange's equations of motion* of the first kind.

APPENDIX 10

Three DOF Open Kinematic Chain – *Mathematica* Program for Dynamics

```
(* three DOF open kinematic chain *)

Off[General::spell]
Off[General::spell1]

Cross [ xx_ , yy_ ] :=
{ xx[[2]] yy[[3]] - xx[[3]] yy[[2]] ,
  xx[[3]] yy[[1]] - xx[[1]] yy[[3]] ,
  xx[[1]] yy[[2]] - xx[[2]] yy[[1]] };

(* kinematics *)

(* transformation matrix from RF1 to RF0 *)
R10 = {{1,0,0},
      {0,Cos[q3[t]],Sin[q3[t]]},
      {0,-Sin[q3[t]],Cos[q3[t]]}};
(* transformation matrix from RF2 to RF1 *)
R21={{Cos[q1[t]],0,-Sin[q1[t]]},
     {0,1,0},
     {Sin[q1[t]],0,Cos[q1[t]]}};

(* angular velocity of link 1 in RF0
expressed in terms of RF1 {i1,j1,k1} *)
w10 = {D[q3[t],t],0,0};

(* angular velocity of link 2 in RF0
expressed  in terms of RF1 {i1,j1,k1} *)
w201 = {D[q3[t],t],D[q1[t],t],0};

(* angular velocity of link 2 in RF0
expressed in terms of RF2 {i2,j2,k2} *)
w20=w201.Transpose[R21];

(* angular acceleration of link 1 in RF0
expressed in terms of RF1 {i1,j1,k1} *)
a10=D[w10,t];

(* angular acceleration of link 2 in RF0
expressed in terms of RF2 {i2,j2,k2} *)
a20=D[w20,t];
```

```
(* position vector of mass center C1 of link 1
in RF0 expressed in terms of RF1 {i1,j1,k1} *)
rC1={0,0,L1};

(* linear velocity of mass center C1 of link 1
in RF0 expressed in terms of RF1 {i1,j1,k1} *)
vC1 =D[rC1,t]+Cross[w10,rC1];

(* linear velocity of joint B in RF0
expressed in terms of RF1 {i1,j1,k1} *)
vB =D[{0,0,LB},t]+Cross[w10,{0,0,LB}];

(* position vector of mass center C2 of link 2
in RF0 expressed in terms of RF2 {i2,j2,k2} *)
rC2={0,0,LB}.Transpose[R21]+{0,0,L2};

(* linear velocity of mass center C2 of link 2 in RF0
expressed in terms of RF2 {i2,j2,k2} *)
vC2 =D[rC2,t]+Cross[w20,rC2];

(* position vector of mass center C3 of link 3 in RF0
expressed in terms of RF2 {i2,j2,k2} *)
rC3=rC2+{0,0,q2[t]};

(* linear velocity of  C32 of link 2
expressed in terms of RF2 {i2,j2,k2}
C32 of link 2 is superposed with C3 of link 3*)
vC32 =vC2+Cross[w20,{0,0,q2[t]}];

(* linear velocity of mass center C3 of link 3 in RF0
expressed in terms of RF2 {i2,j2,k2} *)
vC3 =D[rC3,t]+Cross[w20,rC3];

(* another way of computing vC3 is: *)
vC3'=vC32+D[{0,0,q2[t]},t];
(*vC3-vC3'={0,0,0};*)

(* position vector of mass center R of rigid body RB
in RF0 expressed in terms of RF2 {i2,j2,k2} *)
rR=rC3+{p1,p2,p3};

(* linear velocity of mass center R of rigid body RB
in RF0 expressed in terms of RF2 {i2,j2,k2} *)
vR =D[rR,t]+Cross[w20,rR];

(* linear acceleration of mass center C1 of link 1
in RF0 expressed in terms of RF1 {i1,j1,k1} *)
aC1 =D[vC1,t]+Cross[w10,vC1];

(* linear acceleration of mass center C2 of link 2
in RF0 expressed in terms of RF2 {i2,j2,k2} *)
aC2 =D[vC2,t]+Cross[w20,vC2];
```

```
(* linear acceleration of mass center C3 of link 3
in RF0 expressed in terms of RF2 {i2,j2,k2} *)
aC3 =D[vC3,t]+Cross[w20,vC3];

(* linear acceleration of mass center R of RB
in RF0 expressed in terms of RF2 {i2,j2,k2} *)
aR=D[vR,t]+Cross[w20,vR];
vR=Simplify[(D[rR,t] + Cross[w20,rR])];

(* contact and gravitational forces *)

(* rigid link 1 *)
(* contact force that acts on link 1 at A
in RF0 expressed in terms of RF1 {i1,j1,k1} *)
F01={ F01x , F01y , F01z } ;
(* contact torque that acts on link 1
in RF0 expressed in terms of RF1 {i1,j1,k1} *)
T01={ T01x , T01y , T01z } ;
(* gravitational force that acts on link 1 at C1
in RF0 expressed in terms of RF1 {i1,j1,k1} *)
G1={ -m1 g ,0 , 0 } ;

(* rigid link 2 *)
(* contact force that acts on link 2 at B
in RF0 expressed in terms of RF2 {i2,j2,k2} *)
F12={ F12x , F12y , F12z } ;
(* contact torque that acts on link 2
in RF0 expressed in terms of RF2 {i2,j2,k2} *)
T12={ T12x , T12y , T12x } ;
(* gravitational force that acts on link 2 at C2
in RF0 expressed in terms of RF2 {i2,j2,k2} *)
G2={ -m2 g ,0 , 0 }.Transpose[R21];

(* rigid link 3 *)
(* contact force that acts on link 3 at C3
in RF0 expressed in terms of RF2 {i2,j2,k2} *)
F23={ F23x , F23y , F23z } ;
(* contact torque that acts on link 3
in RF0 expressed in terms of RF2 {i2,j2,k2} *)
T23={ T23x , T23y , T23z } ;
(* gravitational force that acts on link 3 at C3
in RF0 expressed in terms of RF2 {i2,j2,k2} *)
G3={ -m3 g ,0 , 0 }.Transpose[R21];

(* rigid body RB *)
(* gravitational force that acts on RB at R
in RF0 expressed in terms of RF2 {i2,j2,k2} *)
GR={ -mR g ,0 , 0 }.Transpose[R21];
```

```
(*generalized active force*)

F1=D[w10,q3'[t]].T01+                          (*link1*)
   D[vC1,q3'[t]].G1+                           (*link1*)
   D[vB, q3'[t]].Transpose[R21].(-F12)+        (*link1*)
   D[w10,q3'[t]].Transpose[R21].(-T12)+        (*link1*)
   D[w20,q3'[t]].T12+                          (*link2*)
   D[vB, q3'[t]].Transpose[R21].(F12)+         (*link2*)
   D[vC2,q3'[t]].G2+                           (*link2*)
   D[w20,q3'[t]].(-T23)+                       (*link2*)
   D[vC32,q3'[t]].(-F23)+                      (*link2*)
   D[w20,q3'[t]].T23+                          (*link3*)
   D[vC3,q3'[t]].G3+                           (*link3*)
   D[vC3,q3'[t]].F23+                          (*link3*)
   D[vR,q3'[t]].GR;                           (*Body RB*)

F2=D[w10,q1'[t]].T01+                          (*link1*)
   D[vC1,q1'[t]].G1+                           (*link1*)
   D[vB, q1'[t]].Transpose[R21].(-F12)+        (*link1*)
   D[w10,q1'[t]].Transpose[R21].(-T12)+        (*link1*)
   D[w20,q1'[t]].T12+                          (*link2*)
   D[vB, q1'[t]].Transpose[R21].(F12)+         (*link2*)
   D[vC2,q1'[t]].G2+                           (*link2*)
   D[w20,q1'[t]].(-T23)+                       (*link2*)
   D[vC32,q1'[t]].(-F23)+                      (*link2*)
   D[w20,q1'[t]].T23+                          (*link3*)
   D[vC3,q1'[t]].G3+                           (*link3*)
   D[vC3,q1'[t]].F23+                          (*link3*)
   D[vR,q1'[t]].GR;                           (*Body RB*)

F3=D[w10,q2'[t]].T01+                          (*link1*)
   D[vC1,q2'[t]].G1+                           (*link1*)
   D[vB, q2'[t]].Transpose[R21].(-F12)+        (*link1*)
   D[w10,q2'[t]].Transpose[R21].(-T12)+        (*link1*)
   D[w20,q2'[t]].T12+                          (*link2*)
   D[vB, q2'[t]].Transpose[R21].(F12)+         (*link2*)
   D[vC2,q2'[t]].G2+                           (*link2*)
   D[w20,q2'[t]].(-T23)+                       (*link2*)
   D[vC32,q2'[t]].(-F23)+                      (*link2*)
   D[w20,q2'[t]].T23+                          (*link3*)
   D[vC3,q2'[t]].G3+                           (*link3*)
   D[vC3,q2'[t]].F23+                          (*link3*)
   D[vR, q2'[t]].GR;                          (*Body RB*)

(* inertia torques *)

(* inertia torque for link 1
expressed in terms of RF1 {i1,j1,k1} *)
I1={{Ax,0,0},{0,Ay,0},{0,0,Az}};
T1in=-a10.I1-Cross[w10,I1.w10];
```

```
(* inertia torque for link 2
expressed in terms of RF2 {i2,j2,k2} *)
I2={{Bx,0,0},{0,By,0},{0,0,Bz}};
T2in=-a20.I2-Cross[w20,I2.w20];

(* inertia torque for link 3
expressed in terms of RF2 {i2,j2,k2} *)
I3={{Cx,0,0},{0,Cy,0},{0,0,Cz}};
T3in=-a20.I3-Cross[w20,I3.w20];

(* inertia torque for RB
expressed in terms of RF2 {i2,j2,k2} *)
IRB={{D11,D21,D31},{D12,D22,D32},{D13,D23,D33}};
TDin=-a20.IRB-Cross[w20,IRB.w20];

(* generalized inertia forces *)

F1in=D[w10,q3'[t]].T1in+                    (*link1*)
     D[vC1,q3'[t]].(-m1 aC1)+               (*link1*)
     D[w20,q3'[t]].T2in+                    (*link2*)
     D[vC2,q3'[t]].(-m2 aC2)+               (*link2*)
     D[w20,q3'[t]].T3in+                    (*link3*)
     D[vC3,q3'[t]].(-m3 aC3)+               (*link3*)
     D[w20,q3'[t]].TDin+                  (*Body RB*)
     D[vR, q3'[t]].(-mR aR);              (*Body RB*)

F2in=D[w10,q1'[t]].T1in+                    (*link1*)
     D[vC1,q1'[t]].(-m1 aC1)+               (*link1*)
     D[w20,q1'[t]].T2in+                    (*link2*)
     D[vC2,q1'[t]].(-m2 aC2)+               (*link2*)
     D[w20,q1'[t]].T3in+                    (*link3*)
     D[vC3,q1'[t]].(-m3 aC3)+               (*link3*)
     D[w20,q1'[t]].TDin+                  (*Body RB*)
     D[vR, q1'[t]].(-mR aR);              (*Body RB*)

F3in=D[w10,q2'[t]].T1in+                    (*link1*)
     D[vC1,q2'[t]].(-m1 aC1)+               (*link1*)
     D[w20,q2'[t]].T2in+                    (*link2*)
     D[vC2,q2'[t]].(-m2 aC2)+               (*link2*)
     D[w20,q2'[t]].T3in+                    (*link3*)
     D[vC3,q2'[t]].(-m3 aC3)+               (*link3*)
     D[w20,q2'[t]].TDin+                  (*Body RB*)
     D[vR, q2'[t]].(-mR aR);              (*Body RB*)
```

```
(* input data *)

indata={
L1->0.3,L2->0.5,LB->1.1,
p1->0.2,p2->0.4,p3->0.6,
Ax->11,
Bx->7,By->6,Bz->2,
Cx->5,Cy->4,Cz->1,
D11->2,D22->2.5,D33->1.3,
D12->0.6,D21->0.6,
D13->-1.1,D31->-1.1,
D32->0.75,D23->0.75,
m1->87,m2->63,m3->42,mR->50,
g->9.81,
b01->464,
g01->306,
b12->216,
g12->285,
b23->169,
g23->56,
q1ref -> N [Pi/3],
q2ref -> 0.4,
q3ref -> N [ Pi 70/180 ]};
control = {
T01x-> - b01 q3'[t] - g01 (q3[t] - q3ref) ,
T12y-> - b12 q1'[t] - g12 (q1[t] - q1ref) +
       g ( ( m2 L2 + m3 (L2 + q2[t]) +
       mR (L2 + q2[t]+p3) ) Cos [q1[t]] -
       mR p1 Sin [q1[t]] ),

F23z->- b23 q2'[t] - g23 (q2[t] - q2ref) +
        g(m3 + mR) Sin [q1[t]]
 }/.indata;


(* Kane's dynamical equations *)

e1=F1+F1in/.indata;
e2=F2+F2in/.indata;
e3=F3+F3in/.indata;

eqs =
{e1==0,e2==0,e3==0.,
         q3'[ 0 ] == 0.1 ,
         q1'[ 0 ] == 0.2 ,
         q2'[ 0 ] == 0.3 ,
         q1 [ 0 ] == N [ Pi / 6 ] ,
         q2 [ 0 ] == 0.1 ,
         q3 [ 0 ] == N [ Pi / 18 ] };

(* numerical simulation *)

kane = NDSolve [eqs/.control,
              { q1 , q2 , q3 } ,
              { t , 0 , 15 } ] ;
```

```
Plot [Evaluate[q1[ t ]/.kane], {t,0,15},
PlotRange->{All,{0,1.3}},
AxesLabel->{"t[s]","q1[rad]"}]

Plot [Evaluate [ q2 [ t ] /.kane ],{t,0,15},
PlotRange->{All,{0,.6}},
AxesLabel->{"t[s]","q2[m]"}]

Plot[Evaluate[q3 [t]/.kane],{t,0,15},
PlotRange->{All,{0,1.6}},
AxesLabel->{"t[s]","q3[rad]"}]

(* kinetic energy *)

(* kinetic energy of link 1 *)
T1=m1 vC1.vC1/2+ w10.I1.w10/2;
(* kinetic energy of link 1 *)
T2=m2 vC2.vC2/2+ w20.I2.w20/2;
(* kinetic energy of link 1 *)
T3=m3 vC3.vC3/2+ w20.I3.w20/2;
(* kinetic energy of RB *)
TR=mR vR . vR/2+ w20.IRB.w20/2;
(* total kinetic energy *)
T=Expand[T1+T2+T3+TR];

(*=================================================================*)
(* LHS of Lagrange's equations *)
LHS1=D[D[T,q3'[t]],t]-D[T,q3[t]];
LHS2=D[D[T,q1'[t]],t]-D[T,q1[t]];
LHS3=D[D[T,q2'[t]],t]-D[T,q2[t]];
(* Lagrange's equations of motion *)
Lagr1=LHS1-F1/.indata;
Lagr2=LHS2-F2/.indata;
Lagr3-LHS3-F3/.indata;

lageq={Lagr1==0,Lagr2==0,Lagr3==0, q3'[ 0 ] == 0.1 , q1'[ 0 ] == 0.2 ,
         q2'[ 0 ] == 0.3 , q1 [ 0 ] == N [ Pi / 6 ] ,
         q2 [ 0 ] == 0.1 , q3 [ 0 ] == N [ Pi / 18 ] };

(* numerical simulation of Lagrange's eom *)
lagrange = NDSolve [ lageq/.control,
               { q1 , q2 , q3 } ,
               { t , 0 , 15 } ] ;
```

Bibliography

[1] P. Appell, *Traité de Mécanique Rationnelle*, Gautier-Villars, Paris, 1941.
[2] M. Atanasiu, *Mecanica*, EDP, Bucharest, 1973.
[3] I. I. Artobolevski, *Mechanisms in Modern Engineering Design*, MIR, Moscow, 1977.
[4] A. Bedford and W. Fowler, *Dynamics*, Addison-Wesley, Menlo Park, 1999.
[5] A. Bedford and W. Fowler, *Statics*, Addison-Wesley, Menlo Park, 1999.
[6] M. I. Buculei, *Mechanisms*, University of Craiova Press, Craiova, 1976.
[7] M. I. Buculei et al., *Analysis of Mechanisms with Bars*, Scrisul romanesc, Craiova, 1986.
[8] A. G. Erdman and G. N. Sandor, *Mechanisms Design*, Prentice-Hall, Upper Saddle River, 1984.
[9] R. C. Hibbeler, *Engineering Mechanics – Statics and Dynamics*, Prentice-Hall, Upper Saddle River, 1995.
[10] R. C. Juvinall and K. M. Marshek, *Fundamentals of Machine Component Design*, Wiley, New York, 1983.
[11] T. R. Kane, *Analytical Elements of Mechanics*, Vol. 1, Academic, New York, 1959.
[12] T. R. Kane, *Analytical Elements of Mechanics*, Vol. 2, Academic, New York, 1961.
[13] T. R. Kane and D. A. Levinson, "The Use of Kane's Dynamical Equations in Robotics," *MIT International Journal of Robotics Research*, No. 3, pp. 3–21, 1983.
[14] T. R. Kane, P. W. Likins, and D. A. Levinson, *Spacecraft Dynamics*, McGraw-Hill, New York, 1983.
[15] T. R. Kane and D. A. Levinson, *Dynamics*, McGraw-Hill, New York, 1985.
[16] J. T. Kimbrell, *Kinematics Analysis and Synthesis*, McGraw-Hill, New York, 1991.
[17] N. I. Manolescu, F. Kovacs, and A. Oranescu, *The Theory of Mechanisms and Machines*, EDP, Bucharest, 1972.
[18] D. H. Myszka, *Machines and Mechanisms*, Prentice-Hall, Upper Saddle River, 1999.
[19] R. L. Norton, *Machine Design*, Prentice-Hall, Upper Saddle River, 1996.
[20] R. L. Norton, *Design of Machinery*, McGraw-Hill, New York, 1999.
[21] W. C. Orthwein, *Machine Component Design*, West Publishing, St. Paul, 1990.
[22] R. M. Pehan, *Dynamics of Machinery*, McGraw-Hill, New York, 1967.
[23] I. Popescu, *Planar Mechanisms*, Scrisul romanesc, Craiova, 1977.
[24] I. Popescu, *Mechanisms*, University of Craiova Press, Craiova, 1990.
[25] F. Reuleaux, *The Kinematics of Machinery*, Dover, New York, 1963.
[26] I. H. Shames, *Engineering Mechanics – Statics and Dynamics*, Prentice-Hall, Upper Saddle River, 1997.
[27] J. E. Shigley and C. R. Mischke, *Mechanical Engineering Design*, McGraw-Hill, New York, 1989.

[28] J. E. Shigley and J. J. Uicker, *Theory of Machines and Mechanisms*, McGraw-Hill, New York, 1995.
[29] R. Voinea, D. Voiculescu, and V. Ceausu, *Mecanica*, EDP, Bucharest, 1983.
[30] K. J. Waldron and G. L. Kinzel, *Kinematics, Dynamics, and Design of Machinery*, Wiley, New York, 1999.
[31] C. E. Wilson and J. P. Sadler, *Kinematics and Dynamics of Machinery*, Harper Collins, New York, 1991.
[32] Anon., *The Theory of Mechanisms and Machines (Teoria mehanizmov i masin)*, Vassaia scola, Minsc, 1970.

Index